EXTERNAL GALAXIES AND QUASI-STELLAR OBJECTS

INTERNATIONAL ASTRONOMICAL UNION
UNION ASTRONOMIQUE INTERNATIONALE

SYMPOSIUM No. 44

HELD IN UPPSALA, SWEDEN, 10–14 AUGUST 1970

EXTERNAL GALAXIES AND QUASI-STELLAR OBJECTS

EDITED BY

DAVID S. EVANS

University of Texas, Austin, Texas, U.S.A.

ASSISTED BY

DEREK WILLS AND BEVERLEY J. WILLS

D. REIDEL PUBLISHING COMPANY

DORDRECHT-HOLLAND

1972

Published on behalf of
the International Astronomical Union
by
D. Reidel Publishing Company, Dordrecht, Holland

All Rights Reserved
Copyright © 1972 by the International Astronomical Union

Library of Congress Catalog Card Number 77-154736

ISBN 90 277 0199 7

No part of this book may be reproduced in any form, by print, photoprint, microfilm, or any other means, without written permission from the publisher

Printed in The Netherlands by D. Reidel, Dordrecht

EDITORIAL NOTE

The printed text represents the proceedings at Uppsala with the following exceptions: The papers by Einasto, by Pronik and Chuvaev, and by Notni, Oleak and Richter were read by title only. The paper by Shklovsky was read by Woltjer. The paper read by Rogstad and Shostak is not printed. Miss Beverley Harris has now become Mrs Derek Wills. In a number of cases abstracts and figure captions have been drafted by the editors.

D. S. EVANS

ACKNOWLEDGEMENTS

The Editors acknowledge with thanks permission to use the following copyright material:

The Editors of *Nature* for the use of figures by Dr H. C. Arp and by Drs E. M. Burbidge and G. R. Burbidge in the contributions by Dr H. C. Arp and by Dr M. J. Rees.

The Tata Institute of Fundamental Research for the use of a figure by Dr J. R. Shakeshaft in the contribution of Dr M. J. Rees.

The Editors of *The Astronomical Journal* and Dr J. G. Bolton for the use of a figure by the latter in the contribution of Dr M. J. Rees.

Annual Reviews Inc., Palo Alto, California for the use of a figure by Sir Martin Ryle in the contribution of Dr M. J. Rees.

The publishers of *Astrophysical Letters* for permission to reproduce a figure from Vol. **5**, No. 6, pp. 257–258 in the contribution by Dr H. C. Arp.

The Chairman of the Scientific Organizing Committee wishes to thank Mrs Helen Holloway for her most effective assistance in the preparation of the scientific program. Grateful acknowledgement is also made to the IAU Executive for financial support.

SCIENTIFIC ORGANIZING COMMITTEE

M. Schmidt (Chairman), E. M. Burbidge, D. S. Evans, E. B. Holmberg, J. Lequeux,
J. R. Shakeshaft, B. Vorontsov-Veljaminov

LOCAL ORGANIZING COMMITTEE

E. B. Holmberg (Chairman), P. Nilson, B. Gustafsson

INTRODUCTORY REMARKS

As Chairman of the Local Organizing Committee I wish to extend to all Symposium Participants a hearty welcome to Uppsala. I hope that your stay here will be enjoyable, also profitable from a scientific point of view. All the meetings will take place in the main building of the Uppsala university, today used mostly for administration purposes; the number of students has grown to 25 000. In the immediate neighborhood you may find a number of university buildings of historical interest, the oldest of them dating back to the 16th century. As you may know already, the university was founded in 1477, and it thus belongs to the comparatively few European universities that were started in the Middle Ages.

Astronomy has been taught at this university since the beginning. According to lecture notes preserved from the 1480's the three main subjects at that time were theology, philosophy, and astronomy. Among the well-known astronomers I may mention Anders Celsius, the originator of the 100-degree thermometer scale more than 200 years ago. The observatory that he built in 1739 is still standing in the central part of Uppsala. It may be noted that the ångström unit was, in a way, born here at Uppsala; Anders Ångström, associate professor in astronomy and later on professor in physics, was the first spectroscopist to refer his wavelength measurements to the metric system.

It has been possible for us to arrange rooms for you all at the same place, the Waldenstrom Student Hotel. We hope that you will find these student rooms comfortable, and the practical arrangements convenient. If you have any problems that you want to discuss, or any complaints to make, please do not hesitate to talk to me or to any other member of the local staff. We are all recognized by red name badges. Thank you.

ERIK HOLMBERG

TABLE OF CONTENTS

EDITORIAL NOTE	V
THE ORGANIZING COMMITTEES	VI
INTRODUCTORY REMARKS: E. HOLMBERG	VII
LIST OF PARTICIPANTS	XIII

1. S. VAN DEN BERGH / Stellar Populations in Galaxies *(Survey Lecture)* — 1
2. M. S. ROBERTS / The Gaseous Content of Galaxies *(Survey Lecture)* — 12
3. J. EINASTO / Structural and Kinematic Properties of Populations of the Andromeda Galaxy — 37
4. P. W. HODGE / The Population I Content of the Elliptical Companions of M31 — 46
5. M. F. WALKER / Electronographic Photometry of Stars in the Globular Clusters of the Magellanic Clouds (Abstract) — 48
6. V. C. RUBIN and W. K. FORD, JR. / Gas in the Nucleus and Disk of M31 — 49
7. B. T. LYNDS / Distribution of Dust and H II Regions of Spiral Galaxies — 56
8. I. PRONIK and K. CHUVAEV / Spiral Arm Patches in Sc and SBc Galaxies — 62
9. L. SEARLE / Observations of H II Regions in Sc Galaxies (Abstract) — 66
10. R. D. DAVIES / The Neutral Hydrogen Distribution in Spiral and Irregular Galaxies — 67
11. M. GUÉLIN and L. WELIACHEW / Absorption by Neutral Hydrogen in the Irregular Galaxy M82 (Abstract) — 74
12. Y. TERZIAN / Thermal Radio Emission from Normal Galaxies — 75
13. P. NOTNI, H. OLEAK, and G.-M. RICHTER / Radio-Emission from Supernovae Remnants in Distant Galaxies — 82
14. I. R. KING and R. MINKOWSKI / Mass-Luminosity Ratios and Sizes of Giant Elliptical Galaxies — 87
15. R. J. DICKENS and J. V. PEACH / On the Mass-to-Light Ratios for Double Galaxies — 89
16. W. C. SASLAW / Do Galaxies Evolve along the $R-\omega$ Sequence? — 93
17. W. W. MORGAN / Classification of Compact Objects: QSS, QSOs, N-type and Compact Galaxies, Seyfert and Galactic Nuclei *(Survey Lecture)* — 97
18. L. RICHTER, N. B. RICHTER, and P. SCHNELLER / Astrophysical Statistics of 745 Compact Galaxies near the Galactic North Pole — 104
19. E. M. BURBIDGE / Optical Spectra of Compact Objects *(Survey Lecture)* — 109
20. C. R. LYNDS / The Absorption-Line QSOs — 127
21. J. B. OKE / Spectral Energy Distributions of Nuclei of Peculiar Galaxies *(Survey Lecture)* — 139

22. J. D. SCARGLE, L. J. CAROFF, and P. D. NOERDLINGER / The Physcial Properties of the Absorption Envelopes of Two QSOs — 151
23. R. WEYMANN and R. CROMWELL / The Balmer lines in the Seyfert Galaxies NGC 5548 and NGC 4151 — 155
24. E. Y. KHACHIKIAN / Physical Conditions in Seyfert Type Galaxies Markarian 9, 10 and 42 — 160
25. G. NEUGEBAUER / Infrared Radiation from Compact Objects (Abstract) (Survey Lecture) — 163
26. T. D. KINMAN / The Relation between the Optical and Centrimetric Polarized Emission in BL Lac and other QSOs (Abstract) — 164
27. A. G. PACHOLCZYK / Activity in the Nuclei of Seyfert Galaxies in the Visual and Infrared — 165
28. R. J. ANGIONE and H. J. SMITH / Optical Variability of Twenty-Two Quasi-Stellar Objects — 171
29. J. TERRELL and K. H. OLSEN / Can the Optical Fluctuations of 3C 273 be Random? — 179
30. D. ALLOIN, Y. ANDRILLAT, and S. SOUFFRIN / A Study of the Continua of the Nuclei of Galaxies — 188
31. K. I. KELLERMANN / Radio Emission from Compact Objects (Survey Lecture) — 190
32. H. P. PALMER / The Angular Structures of Some Compact Sources — 214
33. G. K. MILEY and G. H. MACDONALD / The Radio Structure of Quasars — 216
34. R. D. EKERS / Compact Radio Sources in the Nuclei of Elliptical Galaxies — 222
35. G. D. NICOLSON / Intensity Variations in Extragalactic Radio Sources at 13 cm (Abstract) — 224
36. V. A. EFANOV, I. G. MOISEEV, H. M. TOVMASJAN, V. B. SHTEINSHLEGER, and V. I. ZAGATIN / Observations of Variable Radio Sources at 8.2 mm — 225
37. E. E. EPSTEIN / Variable Radio Sources: Comparison of Observations with the Adiabatic Spherical Expansion Source Model (Abstract) — 227
38. A. T. MOFFET, J. GUBBAY, D. S. ROBERTSON, and A. J. LEGG / High-Resolution Observations of Variable Radio Sources — 228
39. R. G. CONWAY / Polarization of Quasars — 230
40. B. J. HARRIS / QSOs and Radio Galaxies – Their Spectra and Time Variations at Radio Frequencies — 232
41. M. S. LONGAIR / M87 and the X-Ray Emission from Compact Objects — 249
42. G. WLÉRICK, G. LELIÈVRE, et P. VÉRON / Identification Optique et Photométrie de Radiosources du Catalogue 3 CR — 251
43. G. G. POOLEY / Optical Identifications of Compact Radio Sources (Abstract) — 258
44. W. A. DENT / Attempts to Detect Neutral Hydrogen in Compact Objects — 259
45. J. HEIDMANN / Neutral Hydrogen in Compact Galaxies — 264
46. B. M. LEWIS / Seyfert Galaxies — 267
47. R. J. ALLEN, B. F. DARCHY, and R. LAUQUÉ / Radio Observations of Neutral Hydrogen in Four Seyfert Galaxies (Abstract) — 269

48. M. H. COHEN / 21-cm Absorption in BL Lac (Abstract) — 271
49. I. S. SHKLOVSKY / The Law of Momentum Conservation and Some Problems of Metagalactic Astronomy — 272
50. L. WOLTJER / Theoretical Considerations of Compact Objects *(Survey Lecture)* — 277
51. W. H. MCCREA / The Large-Scale Variations of Quasi-Stellar Objects — 283
52. D. F. FALLA and A. EVANS / A 'Single Electron' Synchrotron Radiation Model and the Quasi-Stellar Objects — 285
53. L. M. OZERNOY / A Probable Mechanism of Repeated Explosions of Compact Objects — 290
54. P. KAFKA / On the Evolution of Quasars and their Remnants — 296
55. R. V. WAGONER and E. E. SALPETER / Supermassive Disks — 300
56. A. ELVIUS / A Matter-Antimatter Model for Quasi-Stellar Objects — 306
57. G. STEIGMAN and P. A. STRITTMATTER / Matter-Antimatter Annihilation as an Energy Source in Seyfert Galaxies (Abstract) — 311
58. J. L. SĚRSIC / Transient Annular Structures in Exploding Galaxies (Abstract) — 313
59. J. V. PEACH / Cosmological Information from Galaxies and Radio Galaxies *(Survey Lecture)* — 314
60. G. O. ABELL / Problems Concerning the Extragalactic Distance Scale — 341
61. G. DE VAUCOULEURS / The Velocity-Distance Relation and the Hubble Constant for Nearby Groups of Galaxies — 353
62. W. G. TIFFT / Rapid Evolution of Galactic Nuclei — 367
63. C. R. LYNDS / The Radial Velocities of Galaxies near NGC 7331 — 376
64. H. C. ARP / Ejection of Small Compact Galaxies from Larger Galaxies — 380
65. W. A. BAUM / The Diameter-Redshift Relation — 393
66. J.-E. SOLHEIM and B. M. TINSLEY / Analysis of the Magnitude-Redshift Relations Including Possible Effects of Evolution — 397
67. J.-E. SOLHEIM / A Combined Cosmological Test — 401
68. J. PACHNER / Notes on Cosmology — 404
69. M. J. REES / Cosmological Evidence from QSOs and Radio Galaxies *(Survey Lecture)* — 407
70. A. H. BRIDLE and M. M. DAVIS / The $N(S)$ Relationship at 1400 MHz — 437
71. I. I. K. PAULINY-TOTH, K. I. KELLERMANN, and M. M. DAVIS / Number Counts and Spectral Distribution of Radio Sources at Centimeter Wavelengths — 444
72. A. BRACCESI / Statistical Properties of QSOs — 453
73. M. ROWAN-ROBINSON / Radio Source-Counts and Cosmology — 458
74. V. PETROSIAN / Redshift Distribution of Quasi-Stellar Objects and the Radio Source Counts — 464
75. M. S. LONGAIR / The Luminosity-Volume Test for Quasi-Stellar Objects (Abstract) — 470
76. R. C. ROEDER / Selection of Cosmological Models Using QSOs — 471

77. J. SILK / Concerning the Primordial Abundance of Helium in Quasi-Stellar Sources 474
78. J. M. BARNOTHY and M. F. BARNOTHY / Ghost Images in a Universe with Inhomogeneous Mass Distribution 478
79. G. R. BURBIDGE / Intergalactic Matter and Radiation *(Survey Lecture)* 492
80. E. K. CONKLIN / Observations of Large-Scale Anisotropy in the 3 K Background Radiation (Abstract) 518
81. K. BRECHER / Galactic and Metagalactic Magnetic Fields 520
82. F. HOYLE / General Review of Cosmological Theories *(Survey Lecture)* 526

INDEX OF NAMES 532

INDEX OF SUBJECTS 541

LIST OF PARTICIPANTS

G. O. Abell, Los Angeles, Calif., U.S.A.
H. O. G. Alfvén, Stockholm, Sweden
R. J. Allen, Groningen, Netherlands
M. A. Arakeljan, Bjurakan, Armenia, U.S.S.R.
H. C. Arp, Pasadena, Cal., U.S.A.
J. Bahcall, Princeton, N.J., U.S.A.
C. B. Barbieri, Padua, Italy
J. M. Barnothy, Evanston, Ill., U.S.A.
M. F. Barnothy, Evanston, Ill., U.S.A.
W. A. Baum, Flagstaff, Ariz., U.S.A.
L. P. Bautz, Evanston, Ill., U.S.A.
J. Bergeron, Paris, France
F. Bertola, Padua, Italy
A. Braccesi, Bologna, Italy
K. Brecher, La Jolla, Calif., U.S.A.
A. H. Bridle, Kingston, Ont., Canada
E. M. Burbidge, La Jolla, Calif., U.S.A.
G. R. Burbidge, La Jolla, Calif., U.S.A.
G. J. Carranza, Cordoba, Argentina
A. Cavaliere, Cambridge, Mass., U.S.A.
M. H. Cohen, Pasadena, Calif., U.S.A.
P. Collinder, Uppsala, Sweden
E. K. Conklin, Tucson, Ariz., U.S.A.
R. G. Conway, Jodrell Bank, Great Britain
N. Dallaporta, Padua, Italy
R. D. Davies, Jodrell Bank, Great Britain
M. M. Davis, Green Bank, W. Va., U.S.A.
M. H. J. Demoulin, Paris, France
W. A. Dent, Amherst, Mass., U.S.A.
A. de Vaucouleurs, Austin, Texas, U.S.A.
G. de Vaucouleurs, Austin, Texas, U.S.A.
R. J. Dickens, Herstmonceux, Great Britain
S. D'Odorico, Asiago, Italy
R. D. Ekers, Pasadena, Calif., U.S.A.
A. M. Elvius, Stockholm, Sweden
T. Elvius, Lund, Sweden
E. E. Epstein, Los Angeles, Calif., U.S.A.

D. S. Evans, Austin, Texas, U.S.A.
D. F. Falla, Aberystwyth, Great Britain
J. E. Felten, Cambridge, Great Britain
S. C. B. Gascoigne, Canberra, Australia
O. Godart, Louvain, Belgium
M. Grewing, Bonn, W. Germany
K. Gyldenkerne, Copenhagen, Denmark
B. J. Harris, Columbus, Ohio, U.S.A.
C. Hazard, Cambridge, Great Britain
O. Heckmann, Hamburg, W. Germany
D. S. Heeschen, Charlottesville, Va., U.S.A.
J. Heidmann, Meudon, France
R. W. Hobbs, Greenbelt, Md., U.S.A.
P. W. Hodge, Seattle, Wash., U.S.A.
D. E. Hogg, Charlottesville, Va., U.S.A.
E. Holmberg, Uppsala, Sweden
F. Hoyle, Cambridge, Great Britain
J. A. Högbom, Westerbork, Netherlands
O. N. Izakova, Moscow, U.S.S.R.
J. C. Jackson, Falmer, Sussex, Great Britain
M. N. Joshi, Bombay, India
P. Kafka, München, W. Germany
K. I. Kellermann, Green Bank, W. Va. U.S.A.
S. Kenderdine, Cambridge, Great Britain
E. E. Khachikian, Bjurakan, Armenia, U.S.S.R.
I. R. King, Berkeley, Calif., U.S.A.
T. D. Kinman, Tucson, Ariz., U.S.A.
M. M. Komesaroff, Sydney, Australia
B. Z. Kozlovsky, Tel Aviv, Israel
T. Kuznetsova, U.S.S.R.
G. Larsson-Leander, Lund, Sweden
B. M. Lasker, Tucson, Ariz., U.S.A.
J. Lequeux, Meudon, France
B. M. Lewis, Jodrell Bank, Great Britain
C. C. Lin, Cambridge, Mass., U.S.A.
U. Lindoff, Lund, Sweden
J. L. Locke, Ottawa, Ontario, Canada
L. O. Lodén, Uppsala, Sweden
M. S. Longair, Cambridge, Great Britain
B. T. Lynds, Tucson, Ariz., U.S.A.
C. R. Lynds, Tucson, Ariz., U.S.A.
C. D. Mackay, Cambridge, Great Britain
W. H. McCrea, Falmer, Sussex, Great Britain

G. C. McVittie, Urbana, Ill., U.S.A.
T. K. Menon, Honolulu, Hawaii, U.S.A.
J. K. Merkelijn, Leiden, Netherlands
G. K. Miley, Charlottesville, Va., U.S.A.
R. Minkowski, Berkeley, Calif., U.S.A.
A. T. Moffet, Pasadena, Calif., U.S.A.
G. M. Monnet, Marseille, France
W. W. Morgan, Willams Bay, Wisc., U.S.A.
G. F. Moseley, Austin, Texas, U.S.A.
J. V. Narlikar, Cambridge, Great Britain
Y. Ne'eman, Tel Aviv, Israel
G. Neugebauer, Pasadena, Calif., U.S.A.
J. Neyman, Berkeley, Calif., U.S.A.
G. D. Nicolson, Johannesburg, South Africa
P. Noerdlinger, Socorro, N.M., U.S.A.
P. Notni, Babelsberg, D.D.R.
T. Oja, Uppsala, Sweden
J. B. Oke, Pasadena, Calif., U.S.A.
H. Oleak, Babelsberg, D.D.R.
J. H. Oort, Leiden, Netherlands
L. M. Ozernoy, Moscow, U.S.S.R.
J. Pachner, Regina, Sask., Canada
A. G. Pacholczyk, Tucson, Ariz., U.S.A.
F. Pacini, Ithaca, N.Y., U.S.A.
H. P. Palmer, Jodrell Bank, Great Britain
M. G. Pastoriza, Cordoba, Argentina
I. I. K. Pauliny-Toth, Bonn, W. Germany
J. V. Peach, Oxford, Great Britain
P. J. E. Peebles, Princeton, N.J., U.S.A.
M. Peimbert, Mexico, D. F., Mexico
V. Petrosian, Stanford, Calif., U.S.A.
P. Pishmish, Mexico, D.F., Mexico
T. Pjatunina, U.S.S.R.
G. G. Pooley, Cambridge, Great Britain
K. H. Prendergast, New York, N.Y., U.S.A.
V. I. Pronik, Crimea, U.S.S.R.
E. Raimond, Westerbork, Netherlands
M. J. Rees, Cambridge, Great Britain
N. B. Richter, Tautenburg, D.D.R.
M. S. Roberts, Charlottesville, Va., U.S.A.
R. C. Roeder, Toronto, Ontario, Canada
D. H. Rogstad, Groningen, Netherlands
M. Rowan-Robinson, London, Great Britain

V. C. Rubin, Washington, D. C., U.S.A.
E. E. Salpeter, Ithaca, N.Y., U.S.A.
W. C. Saslaw, Berkeley, Calif., U.S.A.
P. A. G. Scheuer, Cambridge, Great Britain
M. Schmidt, Pasadena, Calif., U.S.A.
D. W. Sciama, Cambridge, Great Britain
E. L. Scott, Berkeley, Calif., U.S.A.
L. Searle, Pasadena, Calif., U.S.A.
G. A. Seielstad, Pasadena, Calif., U.S.A.
J. L. Sěrsic, Cordoba, Argentina
G. Setti, Bologna, Italy
J. R. Shakeshaft, Cambridge, Great Britain
J. I. Silk, Princeton, N.J., U.S.A.
M. Simon, Stony Brook, N.Y., U.S.A.
U. Sinnerstad, Stockholm, Sweden
H. J. Smith, Austin, Texas, U.S.A.
J.-E. Solheim, Oslo, Norway
S. Souffrin, Paris, France
R. Stabell, Oslo, Norway
G. Stanley, Pasadena, Calif., U.S.A.
G. Steigman, Cambridge, Great Britain
J. M. Sutton, Cambridge, Great Britain
B. Svcnonius, Gothenburg, Sweden
G. Swarup, Bombay, India
J. Terrell, Los Alamos, N.M., U.S.A.
Y. Terzian, Ithaca, N.Y., U.S.A.
W. G. Tifft, Tucson, Ariz., U.S.A.
G. M. Tovmasjan, Bjurakan, Armenia, U.S.S.R.
A. J. Turtle, Sydney, Australia
S. van den Bergh, Richmond Hill, Ontario, Canada
H. van der Laan, Leiden, Netherlands
H. van Woerden, Groningen, Netherlands
P. Véron, Meudon, France
N. Visvanathan, Cambridge, Mass., U.S.A.
R. V. Wagoner, Ithaca, N.Y., U.S.A.
M. F. Walker, Santa Cruz, Calif., U.S.A.
Å. Wallenquist, Uppsala, Sweden
B. L. Webster, Herstmonceux, Great Britain
L. N. Weliachew, Pasadena, Calif., U.S.A.
B. E. Westerlund, Santiago, Chile
R. Weymann, Tucson, Ariz., U.S.A.
A. E. Whitford, Santa Cruz, Calif., U.S.A.
D. Wills, Austin, Texas, U.S.A.

G. Wlérick, Meudon, France
A. M. Wolfe, La Jolla, Calif., U.S.A.
L. Woltjer, New York, N.Y., U.S.A.
J. D. Wray, Evanston, Ill., U.S.A.
J. P. Wright, Washington, D.C., U.S.A.

STELLAR POPULATIONS IN GALAXIES

SIDNEY VAN DEN BERGH
David Dunlap Observatory, Toronto, Ontario, Canada

Abstract. There are significant differences among old stellar systems which may partly be a function of luminosity. One such difference is the presence of nuclei in many high luminosity spiral galaxies but their absence in low luminosity irregular systems. Other differences are seen in the spectroscopic and photometric properties.

1. Nuclei of Galaxies

At first sight the data on old stellar systems suggest that all such objects have very similar morphological, spectroscopic and photometric characteristics. On closer inspection a number of subtle differences become apparent. The major morphological difference between dwarf spheroidal systems on the one hand and giant ellipticals on the other is that giants have nuclei and dwarfs do not. It should, however, be emphasized that not all giant ellipticals have *semi-stellar* nuclei (King and Minkowski, 1966). The dividing line between galaxies with and without nuclei appears to occur at $M_V \simeq -15$. Among the dwarf companions of the Andromeda Nebula NGC 185 ($M_V = -15.2$) has no nucleus, NGC 147 ($M_V = -14.9$) has a faint nucleus and NGC 221 = M32 ($M_V = -16.4$) has a well-developed bright semi-stellar nucleus. The data listed in Table I show that this relation between absolute luminosity and the existence of a nucleus is not confined to elliptical galaxies. The table shows that late-type galaxies of high luminosity are predominantly spirals (which have nuclei) whereas most late-type objects of low luminosity are irregulars which, of course, do not have nuclei.

TABLE I

Relative frequency of spiral and irregular galaxies at different magnitude levels (Van den Bergh, 1966)

$\langle M_B \rangle$ [a]	Spirals [b]	Irregulars	$N(\text{Ir})/N(\text{Sp})$
−20.2	56	0	0.00
−19.4	112	1	0.01
−18.2	66	4	0.06
−17.3:	26	5	0.19
−15::	22	64	2.91

[a] $H = 100$ km s^{-1} Mpc^{-1} assumed.
[b] Barred spirals omitted.

Observations by de Vaucouleurs (1961) and by Tifft (1969) show that the cores of giant E galaxies are usually redder than their envelopes. McClure's (1969) observations of the intermediate-band cyanogen index $C(41-42)$ show that this colour gradient must be due mainly to a change in stellar population rather than to a concentration of dust in the nuclei of E galaxies. McClure's observations of the variation of cyanogen strength within the central regions of M31 and M81 are plotted in Figure 1. These

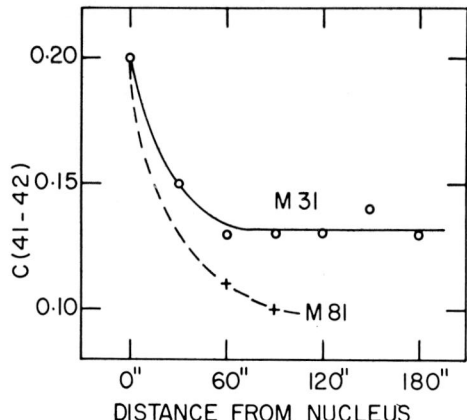

Fig. 1. Variation of cyanogen strength as a function of distance from the nuclei of M31 and M81.

results show that cyanogen is strongly enhanced within a radius of 60" (200 pc) of the nucleus of M31. Possibly this cyanogen enhancement is in some way related to the steep radial abundance gradient that Peimbert (1968) finds for the nitrogen in the interstellar gas in spiral galaxies. Alternatively it might be assumed that the giant stars in the nuclei of galaxies are somehow related to such nearby cyanogen stars as μ Leonis and α Serpentis (Spinrad, 1966).

2. Metal Abundance and Galaxy Luminosity

Baum (1959) and de Vaucouleurs (1961) have pointed out that the integrated colours of old stellar systems are strongly correlated with their intrinsic luminosities. This effect is illustrated in Figure 2 which shows a plot (McClure and van den Bergh, 1968b) of the reddening-free parameter $Q = U - B - 0.72 (B - V)$ vs brightness for E and S0 galaxies in the Virgo Cluster. The figure shows that the most luminous E galaxies are much redder than are fainter ellipticals. The relatively tight correlation between Q and the luminosity of galaxies provides a powerful new tool for the study of the isotropy of the Hubble flow for nearby galaxies.

Available observations (see Table II) show that galaxies of the same absolute magnitude have similar Q values in clusters and in the general field. This result may be interpreted in two different ways: Either (1) the evolutionary history of an E galaxy is independent of the environment in which it is formed i.e. a galaxy is essentially insulated from its neighbours once star formation and heavy element enrichment begin or (2) field E galaxies have escaped from clusters. In connection with the latter

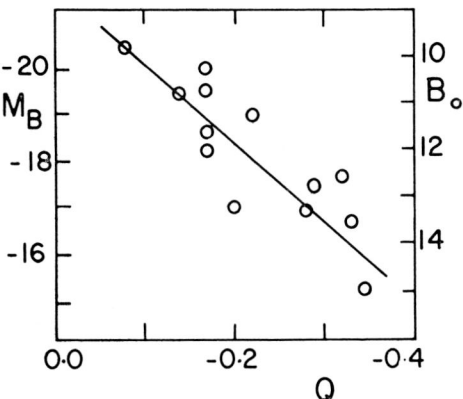

Fig. 2. Plot of the reddening-free parameter Q vs luminosity for E and S0 galaxies in the Virgo Cluster.

TABLE II

Colours of cluster and of non-cluster E galaxies
(McClure and Van den Bergh, 1968b)

Environment	$\langle Q \rangle$[a]	n_{gal}
In rich clusters	-0.16 ± 0.01	23
In poor clusters	-0.18 ± 0.01	10
Field galaxies	-0.18 ± 0.01	27

[a] All data reduced to $\langle M_B \rangle = -19.6$.

possibility it should be emphasized (Van den Bergh, 1962a) that 76% of all E galaxies are members of clusters. This observation implies that most clusters containing elliptical galaxies must be stable over time-scales comparable to the lifetime of the Universe.

A. DWARF SPHEROIDAL GALAXIES

Baum (1959) has pointed out that the integrated colours of dwarf ellipticals and dwarf spheroidal galaxies with $-15 < M_V < -8$ resemble those of metal-poor halo globular clusters. This observation, in conjunction with available colour-magnitude diagrams for the Draco (Baade and Swope, 1961), Sculptor (Hodge, 1965) and Ursa Minor (Van Agt, 1967) systems makes it virtually certain that the faintest dwarf systems are exceedingly metal-poor. This low metal abundance in dwarf spheroidal systems can probably be understood in terms of the low escape velocity from such objects. Presumably this low escape velocity permitted most of the interstellar matter to escape before a second generation of stars could be formed from enriched interstellar gas.

It has been known for some time (Arp, 1955; Sandage and Wallerstein, 1960) that the red giants in metal-poor galactic globular clusters are brighter than are those in metal-rich clusters. A similar correlation between metal abundance and red giant luminosity appears to exist in the globular clusters associated with the Andromeda Nebula (Van den Bergh, 1969). This suggests that a significant spread in the metal abundance of stars in dwarf spheroidal galaxies would manifest itself as a luminosity dispersion in the red giant branch of such a system. Such a dispersion is not seen in Baade and Swope's (1961) observations of the Draco System ($M_V = -8.6$). Spectroscopic observations (Van den Bergh, 1969) do, however, show evidence for abundance differences among the globular clusters associated with the Fornax dwarf system ($M_V = -13.6$). This suggests that 'second generation' stars can be formed in systems with masses larger than approximately $1 \times 10^6 \mathcal{M}_\odot$. For elliptical galaxies with masses larger than $10^9 \mathcal{M}_\odot$ the contribution of 'first generation' stars to the total integrated light is swamped by subsequent generations of metal-rich stars.

B. ELLIPTICAL GALAXIES OF INTERMEDIATE LUMINOSITY

The elliptical galaxy M32 = NGC 221, for which $M_V = -16.4$, is the best observed early-type galaxy of intermediate luminosity. McClure and Van den Bergh (1968a) find that wide-band UBV photometry and intermediate-band photometry by McClure and by Wood (1966) are well represented by a population model consisting of a dwarf-enriched intermediate abundance globular cluster model. Their model, which is compared with observations in Table III, contains the following contributions to V light: globular cluster 47.5%; G8V 25%; K2V 15%; K6V 8% and M1V 4.5%.

TABLE III

Comparison of the observations of M32 with a dwarf-enriched globular cluster model
(McClure and Van den Bergh, 1968a)

Colour index	Observed[a]	O–C
$C(35-38)$	$-0^m.46 \pm 0.03$	$-0^m.01$
$C(38-41)$	$+0.73 \pm 0.02$	$+0.02$
$C(41-42)$	$+0.08 \pm 0.03$	0.00
$C(42-45)$	$+0.78 \pm 0.03$	$+0.01$
C_{35}	$+2.21 \pm 0.06$	0.00
C_{41}	$+1.09 \pm 0.02$	-0.01
C_{47}	$+0.47 \pm 0.02$	$+0.02$
C_{55}	$+0.04 \pm 0.00$	0.00
C_{60}	-0.16 ± 0.01	-0.01
C_{67}	-0.34 ± 0.03	0.00
C_{73}	-0.54 ± 0.22	$+0.09$
L_{52}	$+0.07 \pm 0.01$	-0.02
L_{59}	-0.10 ± 0.02	-0.04
L_{62}	$+0.06 \pm 0.01$	-0.02
L_{66}	-0.05 ± 0.01	-0.01
L_{71}	-0.04 ± 0.05	0.00
$U-B$	$+0.45$	$+0.06$
$B-V$	$+0.83$	0.00

[a] Colours corrected for a reddening of $E_{B-V} = 0.10$.

In the model observations of the intermediate-abundance globular cluster NGC 6356 (Morgan Class VI, $Q = -0.24$, $Sp = G5$) were used to provide the globular cluster contribution to the integrated light of M32. It should be emphasized that *a satisfactory fit to the observations cannot be obtained by adding main sequence stars to a metal-poor globular cluster base*. Early attempts to synthesize the stellar content of M32 by adding red dwarfs to a metal-poor globular cluster base (Roberts, 1956; Baum, 1959) have previously been criticized by Spinrad (1962) because elliptical galaxies have stronger metal lines and weaker hydrogen lines than do metal-poor globular clusters.

From spectrum scans with the Lick 120-in. telescope Spinrad (1970) has obtained evidence for a small radial composition gradient within M32. Spinrad also finds the nuclear region of this small galaxy to be somewhat redder and stronger lined than its outer regions.

C. GIANT ELLIPTICAL GALAXIES

The galaxy M86 = NGC 4406 in the Virgo Cluster is a good example of a giant elliptical. For $H = 100$ km s^{-1} Mpc^{-1} M86 has an absolute magnitude $M_V = -21.0$. The 'redshift' of M86 is -292 km s^{-1} so that no K corrections need to be applied to the observed colours. Observations show that M86 is typical of giant ellipticals in all five intermediate-band colours of the DDO system. It has also been observed on the UBV system and on Wood's 12-colour system.

Two population models that fit the observations reasonably well are given in Table IV. The colours obtained from these models are compared with observations in Table V. The only significant discrepancy between the observed and the computed colours occurs for the DDO intermediate-band index $C(41-42)$, which measures the strength of the cyanogen band at $\lambda\,4216$. A similar discrepancy is also found for the nuclear region of M31. The only way to remove this discrepancy in M86 (and in M31) is to replace the normal K giants in the model by 'cyanogen rich' stars similar to μ Leonis and α Serpentis.

TABLE IV

Population models for the giant elliptical M86
(McClure and Van den Bergh, 1968a)

Star type	Percent contribution to V light	
	Model A	Model B
F5V	2	0
G0V	10.5	10.5
G8V	13.5	13.5
K2V	17	17
K6V	12.5	12.5
M1V	3	3
M5V	3	3
G8IV	3	3
K0III	33	33
K5III	2.5	2.5
metal-poor globular cluster	0	2

The two percent contribution of F5V stars to the total visual light of model A was found to be necessary to obtain a good fit to the continuum colours. Such stars are, however, somewhat earlier in spectral type than the main sequence cut-off expected for a metal-rich stellar population with an age $\sim 1 \times 10^{10}$ yr. In model B the F5V stars have been replaced by a small contribution from a metal-poor globular cluster type

TABLE V

Comparison of population models of M86 with observations
(McClure and Van den Bergh, 1968a)

Colour index	Observed colour[a]	Model A O–C	Model B O–C
$C(35-38)$	-0.49 ± 0.04	-0.02	-0.03
$C(38-41)$	$+0.80 \pm 0.04$	-0.01	-0.01
$C(41-42)$	$+0.14 \pm 0.02$	$+0.06$	$+0.06$
$C(42-45)$	$+0.85 \pm 0.00$	$+0.01$	$+0.01$
C_{35}	$+2.49 \pm 0.03$	$+0.01$	0.00
C_{41}	$+1.24 \pm 0.02$	$+0.01$	0.00
C_{47}	$+0.48 \pm 0.01$	-0.03	-0.03
C_{55}	$+0.05 \pm 0.01$	0.00	$+0.01$
C_{60}	-0.18 ± 0.01	0.00	-0.01
C_{67}	-0.37 ± 0.02	$+0.01$	$+0.01$
C_{73}	-0.70 ± 0.06	$+0.03$	$+0.04$
L_{52}	$+0.10 \pm 0.01$	$+0.01$	$+0.01$
L_{59}	-0.05 ± 0.01	$+0.01$	$+0.02$
L_{62}	$+0.07 \pm 0.01$	-0.01	-0.01
L_{66}	-0.07 ± 0.02	-0.01	0.00
L_{71}	0.00 ± 0.06	$+0.01$	$+0.01$
$U-B$	$+0.54$	0.00	-0.02
$B-V$	$+0.88$	-0.05	-0.07

[a] Observed colours corrected for $E_{B-V} = 0.06$.

population. The data in Table V show that model B, in which the globular cluster population gives a two percent contribution to the total light, gives an acceptable fit to the observations. It should be emphasized that *the observed and the computed* $C(35-38)$, $C(38-41)$ *and* $U-B$ *colours can no longer be brought into agreement if more than two percent of the visual light in M86 is contributed by a metal-poor globular cluster population.* In an independent investigation Spinrad (1970) finds that population models with a few late F stars give a somewhat better fit to his spectrum scans than do models with a small metal-poor globular cluster contribution. Perhaps this indicates that the halo population (including globular clusters) is moderately metal-rich in giant elliptical galaxies. A crucial test of this hypothesis will be provided by photometry of globular clusters associated with giant elliptical galaxies. Such a test is currently being undertaken by Racine and Hanes, who are using 200-in. plates to do photographic photometry of the globular clusters associated with M87.

3. The Mass-to-Light Ratio in Elliptical Galaxies

The mass-luminosity relation for dust-poor stellar systems has been discussed by Poveda (1961). His investigation shows that there is a progressive increase in the mass-to-light ratio from globular clusters through dwarf ellipticals to giant E galaxies. The models for M32 and for M86, that were discussed in the previous section, are consistent with Poveda's results. Both models for M86 give $\mathcal{M}/L_B \simeq 13$, which is significantly higher than the value $\mathcal{M}/L_B \simeq 5$ that is obtained for our model of M32. It should, however, be emphasized that $\mathcal{M}/L_B$ could be increased significantly in all models by adding large numbers of M8V stars which do not make a significant contribution to the total V luminosity of a galaxy.

4. Stellar Populations in the Galactic Disk

Observations of the frequency with which main sequence stars of different metal abundance occur near the Sun (Van den Bergh, 1962b) show that the rate of heavy element enrichment in the Galaxy has decreased much faster than the rate of star formation. This observation suggests that the stellar birthrate function initially contained more massive stars (which rapidly become supernovae) than does the present luminosity function of star formation. Rather similar arguments have also been used by Schmidt (1963) to show that the birthrate of massive stars has decreased faster than has the birthrate of low-mass stars. It is not yet clear how this conclusion is to be reconciled with observations of the stellar luminosity function near the Sun.

Comparison of the mean luminosity function of star clusters (Van den Bergh, 1961)

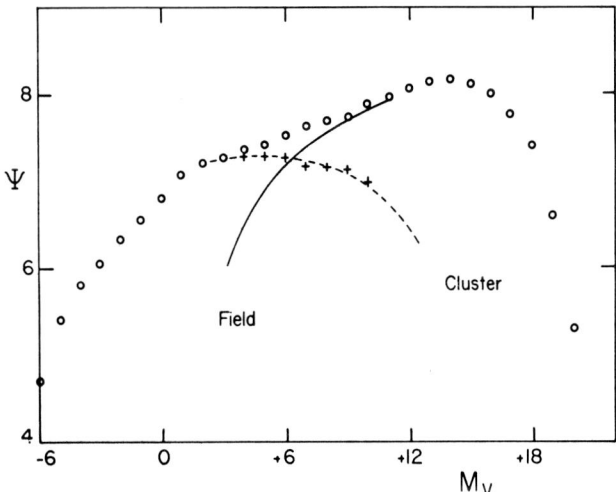

Fig. 3. This figure suggests that the stellar birthrate function near the Sun might consist of two distinct components.

with the stellar birthrate function $\psi(M_V)$ (Limber, 1960) shows that open clusters contain fewer faint stars than does the luminosity function of field stars near the Sun. This discrepancy appears significant even though rather large differences exist between the luminosity functions of different clusters. This observation shows that the galactic disk cannot have been populated entirely by disintegrating open clusters. In fact Figure 3 suggests that the stars in the galactic disk near the Sun may constitute a mixture of two components: (1) a population derived from the disintegration of open clusters and (2) a 'field star' population. This result might be explained by the *ad hoc* hypothesis that the luminosity function of star formation in zones of positive energy (associations) contains more faint stars than it does in regions of negative energy (clusters). Alternatively it might be assumed that the 'field stars' represent a very old constituent of the galactic disk population. In the latter case the 'field star' component of the galactic disk might be related to the faint star component that contributes most of the mass in giant ellipticals.

The hypothesis that 'field stars' constitute an old component of the galactic disk leads to an interesting paradox: *to explain the observed metal abundance of stars in the galactic disk it is necessary to postulate that the stellar birthrate function initially contained more massive stars than it does now. On the other hand the luminosity function of disk stars suggests that the stellar birthrate function initially contained more low-mass stars than it does now.* Perhaps the early history of the Galaxy was much more complex than the situation envisaged in currently fashionable models for galactic evolution.

5. Differences Among Old Metal-Poor Populations

Detailed observations of individual stars of Population II are only possible in the Galaxy, in the Magellanic Clouds and in nearby dwarf spheroidal systems. The rather

TABLE VI

Colour-magnitude diagrams of globular clusters in the Magellanic Clouds (Van den Bergh, 1968)

Cluster	Horizontal branch*	B − V colour at tip of giant branch
SMC NGC 121	R	1.9
NGC 339	R	2.2
NGC 361	–	2.1
NGC 419	–	2.7
L1	R	2.4
L8 = K3	R	2.2
LMC NGC 1466	B	1.6
NGC 1783	R?	2.4
NGC 1841	B	1.7
NGC 1846	–	2.6
NGC 1978	–	2.1
NGC 2257	B	1.8

* R – most of stars to the red of variable star gap,
 B – most of stars on the blue side of this gap.

fragmentary data that are so far available hint at a number of significant differences between galactic and extra-galactic Population II stars. Perhaps the most striking differences that have so far been observed (see Table VI) are those between the globular clusters in the Magellanic Clouds and those in the Galaxy. All the globulars in the Small Cloud and about half of those that have so far been observed in the Large Cloud seem to contain a few exceedingly red giant stars with colour indices in the range $1.9 \leqslant B-V \leqslant 2.7$. Such very red stars are not known to occur in galactic globular clusters. It might be speculated that these very red stars in the Cloud clusters are mild carbon stars. If this is indeed the case they might possibly represent objects in the helium shell burning phase that have managed to mix carbon from their inert carbon-oxygen cores up to their surfaces. In his study of the Sculptor system Hodge (1965) found two very red giant stars that are possibly similar to those that occur in the Small Cloud clusters. This observation suggests that some dwarf spheroidal galaxies may contain a stellar population similar to that which is encountered in the Small Cloud globulars.

An additional argument for differences between Population II stars in the Galaxy and in dwarf spheroidal systems is provided by Van Agt's (1967) observations of BL Herculis stars in dwarf spheroidals. At a given period he finds these variables to be almost a magnitude brighter in dwarf spheroidals than they are in galactic globular clusters.

Tifft's (1963) observations of the $1^{d}\!.43$ variable in the Small Cloud cluster NGC 121 gives a distance modulus that is in excellent agreement with that obtained from classical Cepheids *if* it is placed on Van Agt's period luminosity relation for BL Herculis stars in dwarf spheroidal systems rather than on that for galactic globulars. This observation tends to strengthen the suspicion that the stars in the Cloud globulars are more closely related to those in dwarf spheroidal systems than they are to galactic Population II.

The properties of RR Lyrae variables provide additional evidence for differences between Population II stars in galactic globular clusters and in dwarf spheroidal galaxies. According to Van Agt and Oosterhoff (1959) the mean periods $\langle P \rangle_{ab}$ of RR Lyrae variables in galactic globulars cluster around $0^{d}\!.55$ and $0^{d}\!.65$. No such clustering is apparent for dwarf spheroidal systems. In fact two of the four dwarf spheroidals that have so far been studied have $\langle P \rangle_{ab} \simeq 0.60$, which places them halfway between Oosterhoff's long-period and short-period groups.

Dwarf spheroidal galaxies are exceedingly fragile objects. Because of their low space densities such objects can never have approached any of the more massive members of the Local Group very closely. It therefore seems improbable that they could have suffered significant contamination from nucleosynthesis in these massive galaxies. This suggests that individual dwarf spheroidal galaxies probably had a rather simple chemical evolution. It is therefore rather surprising that significant differences appear to exist between the colour-magnitude diagrams of different dwarf spheroidal galaxies. Van Agt's (1967) observations show that the Ursa Minor system has a blue horizontal branch. On the other hand the Draco system (Baade and Swope, 1961) and the Leo II system (Swope, 1967) exhibit red horizontal branches similar to those of the intergalactic globular clusters Pal. 3 and Pal. 4 (Burbidge and Sandage, 1958).

Available observations of the colour-magnitude diagrams of galactic globular clusters (White, 1970) show that these clusters (with the possible exception of NGC 362) have similarities (family traits?) that suggest a common ancestry. It is, however, quite obvious (Hartwick, 1968) that the galactic globular clusters do not form a one parameter sequence i.e. their 'family tree' has many branches.

It has been suggested (Van den Bergh, 1965, 1967; Sandage and Wildey, 1967) that the parameters which determine the characteristics of globular cluster colour-magnitude diagrams are helium abundance and metal abundance. An alternative suggestion by Rood and Iben (1968) is that the observed differences between globular clusters are due to differences in age and metal abundance. Rood and Iben argue that the age differences among galactic globular clusters might well exceed the 2×10^8 yr collapse time-scale envisioned by Eggen *et al.* (1962). It turns out, however, that rather extreme assumptions have to be made to obtain age differences as large as 1×10^9 yr. A third possibility, which deserves to be explored in more detail is that the observed differences among Population II objects of similar metal abundance are due to variations in the relative abundances of different species of heavy elements. Such differences might arise if different types of supernovae had differing spatial distributions.

Acknowledgements

I thank Drs. Robert McClure, Allan Sandage and Hyron Spinrad for a number of stimulating discussions.

References

Agt, S. L. T. J. van: 1967, *Bull. Astron. Inst. Neth.* **19**, 275.
Agt, S. L. T. J. van and Oosterhoff, P. T.: 1959, *Leiden Ann.* **21**, 253.
Arp, H. C.: 1955, *Astron. J.* **60**, 317.
Baade, W. and Swope, H. H.: 1961, *Astron. J.* **66**, 300.
Baum, W. A.: 1959, *Publ. Astron. Soc. Pacific* **71**, 106.
Bergh, S. van den: 1961, *Astrophys. J.* **134**, 553.
Bergh, S. van den: 1962a, *Z. Astrophys.* **55**, 21.
Bergh, S. van den: 1962b, *Astron. J.* **67**, 486.
Bergh, S. van den: 1965, *J. Roy. Astron. Soc. Can.* **59**, 151.
Bergh, S. van den: 1966, *Astron. J.* **71**, 922.
Bergh, S. van den: 1967, *Publ. Astron. Soc. Pacific* **79**, 460.
Bergh, S. van den: 1968, *J. Roy. Astron. Soc. Can.* **62**, 145.
Bergh, S. van den: 1969, *Astrophys. J. Suppl.* **19**, 145.
Burbidge, E. M. and Sandage, A. R.: 1958, *Astrophys. J.* **127**, 527.
Eggen, O. J., Lynden-Bell, D., and Sandage, A. R.: 1962, *Astrophys. J.* **136**, 748.
Hartwick, F. D. A.: 1968, *Astrophys. J.* **154**, 475.
Hodge, P. W.: 1965, *Astrophys. J.* **142**, 1390.
King, I. R. and Minkowski, R.: 1966, *Astrophys. J.* **143**, 1002.
Limber, D. N.: 1960, *Astrophys. J.* **131**, 168.
McClure, R. D.: 1969, *Astron. J.* **74**, 50.
McClure, R. D. and Bergh, S. van den: 1968a, *Astron. J.* **73**, 313.
McClure, R. D. and Bergh, S. van den: 1968b, *Astron. J.* **73**, 1008.
Peimbert, M.: 1968, *Astrophys. J.* **154**, 33.
Poveda, A.: 1961, *Astrophys. J.* **134**, 910.
Roberts, M. S.: 1956, *Astron. J.* **61**, 195.

Rood, R. and Iben, I.: 1968, *Astrophys. J.* **154**, 215.
Sandage, A. R. and Wallerstein, G.: 1960, *Astrophys. J.* **131**, 598.
Sandage, A. R. and Wildey, R. L.: 1967, *Astrophys. J.* **150**, 469.
Schmidt, M.: 1963, *Astrophys. J.* **137**, 758.
Spinrad, H.: 1962, *Astrophys. J.* **135**, 715.
Spinrad, H.: 1966, *Publ. Astron. Soc. Pacific* **78**, 367.
Spinrad, H.: 1970, private communication.
Swope, H. H.: 1967, *Publ. Astron. Soc. Pacific* **79**, 439.
Tifft, W. G.: 1963, *Monthly Notices Roy. Astron. Soc.* **125**, 199.
Tifft, W. G.: 1969, *Astron. J.* **74**, 354.
Vaucouleurs, G. de: 1961, *Astrophys. J. Suppl.* **5**, 233.
White, R. E.: 1970, *Astrophys. J. Suppl.* **19**, 343.
Wood, D. B.: 1966, *Astrophys. J.* **145**, 36.

Discussion

Lewis: Have you taken into account the possible explanation offered by 'collapsars' or 'black holes' to the high $\mathscr{M}/L$ ratios of E's etc.; while these need not contribute to the light they give the high $\mathscr{M}/L$ ratio without resort to the filling of E's with dwarf type stars. The existence of these past super-massive stars would provide a natural explanation of considerable abundances of metals from the past history.

Van den Bergh: No black holes are needed *inside* galaxies! Stellar population models that fit existing multicolour observations require mass-to-light ratios similar to those that King and Minkowski obtain from dynamic considerations.

King: The King-Minkowski paper will give evidence that there are not enough collapsars to have any appreciable gravitational effect. Regarding the correlation between Q and M_B, we do not find such a correlation between $\mathscr{M}/L$ and M_B for Virgo giants – we see only a difference between giants and dwarfs. We find the $\mathscr{M}/L$ of M86 to be the *least* extreme in $\mathscr{M}/L$ of all the Virgo giants that we have studied; the others have even *higher* $\mathscr{M}/L$.

Mrs. Rubin: What is the evidence that the strong CN or at least some part of it does not arise in interstellar gas in the galaxies? In some galaxies it appears that Na D-lines may be partly of interstellar origin.

Van den Bergh: I would guess that an unreasonably large amount of interstellar gas would be required to account for the observed cyanogen strength in the nuclei of galaxies.

Tifft: Small aperture four-colour uniform photometry of Virgo ellipticals shows a close colour index correlation with luminosity and supports the Q, M correlation presented by Van den Bergh.

Morgan: In connection with the measures of the CN intensity in giant ellipticals (NGC 4472, etc.) and in the nuclear region of Sb spirals (M31, etc.); have you corrected for the effects of non-CN absorptions of Ca I, Cr I, etc. from K4–early M giants?

Van den Bergh: Yes, our models (see Table IV) do include K5III stars. Furthermore, recent models of the nucleus of M31 presented by Spinrad at the Rome Symposium contain both K and M giants. These models show the same cyanogen deficiency exhibited by our models.

Mrs. Burbidge: Could you comment on the stellar population in globular clusters in M31? How does the high proportion of metal-rich clusters which you found tie in with the nuclear population?

Van den Bergh: Perhaps the average metal abundance of globular clusters will turn out to depend on the luminosity of their parent galaxy. The globulars in the Fornax dwarf system are *all* metal-poor. Most of the globular clusters associated with the Galaxy are metal-poor, but there is also a significant number of relatively metal-rich galactic globular clusters. Finally my recent photometric and spectroscopic observations of the globulars in M31 seem to show that these clusters are, on the whole, richer in metals than are those in the Galaxy. It would be interesting to know if the globular clusters associated with such giant E galaxies as M87 are metal-rich. Perhaps the colour observations by Racine and Hanes, to which I referred previously, will provide an answer to this question.

Ekers: Can you place limits on a possible non-stellar light contribution in the giant ellipticals? To what extent could this affect your analysis of stellar content?

Van den Bergh: Available observations now cover a wide wavelength baseline. This places quite severe constraints on the amount of non-thermal background radiation that can be present in normal elliptical galaxies. Galaxies with known peculiarities, such as M87, were avoided during the David Dunlap Observatory observing program.

THE GASEOUS CONTENT OF GALAXIES

MORTON S. ROBERTS

National Radio Astronomy Observatory,* Green Bank, West Va., U.S.A.

Abstract. A general review is given of the content and distribution of interstellar gas within galaxies. The constancy of the ratio $N(\text{He})/N(\text{H})$, independent of galaxy type (spirals and irregulars), is discussed and the possible mechanisms for this constancy are considered. The helium abundance does not vary across the disk of spirals, although nitrogen and possibly other elements do.

The gross features of the neutral hydrogen distribution in our Galaxy and other systems are described. In spirals, the peak of the radial distribution of H I is located well away from the optical center. This is not the case for irregular-type systems. A possible correlation of the relative location of the maxima of H I and H II distributions with galaxy type is described. Many spirals studied with high enough relative angular resolution show concentrations of H I in their outermost regions. These may be due to hydrogen companions or warps in the hydrogen plane. Hydrogen 'bridges' are described and a new example for the triple system M81–M82–NGC 3077 is given. This latter case may be an extreme example of distortion by companion galaxies of the H I associated with a massive galaxy.

The neutral hydrogen content of a galaxy and its correlation with other integral properties is discussed. The absorption profile due to hydrogen associated with the radio galaxy Centaurus A is given. Comparison of optical and 21-cm measurements of galaxian redshifts shows excellent agreement over the radical velocity range -400 to $+5200$ km s^{-1}. There is, however, a systematic difference between 21 cm and optical redshifts over the range ~ 1200 to ~ 2400 km s^{-1} for optical values based on blue-sensitive spectra. The difference, ~ 100 km s^{-1}, is most likely due to blending of galaxian and night sky H and K absorption lines. The Hubble Constant is derived from a redshift-21 cm flux relation. Values in the range 78 to 109 km s^{-1} Mpc^{-1} are derived. A value of 97 km s^{-1} Mpc^{-1} is favored.

1. Introduction

Interstellar gas represents a significant constituent of most types of galaxies. As an example, one-fourth of the total mass of irregular-type systems is in the form of neutral atomic hydrogen. Assuming a normal abundance of helium – a point to be discussed in some detail below – the gaseous component reaches one-third of the total mass. In our galaxy, this number lies between 5 and 10%.

Properly and completely to describe the gaseous component of a galaxy requires the evaluation of a distribution function giving the particle density of the various constituents, $N(p_i)$, in a volume element as a function of position within the galaxy. Additional descriptive information would include the motions of and the excitation condition within the volume element.

In the optical and radio domains, there are four physical processes that supply information to the observer on $N(p_i)$. These are:

(1) Emission: neutral and ionized gas;
(2) Absorption: gas and electrons;
(3) Dispersion: electrons;
(4) Scattering: electrons (also dust and gas).

---

* Operated by Associated Universities, Inc., under contract with the National Science Foundation.

Additional 'probes' of interstellar matter are available through a study of cosmic rays, X-rays, and γ-rays.

The presently available information does not allow us adequately to evaluate the above distribution function – even in our own galaxy. Nevertheless, a picture of the gaseous content of galaxies is slowly emerging for the nearer, and primarily, 'normal' systems. The complexity of the problem is indicated by the rich composition of the interstellar medium in our own Galaxy. Optical studies have identified interstellar lines of such atoms and molecules as sodium, calcium, titanium, iron, CH, and CN.

From studies at radio wavelengths we have thus far identified: neutral atomic hydrogen (H I), the hydroxyl radical (OH), water (H_2O), ammonia (NH_3), formaldehyde (H_2CO), hydrogen cyanide (HCN), carbon monoxide (CO), the CN radical, and cyano-acetylene (HC_3N). The C^{13} and O^{18} isotopes have been found in several of these substances. The presence of the hydrogen molecule (H_2), long suspected as being present in dark clouds, has been identified by Carruthers (1970) in a spectrum of the star ξ Persei obtained in a rocket flight.

Ionized hydrogen (H II) regions, which we assume to be representative of the interstellar medium in general, are excellent sources for studying a number of elements, especially nitrogen, oxygen, sulfur, and helium. We may expect our knowledge of the inventory of this chemical laboratory to continue to increase in the near future.

The electron density, N_e, of the interstellar medium shows immense variation, at least six orders of magnitude, from $\sim 10^{-2}$ to $\sim 10^4$ cm^{-3}, the latter value holding for high-density H II regions. The low-electron density distribution has been described by Bridle and Venugopal (1969) in terms of a disk whose full thickness at half-intensity is 800 pc.

2. The Chemical Composition of the Gas in Extragalactic Systems

Information regarding the chemical composition of the interstellar material in extragalactic systems is almost wholly based on optically derived data. Qualitatively, the chemical composition of H II regions in other galaxies, as evidenced by low and moderate dispersion spectra, is similar to that in our own Galaxy. Thus, similar lines and line ratios are seen in Orion and in extragalactic H II regions. This similarity was used by Hugh Johnson (1959) in the first determination of an extragalactic helium abundance; that of the 30 Doradus nebulae in the Large Magellanic Cloud. A summary of presently available determinations of the ratio $N(He^+)/N(H^+)$ for different types of galaxies is displayed in Figure 1. The data are from a variety of sources but are on a homogeneous system as far as recombination rates are concerned. The three filled circles at the right of the Sc category all refer to NGC 604, a giant H II region in M33. Their scatter shows the range of uncertainty in such determinations.

The average value of this ratio for the extragalactic systems is 0.087. For the six galactic H II regions shown it is 0.092.

The neutral helium component in H II regions may be allowed for through an ionization correction factor obtained from the abundance ratios of different ions of

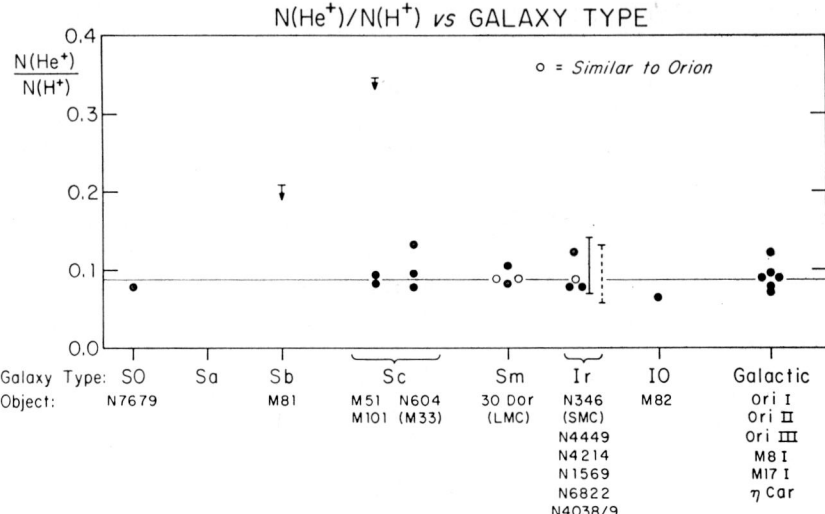

Fig. 1. The ratio $N(He^+)/N(H^+)$ for different galaxy types and for the Galaxy. The H II region or galaxy designation is indicated at the bottom of the figure. The three filled circles to the right of the Sc category refer to three different determinations for the same object, NGC 604 in M33. The solid bar in the irregular type refers to NGC 4214 and represents the limits applied by Mathis (1965) to this ratio. The dashed bar shows the range of $N(He^+)/N(H^+)$ found for different H II regions in the peculiar galaxy NGC 4038/9 by Rubin et al. (1970).

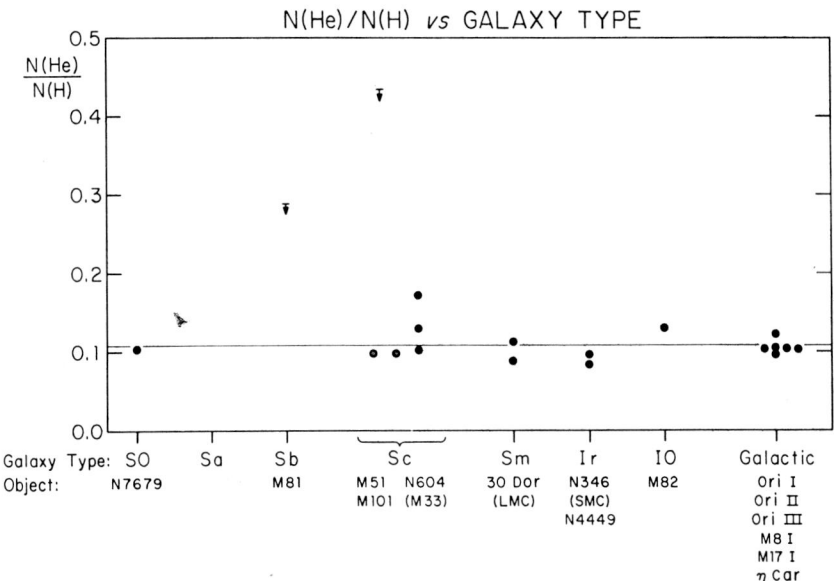

Fig. 2. The ratio $N(He)/N(H)$ for different galaxy types and for the Galaxy. The three filled circles to the right of the Sc category refer to three different determinations for the same object, NGC 604 in M33. The H II region or galaxy designation is indicated at the bottom of the figure.

a particular element such as oxygen or sulphur (Peimbert and Costero, 1969). Figure 2 displays the ratio $N(\text{He})/N(\text{H})$ as a function of structural type. Most of these data are from Peimbert and Spinrad (1970). The large scatter for NGC 604 is again evident. The average value of the helium-to-hydrogen ratio for the extragalactic systems is 0.11; for the galactic H II regions it is the same. This constancy of the helium-to-hydrogen ratio in various spirals and irregular-type systems, regardless of structural type, has been recognized for some time. The more recent data strengthen this conclusion. I shall return to this subject later after discussing the hydrogen content of galaxies.

The question of a variation of helium abundance with distance from the center of a galaxy was first discussed by Schmidt (1962). He obtained spectra of three H II regions in Andromeda at true distances from the center of 25', 70', and 89' and found that the ratio of the D3 line of helium to Hβ was constant to within a factor of two, which was the uncertainty he assigned to his measurements. Schmidt concluded that the helium-to-hydrogen ratio was essentially constant over this range of distances. More recent and more extensive work by Rubin (1970, unpublished) supports this earlier result. She finds *no* systematic variation with radial distance of the helium-to-hydrogen ratio in M31. Searle (1970) reaches a similar conclusion for Sc-type galaxies. This absence of a correlation of helium abundance with radial distance is indirectly supported by the material presented in Figures 1 and 2 since the H II regions involved in the measurements for the various galaxies are at different relative distances from the center.

There does appear to be a systematic variation of nitrogen with respect to hydrogen as a function of radial distance. The variation of [N II]/Hα within a galaxy and among different galaxies has been described by the Burbidges (1962, 1965). Initially, these variations were thought as being possibly due to excitation effects (Burbidge *et al.*, 1963). However, Peimbert's (1968) observations of the [O I], [O II], and [O III] lines, in addition to [N II] and Hα yield the conclusion that the variations are caused by an abundance effect. For the nuclear regions of M51 and M81, Peimbert found, by adopting a solar oxygen-to-hydrogen ratio, an excess of nitrogen of from two to six times the solar value.

Searle (1970) and Rubin (1970) find similar abundance variations in other galaxies.

In summary, we conclude that the helium-to-hydrogen ratio is constant in galaxies of structural types Ir through S0 and is also constant across the disks of these galaxies although there may be local variations. This is not the case for the nitrogen-to-hydrogen ratio which *does* vary with structural type as well as radial distance within a galaxy.

3. The Distribution of Neutral Hydrogen Within Galaxies

The most abundant element, hydrogen, cooperates to a surprising extent by allowing us to study it in a relatively direct and easy manner in both its neutral and excited states. The recombination lines of excited hydrogen can be studied at both optical and radio wavelengths; neutral atomic hydrogen is measurable through its line radiation at 21 cm. With this latter probe, we are able to map the distribution of neutral hydrogen; measure its total content in a galaxy; obtain information on the dynamics within a galaxy and hence its total mass; and finally we can obtain an accurate measure of the systemic velocity of a galaxy.

The material which I will discuss is based on measurements made at a number of observatories, and I shall not attempt to credit the many people involved except when I speak of specific galaxies. Details may be found in a survey by Roberts (1969) of the data available through mid-1969.

For the larger galaxies, filled aperture observations supply sufficient relative resolution to outline the gross features of the H I distribution. These features may be conveniently categorized as: (1) Main body distribution; (2) Extent; (3) Companions; (4) Warp or bending of the plane; (5) Bridges.

Some of these may be interrelated but at present we do not have sufficient information for a large enough sample to do much more than describe such features.

21-cm studies of our Galaxy are a convenient starting point. There are many interesting features in the galactic hydrogen distribution, e.g. the arc-shaped hydrogen concentrations termed 'arms', the high velocity clouds, the extreme flatness of the plane of H I interior to the Sun, and the bending or warp of the plane exterior to the Sun. For comparison with other galaxies, we shall consider only the gross features of the galactic H I distribution. These are: (1) The deficiency of H I in the central region, i.e., a 'ring' of H I; (2) The percentage of the total mass in the form of neutral atomic hydrogen; and (3) The bending of the H I plane in the outer regions of our Galaxy.

The term 'ring' is intended merely as a convenience in describing the average radial variation of the hydrogen surface density (projected onto the plane of the galaxy). Burke (1967) has described the ring as "a disk with a hole in the middle". This is a more descriptive (and more lengthy) phrase since the ring is quite thick in the radial direction. In our Galaxy the radius at maximum density is about 10 kpc and its full width at half intensity is also about 10 kpc. Such a ring-like distribution is not unique to our Galaxy. The half dozen spiral galaxies thus far studied with high enough relative resolution all show a similar distribution.

An interesting and unexplained feature of this ring of neutral hydrogen is its position relative to the location of the principal concentrations of ionized hydrogen, that is, the optical spiral arms. In our Galaxy the giant H II regions, as defined by radio recombination line observations, lie primarily interior to this ring at a distance of ~ 4 to 6 kpc (Reifenstein et al., 1970). The distribution of the thermal continuum radiation also implies a similar central concentration. Presumably, both of these data refer to a prominent spiral arm of our Galaxy. Figure 3, taken from Professor Oort's (1965) Invited Discourse at the Hamburg IAU, shows these features quite clearly.

A similar placing of H I and H II concentrations is found in other spiral galaxies. The prominent, optically-defined spiral features lie interior to the H I ring. An outstanding exception is M31. In this Sb-type galaxy, the optical arms are embedded in the H I ring. The H I ring in M31 is clearly evident in the isometric projection of

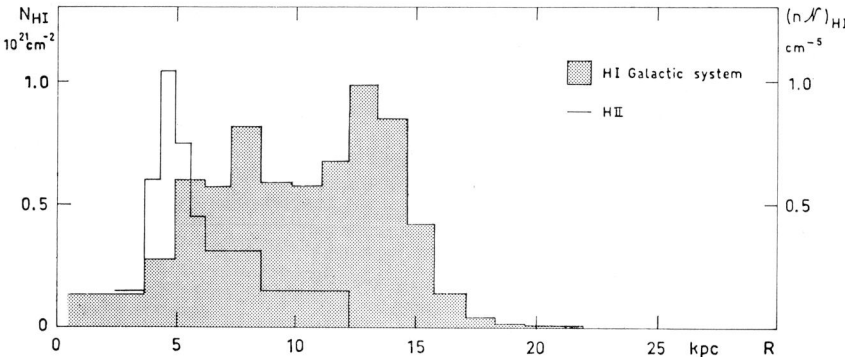

Fig. 3. The radial distribution of neutral and ionized hydrogen in our Galaxy. This figure is from Oort (1965).

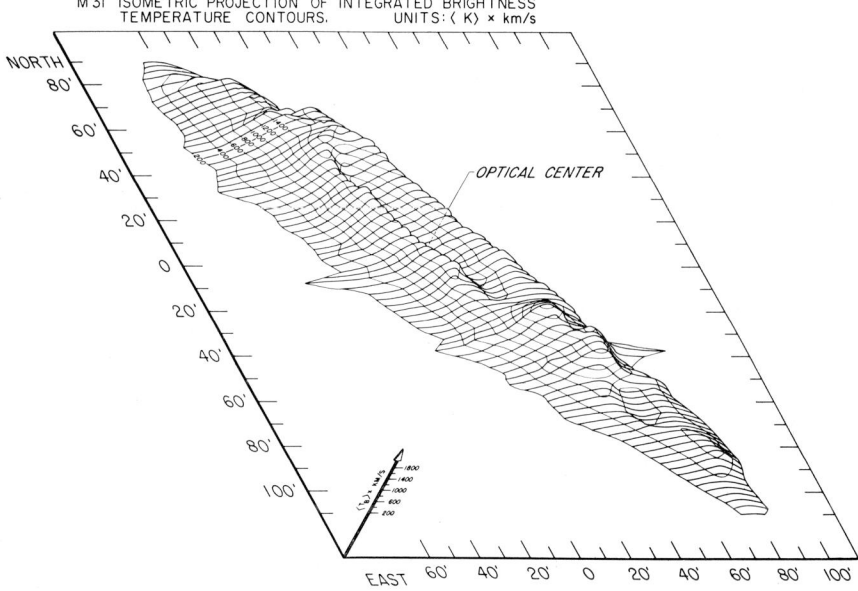

Fig. 4. An isometric projection of 21-cm integrated brightness temperature contours for M31. The lack of a central concentration of neutral hydrogen is clearly evident. These data were obtained with the 300-foot telescope (beamwidth = 10′).

integrated brightness temperature contours shown in Figure 4. M31 is the only Sb system in our rather meagre sample and it is obviously premature to attempt to relate the relative location of neutral and ionized hydrogen to structural type. However, we do have several irregular-type galaxies mapped with sufficient resolution and they do not exhibit a deficiency of H I in their central regions. This additional datum does suggest a relation between the region of star formation and the maximum surface density of H I with type, but more information on earlier-type galaxies is required before this conclusion can be considered well established.

Several clarifying points should be made regarding these features. The hydrogen ring is really defined by the observational quantity of integrated (over velocity), beam-averaged brightness temperature. A minimum in the center of the ring could be due to an observational effect in which much of the H I is concentrated into small regions of high optical depth. Beam-dilution would then give the appearance of a minimum of H I. Another possibility is that much of the hydrogen is in the form of molecular hydrogen. A third possibility is that the hydrogen is ionized. Finally, there may indeed be a relative deficiency of the neutral hydrogen in the central regions of later-type spirals. Regardless of the correct explanation we cannot escape the fact that there is an over-all variation in spirals in the form of amount of hydrogen with respect to distance from the center.

The ring itself has been described as a smooth circular feature. This is a great simplification forced by lack of sufficient resolving power. The ring need not be circular and surely has a large fluctuation in its density structure. Further, the concentration of ionized hydrogen regions interior to the maximum of the ring does *not* imply that there is no neutral hydrogen in these regions. Rather, we may expect local concentrations of neutral hydrogen near and about H II regions, as shown by Orion in our Galaxy (Menon, 1958; C. Gordon, 1970) and the H II regions in the Large Magellanic Clouds (McGee, 1964; McGee and Milton, 1966).

Although there are some H II regions in the ring, an obvious question is: Why haven't the spiral arms formed in this higher density region? Lin (1970) has suggested

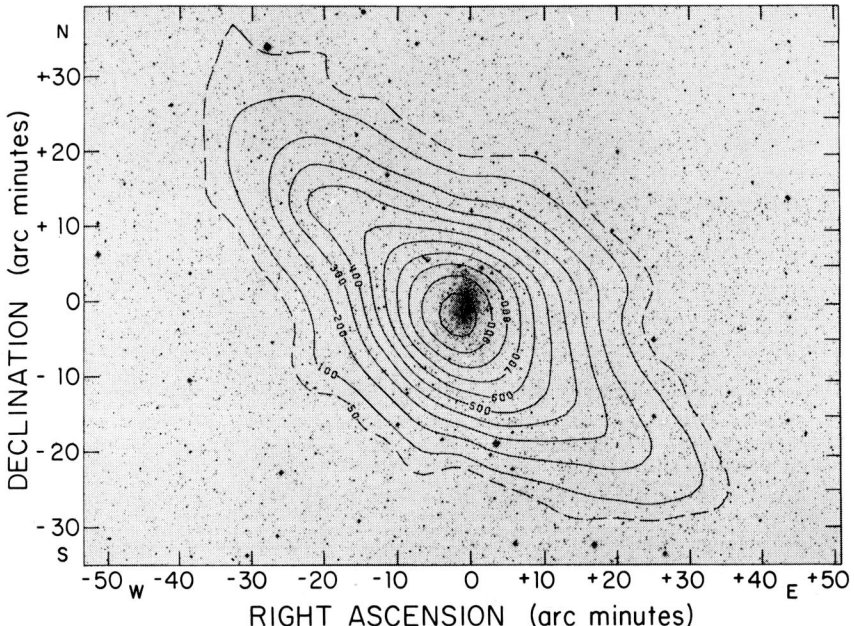

Fig. 5. Contours of 21-cm integrated brightness temperature superposed on a photograph of the irregular-type galaxy NGC 6822. These data were obtained with the 300-foot telescope (beam width = 10′).

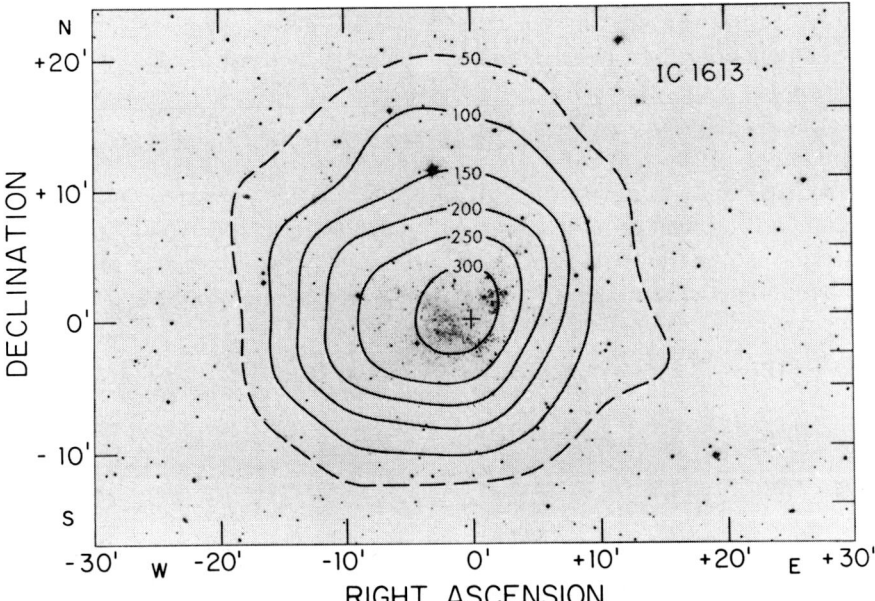

Fig. 6. Contours of 21-cm integrated brightness temperature superposed on a photograph of the irregular-type galaxy IC 1613. These data were obtained with the 300-foot telescope (beam width = 10').

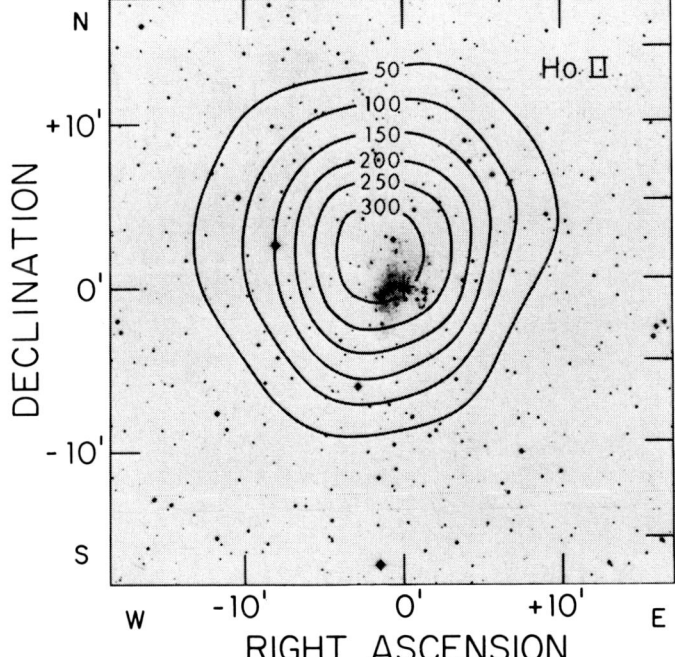

Fig. 7. Contours of 21-cm integrated brightness temperature superposed on a photograph of the irregular-type galaxy Ho II. These data were obtained with the 300-foot telescope (beam width = 10').

that the density wave and its related shock front that he proposes for spiral arm structure does not reach out this far (except in M31). Another possibility would have the *volume* density of H I smaller in the ring (i.e. the ring is thicker in the z-direction) than in the interior portions.

The neutral hydrogen distribution of a galaxy extends beyond its usually measured optical dimensions. Comparably normalized dimensions are significantly larger in neutral hydrogen than those measured at optical wavelengths. An extreme example is the irregular galaxy NGC 6822 shown in Figure 5. Although this galaxy is at a relatively low galactic latitude, 18°, it seems unlikely that as much of the optical image,

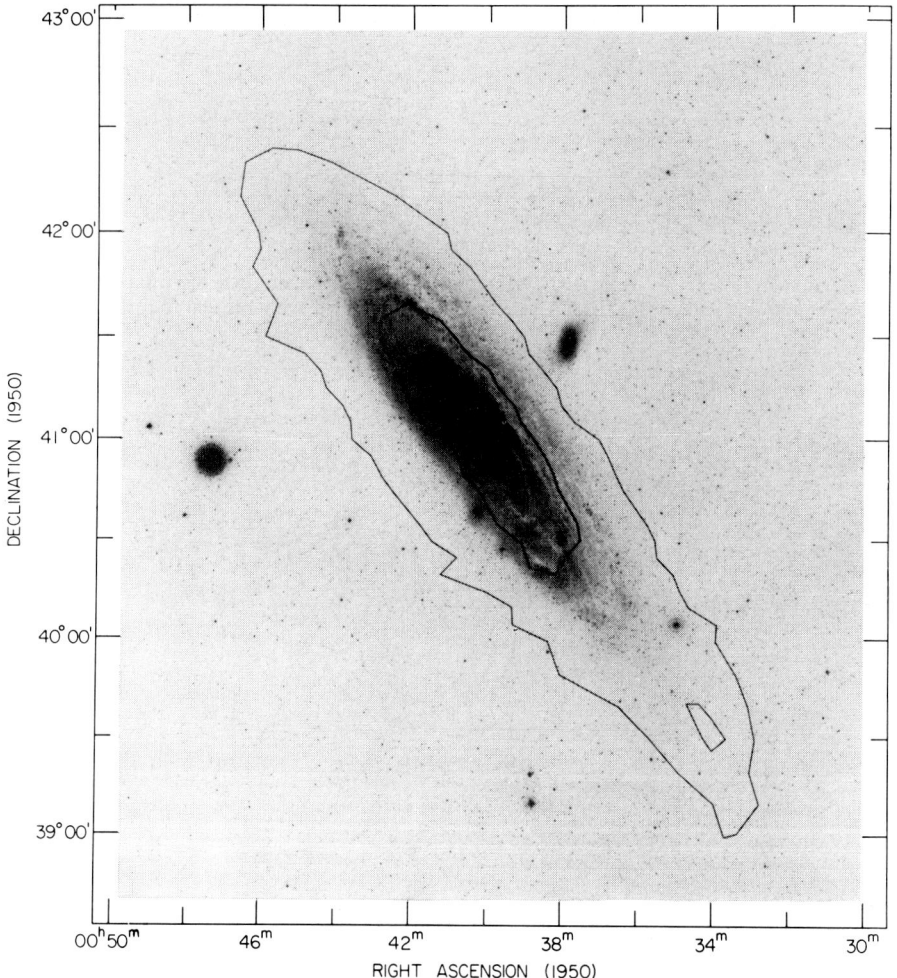

Fig. 8. Contours of 21-cm integrated brightness temperature superposed on a photograph of M31. The inner, heavy contour represents the peak brightness temperature. The outer contour corresponds to 200 K km s^{-1} and is from $\frac{1}{6}$ to $\frac{1}{8}$ the peak contour. The south-preceding feature is represented in the lower right by a contour level of 600 K km s^{-1}. These data were obtained with the 300-foot telescope (beam width = 10′).

as implied by the extent of the 21-cm isophotes, is hidden by galactic obscuration. Of more concern is the low systemic velocity of this system which causes confusion with foreground galactic hydrogen. However, associated with this hydrogen map is a well-ordered radial velocity map showing a range of over 100 km s^{-1}. It seems improbable that we would have such an unusual foreground cloud just in the direction of NGC 6822.

Other, less extreme, examples are shown in Figures 6 and 7. These are also irregular-type systems, IC 1613 and Ho II. For all three cases, the beam-broadened hydrogen isophotes are as shown. These data were obtained with the Green Bank 300-foot telescope which, at this wavelength, has a half-power beam of 10′.

Figure 8 is a superposition of 2 principal H I contour levels on a photograph of M31.

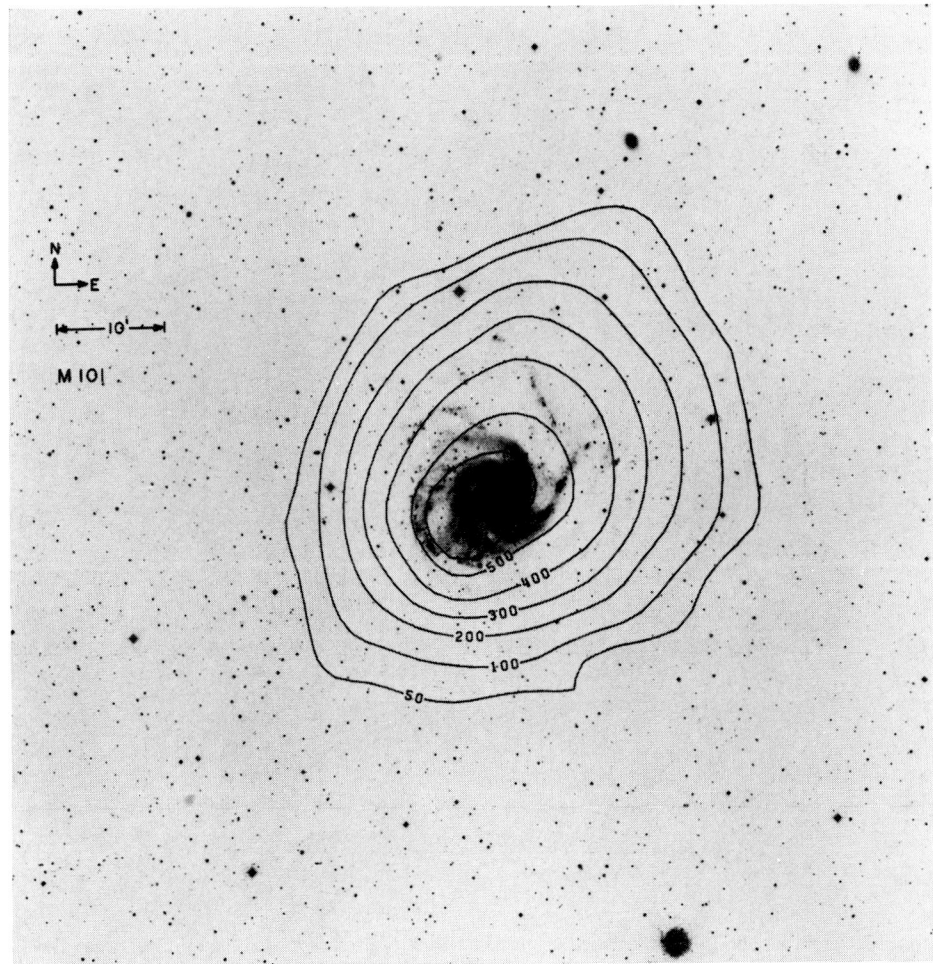

Fig. 9. Contours of 21-cm integrated brightness temperature superposed on a photograph of M101. The asymmetrical H I distribution to the northeast is clearly evident. These data were obtained with the 300-foot telescope (beam width = 10′).

The inner contour locates the peak integrated brightness temperature. The outer contour is $\frac{1}{6}$ to $\frac{1}{8}$ of the average peak level. These data are also broadened by a 10' beam. The south-preceding feature first delineated by Burke *et al.* (1963) can be seen at one end of the major axis. On this relative-resolution scale it stands out almost as a separate feature. For more distant galaxies such a feature would appear only as a smooth extension of the general contours. A possible example of this is shown in

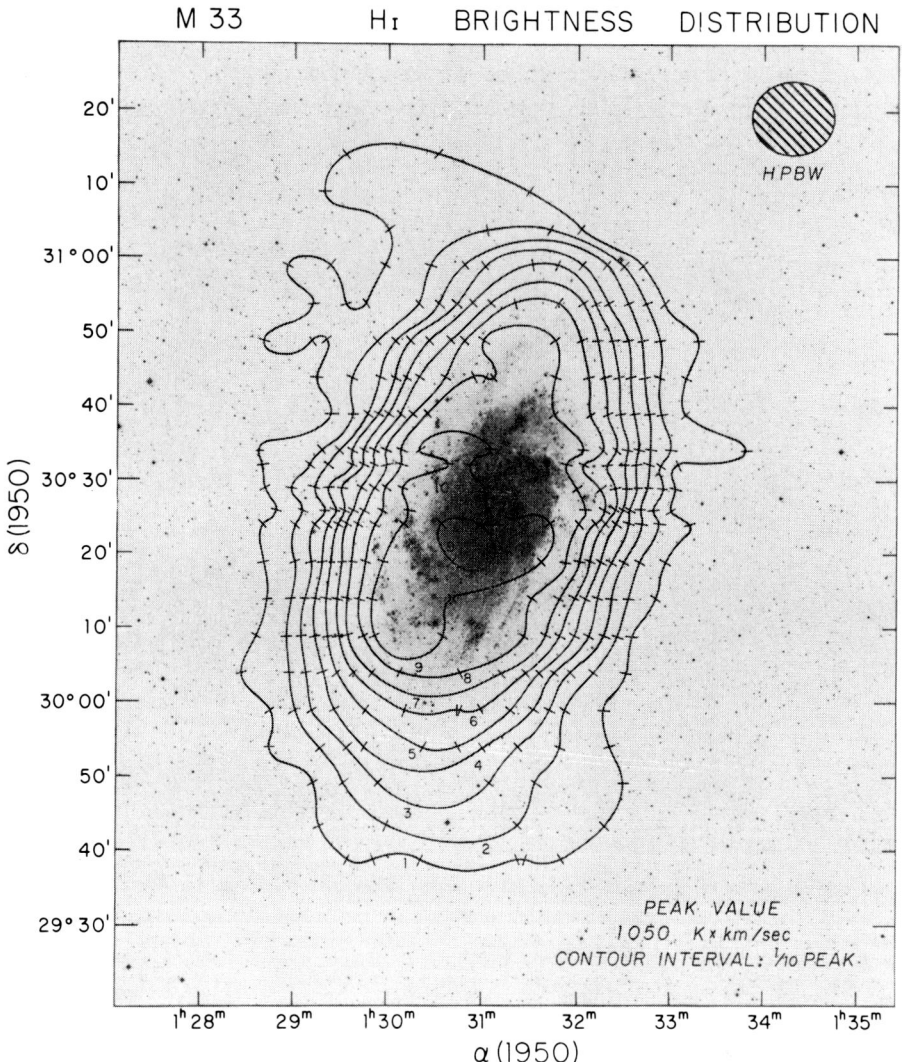

Fig. 10. Contours of 21-cm integrated brightness temperature superposed on a photograph of M33. Data points for constructing the contours are indicated by short lines on the contours. The 'wings' to the northwest (top left) and southeast (bottom right) are clearly evident. These data were obtained by K. Gordon (1970) using the 300-foot telescope. The signal-to-noise ratio has been improved by smoothing. The resultant beam-size is shown in the figure.

Figure 9, which is a map of hydrogen contours superimposed on a photograph of M101. The asymmetrical distortion to the north-east *may* represent beam-smearing of an adjacent hydrogen companion.

Figure 10 is of M33 and shows the relation of the optical image and hydrogen isophotes as derived from 300-foot observations. These data were obtained by K. Gordon (1970). Gordon finds that to comparable levels, the hydrogen extent is approximately twice that given by star counts, B photometry, and the distribution of H II regions.

There is another feature in M33 to which attention is directed. This is the significant deviation of the outer isophotes from the optical major axis. A line connecting these outer features is displaced from the major axis position angle in the direction of rotation of M33. The radial velocity map for M33 shows a displacement in the same sense as the H I contours.

Similar features are seen in other spirals, e.g., NGC 300 (Shobbrook and Robinson, 1967), NGC 2403 (Burns and Roberts, 1970), NGC 5236 (Lewis, 1968). In all cases the shift of the kinematic and H I axes with respect to the optical major axis is in the sense of rotation of the galaxy.

There are several possible explanations for such features:

(a) Tidal distortion by neighboring galaxies;

(b) A warp or bending of the plane;

(c) A non-axisymmetric expansion of the hydrogen in the plane of the galaxy with a stable region on the outskirts of the galaxy where the hydrogen collects;

(d) Hydrogen companions.

The problems of invoking tidal distortion for our Galaxy are well known. For M33, as an example, the situation is even more severe if one wishes to use M31 as the most massive, nearby interacting system (K. Gordon, 1970). As with the Magellanic Clouds and our Galaxy, a near encounter in the orbital motion of M31–M33 could, presumably, be invoked.

A warp in the plane similar to that in our Galaxy will qualitatively explain the features. However, the distortion in the radial velocity field is too great to be accounted for without requiring a large radial streaming in the plane of the galaxy. For M33 such a non-axisymmetric radial motion term is several tens of kilometers per second, increasing outwards. These remarks obviously hold for the third possible explanation, of non-axisymmetric radial motion.

The final possibility would require hydrogen companions in orbits out of the principal plane of the primary galactic system. The expansion term required for the physically connected hydrogen case need not be invoked. Instead, the distortion in the radial velocity map would be attributable to the radial component of the orbital motion of the companions. Alternatively, these companions could be coplanar, and have a high expansion term together with their expected circular motion. [A similar situation exists for the bright optical companion of M51, NGC 5195, where the radial velocity difference of these two systems can only be explained by an expansion term or by an orbit out of the plane of M51 (Roberts and Warren, 1970).]

Of the various explanations, the hydrogen companions appears the most attractive.

In addition, warps in the H I plane of the primary system may also exist and could, possibly, be attributed to these companions.

'Bridges' are the final aspect in the list of hydrogen distribution features. The first described example of such a feature is the link between the Large and Small Magellanic Clouds (Hindman *et al.*, 1963). This link has been studied in some detail by Turner (1970). There are also suggestions that a link exists between our Galaxy and the Magellanic Clouds. Another example is the bridge between NGC 4631 and 4656 (Roberts, 1968; Weliachew, 1969).

Yet another striking example of such hydrogen bridges is found in the triple

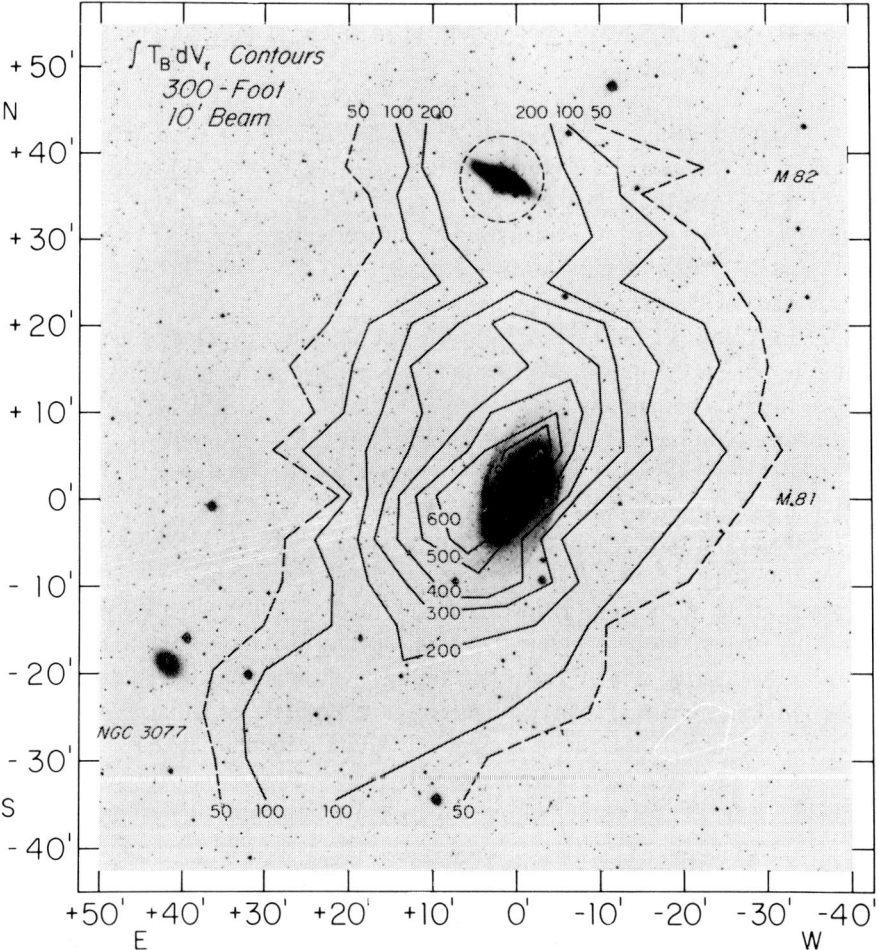

Fig. 11. Contours of 21-cm integrated brightness temperature in the area of the triple system M81–M82–NGC 3077. The photograph is from the red sensitive National Geographic Sky Survey Atlas. Data from the region of M82 (dashed circle), a continuum source, are omitted because of possible gain imbalance between signal and reference frequencies. The integration is over all velocities. Because of low systemic velocity of M81 there is some confusion with foreground galactic hydrogen in the region of M81. The contours include an extrapolation over this velocity range. The other regions are free of confusion; see the velocity map in Figure 13.

system of M81–M82–NGC 3077. (A more appropriate term in this case might be hydrogen streamers.) Figure 11 shows the beam-averaged hydrogen distribution as seen with the 10′ beam of the 300-foot telescope. The hydrogen associated with M81 is clearly distorted in the directions of M82 and NGC 3077. These data were obtained

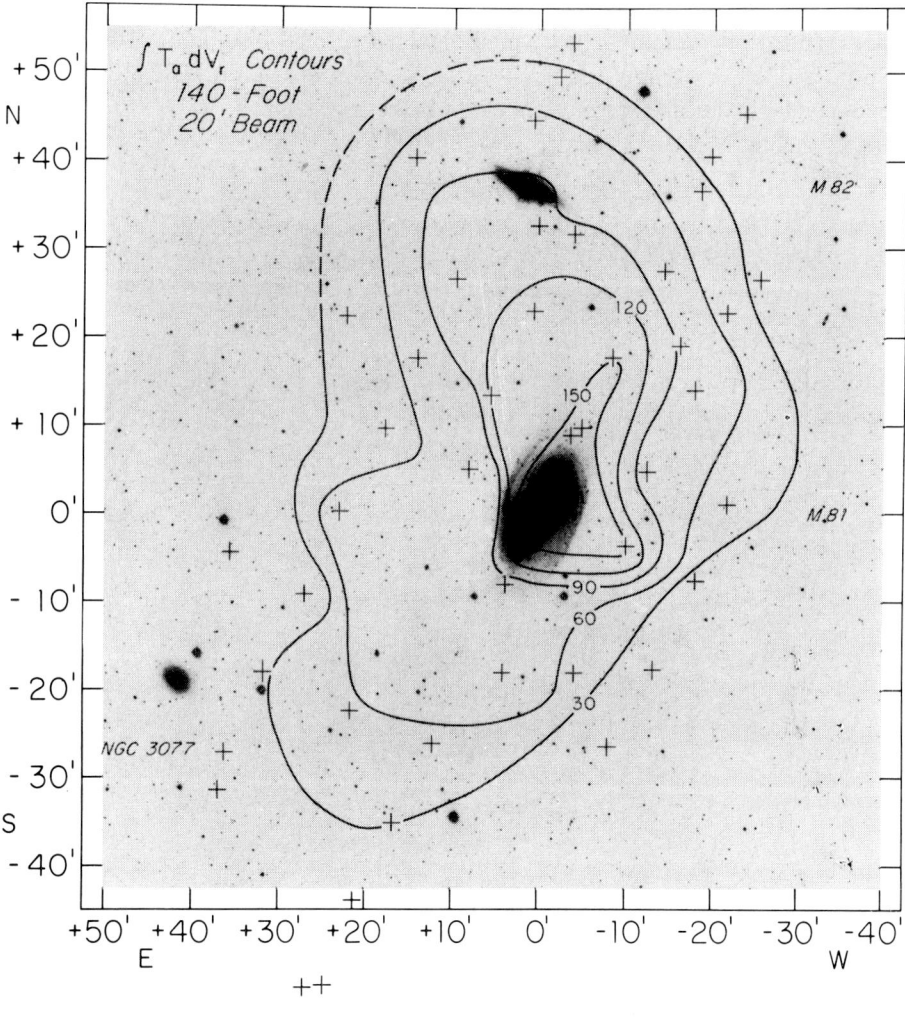

Fig. 12. Contours of 21-cm integrated *antenna* temperature in the of area the triple system M81–M82–NGC 3077. The photograph is from the red sensitive National Geographic Sky Survey Atlas. An autocorrelation receiver was used for these observations and the data near M82 are therefore free of gain imbalance due to the continuum radiation from M82. The points observed are indicated by plus signs. The integration is over all velocities. Because of the low systemic velocity of M81 there is some confusion with foreground galactic hydrogen in the region of M81. The contours include an extrapolation over this velocity range. The other regions are free of confusion; see the velocity map in Figure 13.

with a multichannel receiver and the observations on M82, a radio continuum source, are not included because of possible gain imbalance between the signal and reference frequencies. Figure 12 is a similar map, obtained with the 20' beam of the 140-foot telescope. An autocorrelation receiver, which does not suffer from gain imbalance problems, was used. We see a hydrogen distribution similar to the 300-foot data. On the plane of the sky, M82 is 35 kpc and NGC 3077 is 45 kpc from the center of M81. Figure 13 is a radial velocity map obtained with 10' resolution. It is relatively well-ordered, showing only small distortion towards M82 and NGC 3077 although the kinematic major axis does swing towards these systems.

The systemic radial velocity of M81 is -40 km s^{-1} and observations near this

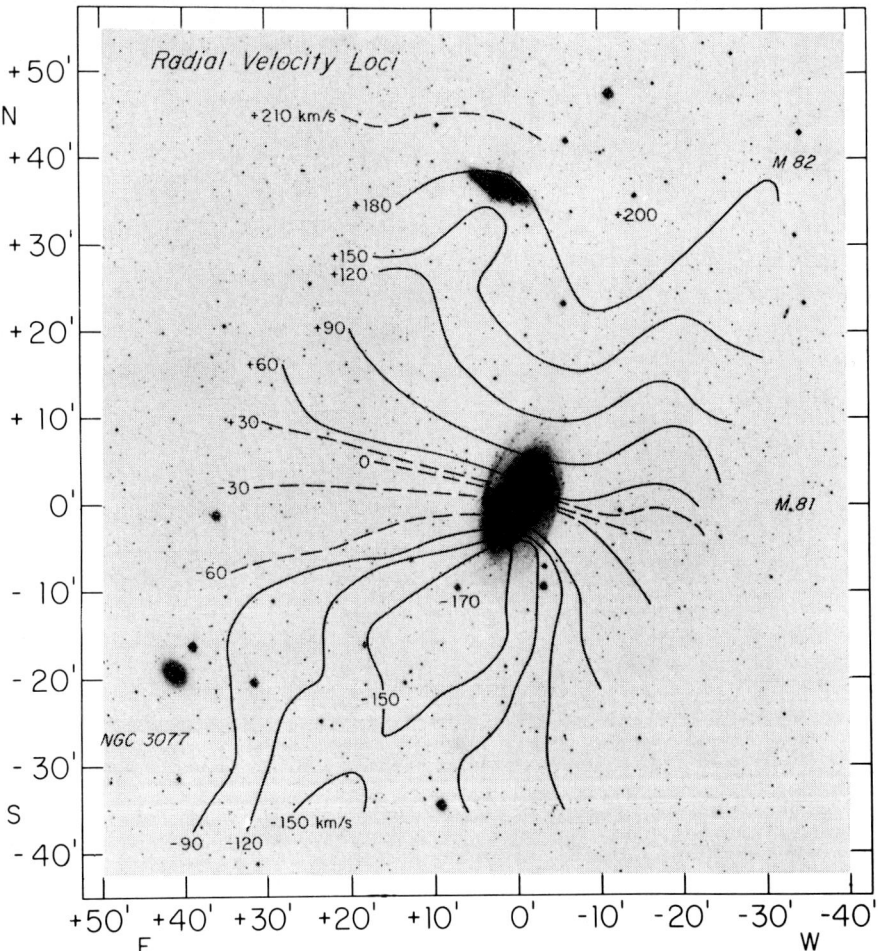

Fig. 13. A radial velocity map (loci of constant radial velocity) in the area of the triple system M81–M82–NGC 3077. The photograph is from the red sensitive National Geographic Sky Survey Atlas. These loci were constructed from the midpoints of velocity profiles obtained with the 300-foot telescope (beam width = 10'). The loci near zero velocity are confused by foreground galactic hydrogen and are therefore uncertain.

velocity are seriously affected by foreground galactic hydrogen. For this reason, the low velocity loci in Figure 13 are uncertain. However, the optically-measured velocity of M82 is well separated from galactic hydrogen and the distorted hydrogen toward this system is free of foreground contamination. The situation is less clear for NGC 3077. Humason *et al.* (1956) give a velocity of -158 km s^{-1} for this system while Demoulin (1969) obtains a velocity of -41 km s^{-1}. The hydrogen data in this latter velocity range are again confused with foreground hydrogen although individual velocity channel data indicate a concentration of hydrogen in the direction of NGC 3077 at a radial velocity of $\sim +10$ km s^{-1} and covering a range of more than ± 30 km s^{-1}. The contours of Figures 11 and 12 do not show this feature because of the conservative extrapolation over the low (foreground) velocity range.*

We conclude that there is little, if any, hydrogen associated with NGC 3077 at a velocity of -158 km s^{-1} although there may be some hydrogen at lower velocities. Similarly, much of the hydrogen previously associated with M82 from single-beam observations (Volders and Högbom, 1961) would appear to refer in large part to M81 hydrogen distorted towards M82. This would also explain the surprisingly large difference between the well-determined optical radial velocity (Mayall, 1960) and the value derived from 21-cm measurements (Volders and Högbom, 1961).

To summarize:

(1) The neutral hydrogen in spirals is not centrally concentrated. Rather, a minimum in the projected H I density occurs near the optical center. The maximum occurs beyond the location of the prominent optical spiral features in Sc-type systems. In M31, an Sb, the arms are approximately coincident with the H I maximum.

In irregular-type galaxies the projected H I density is centrally concentrated.

(2) The hydrogen extent of a galaxy is significantly larger than comparable optical dimensions.

(3) Many (possibly all) spirals show distortions in their hydrogen distribution and radial velocity map. There are a variety of possible explanations including hydrogen companions and/or a warp in the plane.

(4) Hydrogen bridges or links have been found between galaxies whose separation is of the order of a few tens of kiloparsecs.

4. The Hydrogen Content of Galaxies

One of the clearest correlations with structural type of a quantitatively measured parameter is obtained with the neutral hydrogen content of a galaxy. Using photographic luminosity as a normalizing factor, Heidmann (1961) found such a correlation in the sense that the ratio M_H/L increased for later-type systems. Heidmann used data for 10 galaxies. The number of galaxies for which such data are available has been increased almost tenfold and the results are shown in Figure 14. Bottinelli *et al.* (1970) have recently determined an average M_H/L value for 4 S0 galaxies which fits nicely at the lower left of Figure 14 at a value a bit less than one half of the Sb value.

* See note added in proof, p. 34.

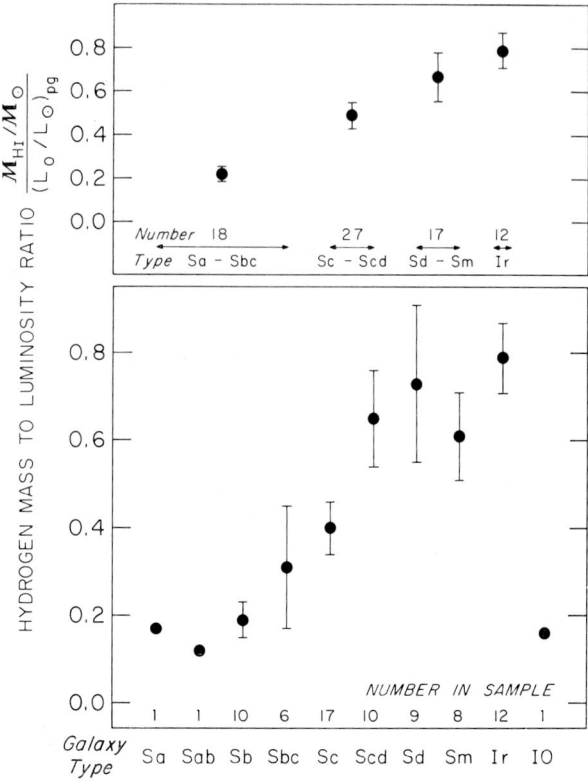

Fig. 14. The ratio: hydrogen mass to photographic luminosity for individual structural types (lower panel) and for broader type intervals (upper panel). The error bars represent the standard deviation from the mean. Both the hydrogen mass and luminosity are corrected for inclination effects.

The ratio of hydrogen mass-to-luminosity is a distance independent quantity so that this great uncertainty in extragalactic work is not present in these data. There is nevertheless a large range in M_H for a given type and luminosity and conversely. This is shown in Figure 15, a plot of M_H vs L. Some of this scatter reflects observational errors and the use of statistical corrections for a face-on orientation for the luminosity and hydrogen mass (Roberts, 1969). However, part of the range in M_H/L values for a given type appears to be real and is correlated with the color of the galaxy as shown in Figure 16. Shown here is the color excess, $C_0 - \langle C_0 \rangle_{type}$, of a galaxy vs the normalized 'hydrogen excess' $[M_H/L - \langle M_H/L \rangle_{type}]/\langle M_H/L \rangle_{type}$. The 'error bars' in this figure show the *full range* in color at a particular abscissa value. This color dependence within a structural class appears to be another aspect of the general relation between M_H/L with type and therefore with color.

Another correlation is found between the fractional hydrogen content of a galaxy and its structural type. This is shown in Figure 17. A detailed discussion of these integral properties of galaxies is given by Roberts (1969).

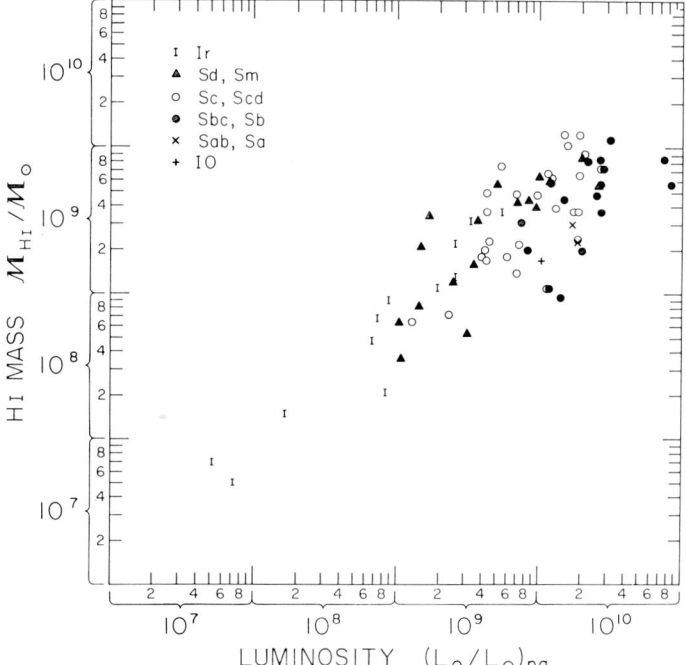

Fig. 15. The hydrogen mass-photographic luminosity relation for spiral and irregular-type galaxies. Both the hydrogen mass and luminosity are corrected for inclination effects.

The fractional hydrogen content of our Galaxy is about 5% (Kerr and Westerhout, 1965). The earlier value of 1 to 2% (Van de Hulst *et al.*, 1957) has been significantly increased by a more detailed analysis.

These data on the hydrogen content of galaxies supply another aspect to the problem of the constancy of the helium-to-hydrogen abundance ratio discussed earlier. We find galaxies of the same structural type differing by a factor of ~ 10 in their hydrogen content, and therefore, presumably, in their helium content. We also find a similar range, going from one morphological class to another, in the percentage of the total mass of a galaxy that is in the form of neutral hydrogen and therefore in helium. We assume here that the helium is well mixed with the hydrogen throughout the total extent of the hydrogen. This assumption is necessary since the helium-to-hydrogen ratio is determined only from H II regions which, as we saw earlier, are not uniformly distributed throughout the region of neutral hydrogen.

Of the two usual proposals invoked to account for the helium abundance: (i) primordial; or (ii) nucleosynthesis in stars, the former would appear the more attractive in terms of the different amounts of helium required in different galaxies. If we wish to invoke the nucleosynthesis explanation, we have the constraint of relating several parameters to the uncycled primordial plus cycled hydrogen (i.e., the present interstellar H I). These parameters are the primordial mass function, the rate (and type)

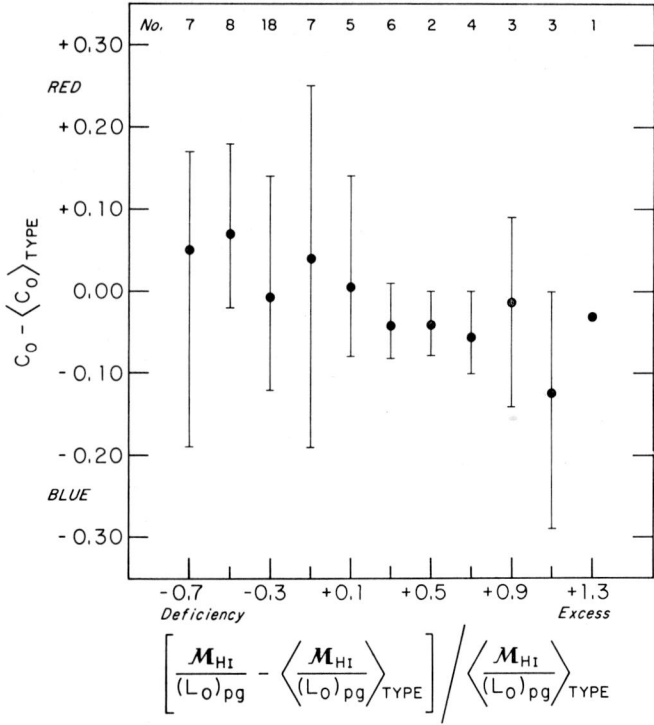

Fig. 16. The color excess of a galaxy *vs* the normalized hydrogen 'excess'. The excess in both parameters is with respect to the mean value for that type. The error bars represent the *full* range in color excess values. The hydrogen 'excess' values are averaged over intervals of 0.2.

of later star formation and the amount and composition of the gas lost by evolving stars. In essence we require that the hydrogen and stellar content of a galaxy be related.

We have seen that the hydrogen content of a galaxy is well correlated with such galaxian properties as luminosity, total mass, color, and structural type. These are, in various ways, related to the stellar content of a galaxy and so the general requirement of a hydrogen-star relationship for the nucleosynthesis explanation is satisfied. However, the detailed relationships among the quantities listed above are not available without an adequate theory of star and galaxy formation and evolution. The approach may possibly be inverted and the relationships derived or at least indicated from the hydrogen and helium data.

5. Neutral Hydrogen in Elliptical and Radio Galaxies

Only one positive detection of 21-cm line radiation from an elliptical galaxy, NGC 4472, has been reported. Robinson and Koehler (1965) derived a hydrogen mass of $3 \times 10^8 M_\odot$, $M_H/L = 0.007$ and 0.02% for the fractional mass content.

Hydrogen in absorption has been detected in the radio galaxy NGC 5128, Centaurus A (Roberts, 1970). This system is particularly favorable for such observations since

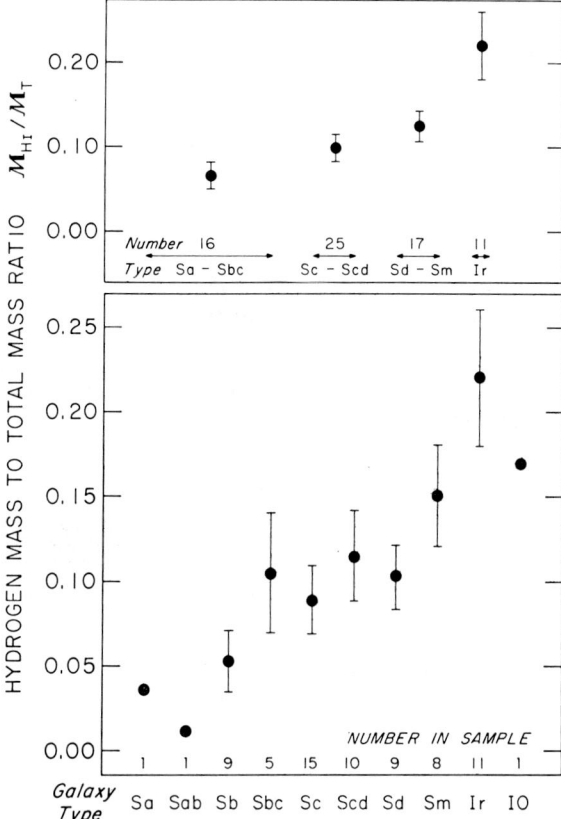

Fig. 17. The ratio: hydrogen mass to total mass for individual structural types (lower panel) and for broader type intervals (upper panel). The error bars represent the standard deviation from the mean.

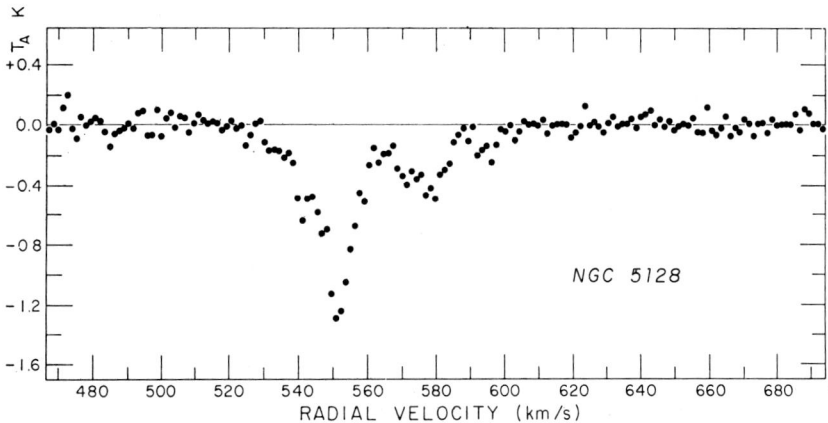

Fig. 18. Absorption velocity profile for Centaurus A, NGC 5128, obtained with the 140-foot telescope (Roberts, 1970).

it is the only strong radio source in which much of the radio radiation comes from within the optical image of the galaxy. Further, the two central radio sources lie approximately on the rotation axis of the system so that the hydrogen seen projected on these sources is not subject to a rotational velocity spread.

The absorption feature, Figure 18, has at least 5 components. Its overall extent, at zero level, is ~ 80 km s^{-1}. The midpoint radial velocity is 563 km s^{-1}. This may be compared with the optically derived systemic velocity of 540 km s^{-1}. This latter value is based on the average of several emission line measurements. Absorption lines give a systemic velocity of 445 km s^{-1}.

Little can be said about the optical depth of the absorbing features because their geometry is unknown. Clark (private communication) has pointed out that within the framework of several reasonable assumptions the absorption lifetime is $\sim 10^3$ yr, which is presumably unimportant with respect to the collision lifetime.

6. Redshifts of Galaxies

The previous discussion has dealt with observational material which yields information on the gaseous component of galaxies. The presence of such gas and its resultant spectral line radiation can also be used to study the motions within a galaxy as well as the systemic motion of a galaxy as a whole. A fundamental system of radial velocities, one essentially free of line blends, night sky emissions, and uncertain rest wavelengths, can be established with 21-cm measurements. Such data can also be used to test the Doppler expression over a wavelength range of half a million by comparing optical and 21-cm redshifts. We would expect a plot of these two quantities to have a slope of one and an intercept of zero. Such a comparison is shown in Figure 19. Because of blending with night sky H and K lines, those optical data taken in the blue part of the spectrum are found to be systematically too large by ~ 100 km s^{-1} over the velocity range ~ 1200 km s^{-1} to ~ 2400 km s^{-1}. These data are shown as an insert in Figure 19. Only red optical data are used for the full figure.

The mean of the two least-squares regressions for 130 velocity pairs over the range -400 to $+5200$ km s^{-1} gives

$$V(\text{OPT}) = -0.71 + 1.000 \ V(\text{21-cm}).$$

We may conclude that the form of the Doppler expression is well confirmed over this velocity range.

A Hubble diagram similar to the classical $(m, c\Delta\lambda_0/\lambda_0)$ may be constructed from the 21-cm data using 21-cm flux density in place of apparent magnitude. Such a diagram with 117 entries is shown in Figure 20; the line shown is one with the expected slope of $\frac{1}{2}$. The Hubble constant can be evaluated from this diagram given a 'typical' galaxian hydrogen mass. Using the M81 group which has 7 H<small>I</small> measurements, we obtain values of H of 83 to 109 km s^{-1} Mpc^{-1} from solutions for different velocity intervals, i.e. all, >500 km s^{-1}, and >1000 km s^{-1}. Similar solutions for the

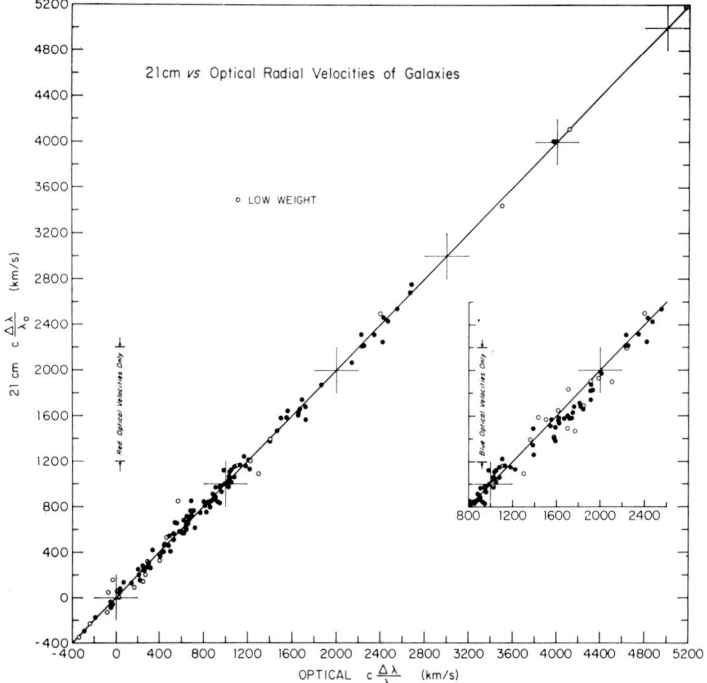

Fig. 19. Comparison of extragalactic redshifts measured at 21 cm and at optical wavelengths. The diagonal line has a slope of one and a zero intercept. *Insert graph* illustrates the systematic difference between 21-cm and optical redshifts over the range ~ 1200 to ~ 2400 km/s^{-1} for optical values based on blue-sensitive spectra. The full graph has values in this velocity range based on red-sensitive spectra (primarily in the Hα region) and shows no systematic differences. The discrepancy with the blue optical data is most likely due to blends of night sky and galaxian H and K lines. The 21-cm data are from a variety of sources but are primarily based on Green Bank and Nançay observations.

Sculptor group, which has 5 H I measurements, gives values of 78 to 103 km s^{-1} Mpc^{-1}. The solutions for $V > 500$ are to be favored since it has a relatively large number of galaxies, 85, and is presumably free of non-isotropic effects in the velocity field (de Vaucouleurs, 1966). The Hubble constant values for these solutions are 100 (M81 group) and 94 (Sculptor group) km s^{-1} Mpc^{-1}, giving an unweighted mean of 97 km s^{-1} Mpc^{-1}.

Acknowledgements

I wish to express my appreciation and thanks to colleagues with whom I have had valuable conversations regarding various topics covered in this review as well as those who have sent me material in advance of publication. Special acknowledgements are due to Drs. E. Churchwell, R. Hjellming, G. Huguenin, V. Rubin, and L. Searle. I am also indebted to Mr. A. H. Rots for his careful analysis and aid in preparing several of the figures and in carrying through the computations of the Hubble Constant from the 21-cm data.

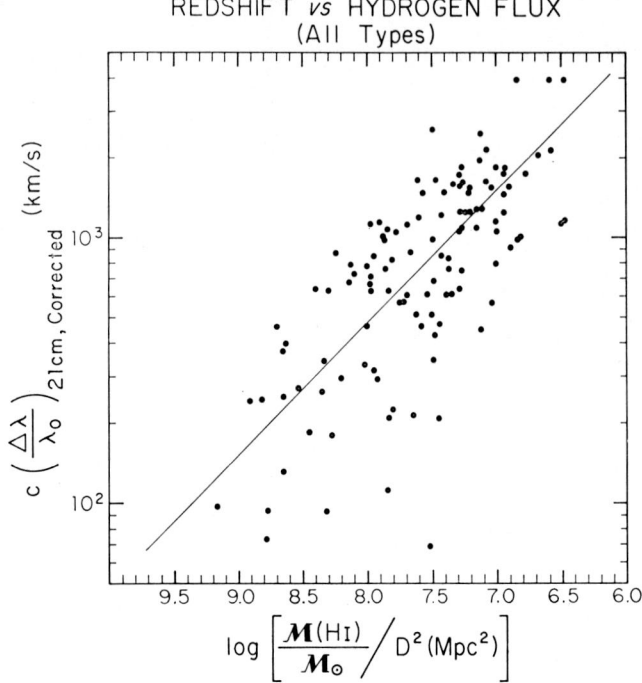

Fig. 20. The 'redshift-magnitude' relation based wholly on 21-cm redshifts and magnitudes (i.e. hydrogen flux densities). The diagonal line has the expected slope of one half. To compare with its optical counterpart see the lower part of Figure 10 in Humason *et al.* (1956).

Note to page 27 added in proof. Dr F. Bertola informs me that he obtains a velocity of -10 km s^{-1} for NGC 3077. This value is based on three spectra (80 Å/mm) obtained with an image tube at the 200-in. telescope. This new, low systemic velocity implies that the concentration of neutral hydrogen seen in the direction of NGC 3077 is most likely associated with NGC 3077 rather than being foreground galactic hydrogen. Thus the contours in Figures 11 and 12 should extend to and envelop NGC 3077.

References

Bottinelli, L., Chamaraux, P., Gouguenheim, L., and Lauqué, R.: 1970, *Astron. Astrophys.* **6**, 453.
Bridle, A. and Venugopal, V.: 1969, *Nature* **224**, 545.
Burbidge, E. M. and Burbidge, G. R.: 1962, *Astrophys. J.* **135**, 694.
Burbidge, E. M. and Burbidge, G. R.: 1965, *Astrophys. J.* **142**, 634.
Burbidge, G. R., Gould, R. J., and Pottasch, S. R.: 1963, *Astrophys. J.* **138**, 945.
Burke, B. F.: 1967, in H. van Woerden (ed.), 'Radio Astronomy and the Galactic System', *IAU Symp.* **31**, 185.
Burke, B. F., Turner, K. C., and Tuve, M. A.: 1963, in Annual Report of the Director, Dept. of Terrestrial Magnetism, 1962–63, Carnegie Institute of Washington, Washington, D.C., p. 289.
Burns, R. and Roberts, M.: 1970, in preparation.
Carruthers, G. R.: 1970, *Astrophys. J. Letters* **161**, L81.
Demoulin, M. T. H.: 1969, *Astrophys. J.* **157**, 81.
Gordon, C.: 1970, *Astron. J.* **75**, 914.
Gordon, K.: 1970, *Astrophys. J.* in press.

Heidmann, J.: 1961, *Bull. Astron. Inst. Neth.* **15**, 314.
Hindman, J. V., Kerr, F. J., and McGee, R. X.: 1963, *Australian J. Phys.* **16**, 570.
Humason, M. L., Mayall, N. U., and Sandage, A. R.: 1956, *Astron. J.* **61**, 97.
Johnson, H. M.: 1959, *Publ. Astron. Soc. Pacific* **71**, 425.
Kerr, F. J. and Westerhout, G.: 1965, in G. P. Kuiper and B. M. Middlehurst (eds.), *Galactic Structure*, University of Chicago Press, Chicago, p. 167.
Lewis, B. M.: 1968, *Proc. Astron. Soc. Australia* **1**, 104.
Lin, C. C.: 1970, *Invited Discourse*, 14th General Assembly IAU.
McGee, R. X.: 1964, *Australian J. Phys.* **17**, 515.
McGee, R. X. and Milton, J. A.: 1966, *Australian J. Phys.* **19**, 343.
Mathis, J. S.: 1965, *Publ. Astron. Soc. Pacific* **77**, 90.
Mayall, N. U.: 1960, *Ann. Astrophys.* **23**, 344.
Menon, T. K.: 1958, *Astrophys. J.* **127**, 28.
Oort, J. H.: 1965, *Trans. IAU*, **12 A**, 789, Figure 2.
Peimbert, M.: 1968, *Astrophys. J.* **154**, 33.
Peimbert, M. and Costero, R.: 1969, *Bol. Obs. Tonantzintla Tacubaya* **5**, 3.
Peimbert, M. and Spinrad, H.: 1970, *Astrophys. J.* **159**, 809.
Reifenstein, E. C. III, Wilson, T. L., Burke, B. F., Mezger, P. G., and Altenhoff, W. J.: 1970, *Astron. Astrophys.* **4**, 357.
Roberts, M. S.: 1968, *Astrophys. J.* **151**, 117.
Roberts, M. S.: 1969, *Astron. J.* **74**, 859.
Roberts, M. S.: 1970, *Astrophys. J. Letters* **161**, L9.
Roberts, M. S. and Warren, J. L.: 1970, *Astron. Astrophys.* **6**, 165.
Robinson, B. J. and Koehler, J. A.: 1965, *Nature* **208**, 993.
Rubin, V. C., Ford, W. K., Jr., and D'Odorico, S.: 1970, *Astrophys. J.* **160**, 801.
Schmidt, M.: 1962, in J. Sahade (ed.) 'Symposium on Stellar Evolution', Astronomical Observatory, La Plata, Argentina, p. 61.
Searle, L.: 1971, this volume, p. 66.
Shobbrook, R. R. and Robinson, B. J.: 1967, *Australian J. Phys.* **20**, 131.
Turner, K. C.: 1970, in Annual Report of the Director, Department of Terrestrial Magnetism, 1968–69, Carnegie Institute of Washington, Washington, D.C., p. 366.
Van de Hulst, H. C., Raimond, E., and Van Woerden, H.: 1957, *Bull. Astron. Inst. Neth.* **14**, 1.
Vaucouleurs, G. de: 1966, in G. Barbera (ed.) 'Galileo Symp. Cosmology', Florence, Italy, p. 37.
Volders, L. and Högbom, J. A.: 1961, *Bull. Astron. Inst. Neth.* **15**, 307.
Weliachew, L.: 1969, *Astron. Astrophys.* **3**, 402.

Discussion

Heidmann: I would like to add a few words to Morton Roberts' talk. Miss Bottinelli just completed her thesis on the large-scale distribution of neutral hydrogen in small galaxies. She obtained the effective H I diameter for 35 galaxies, from lenticulars to irregulars; this is the diameter containing half of the total neutral hydrogen. She found that the H I diameter to optical Holmberg diameter ratio increases from one-half to one from Sb to Magellanic irregulars. Thus, the H I is quite widely distributed and is more so for late types.

She found that one-third of these galaxies have an asymmetrical H I distribution with a tendency for asymmetrical H I distributions to be associated with asymmetrical optical distributions. This may be related to the existence of H I companions pointed out by Roberts.

From the H I diameter, the true mean projected H I density can be deduced. There is practically no correlation of this true density with morphological type; its mean value is 1.3×10^{-3} gm cm^{-2}.

I would like to make a comment about Roberts' relation between color excess and M_H/L excess. This may be related to the dependence of color index on morphological type and on intrinsic luminosity given in Gouguenheim's thesis (*Astron. Astrophys.* **3**, 281, 1969). This may lead again to the fact that the main physical properties of galaxies are determined by type and luminosity as shown for optical diameters by myself (*Astrophys. Letters* **3**, 19, 1969) and for hydrogen masses and total masses by N. Heidmann (*Astrophys. Letters* **3**, 153, 1969).

Arp: The results on the M81 system are exciting because they relate the M81 system to a class of galaxies where companions exist roughly on either side of a large galaxy. In many cases the companions are connected by luminous filaments to the main galaxy. Heretofore, no luminous connections have been observed between M81 and its two companions, but the radio observations just presented by Roberts indicate now that material in M81 extends in both directions towards the two peculiar companions, M82 and NGC 3077.

The importance of the association of *two* companions lies in the explanation for the cause of the connections. If it were hypothesized that the connection was the result of a perturbation caused by an encounter between the main galaxy and an orbiting companion, then it would be extremely unlikely that two such encounters would take place simultaneously on opposite sides of the main galaxy. Therefore, we must look elsewhere for an explanation of the fact that there is material linking the main galaxy and the companions.

Heidmann: In connection with comments on the M81 system, I would like to point out that Bottinelli found no relation between asymmetrical H_I distributions and companion galaxies.

Mrs. Burbidge: I was surprised by the small velocity range given by the H_I in NGC 5128; in the gas in the dust lane we found a considerably larger range due to rotation, from the $H\alpha$ line.

Roberts: The 21-cm absorption arises from in front of the two prominent radio sources located within the optical image of NGC 5128. These sources lie near the axis of rotation of this galaxy which is approximately perpendicular to the dust lane. Thus, essentially no rotational component would be expected in the 21-cm velocity profile.

Ekers: Perhaps there is a more widespread class of H_I clouds without optical counterparts of which only 'companions' have been found while observing known galaxies.

Lewis: The velocity field of NGC 300 can be satisfactorily fitted by a single circular symmetric rotation curve and optimized to fit the most appropriate major axis position angle. This is found to agree closely with the major axis defined by the integrated antenna temperature contours. Likewise, the results from the velocity field of NGC 5236 agree well with fitting an elliptic rotation curve to the data and treating as above. In this case the major axis of the outermost optical isophotes of Sĕrsic and from the integrated antenna temperature contours all agree. Thus, it is not necessary to postulate any perturbations of the velocity field by companion galaxies. Indeed this is the conclusion I have also drawn from the M81–82 velocity field shown by Dr. Roberts.

STRUCTURAL AND KINEMATIC PROPERTIES OF POPULATIONS OF THE ANDROMEDA GALAXY

J. EINASTO

W. Struve Astrophysical Observatory, Tartu, Estonia, U.S.S.R.

Abstract. New observational data (Spinrad, 1970; Van den Bergh, 1970; Rubin and Ford, 1970) are used to determine structural and kinematic parameters of the nucleus, the subsystem of globular clusters, and interstellar hydrogen in M31.

The mass derived for the nucleus from the new spectrophotometric data is in good agreement with the virial mass $6 \times 10^8 \mathcal{M}_\odot$. Model calculations show that there is no appreciable exchange of stars between the nucleus and the bulge. The rotation energy of the nucleus is only 7.5% of the total kinetic energy; the central density is $2 \times 10^6 \mathcal{M}_\odot \, \mathrm{pc}^{-3}$.

The mean radius of the subsystem of globular clusters is 4.5 kpc. This indicates that the subsystem of old stars is not identical with the spheroidal component of the galaxy, whose mean radius is only 1 kpc. Radial velocity dispersion of globular clusters is only half of that of the nucleus. This shows a strong dependence of the velocity dispersion on distance to the center of the galaxy and a bias in mass determination of a galaxy from velocity dispersion near the nucleus.

On the basis of data on rotation two mass distribution models have been found, differing from each other in respect of the mass concentration to the center. Spectrophotometric data on the stellar content of the bulge are urgently needed to solve the mass distribution problem.

1. Introduction

To construct a meaningful physical theory of the structure and evolution of a galaxy one needs reliable data on parameters, describing the spatial and kinematic structure of the galaxy and its subsystems of different ages. As such parameters one can adopt: the mass of the subsystem $\mathcal{M}$, its mean radius, a_0, the axial ratio of equidensity ellipsoids, ε (supposing equidensity surfaces of subsystems to be ellipsoids of rotational symmetry and constant axial ratio), and suitable structural parameters determining the degree of concentration of the mass to the center of the system. As concentration parameters dimensionless quantities K_i may be used (Einasto, 1968). In the present paper, however, we use for this purpose model-parameters N, x_0 of the modified exponential density law (Einasto, 1970).

In a series of papers (Einasto, 1969; Einasto and Rümmel, 1970a, b, c) we have studied the structure of the Andromeda galaxy and its subsystems and obtained preliminary values for descriptive parameters. Last year new observational data have become available for the nucleus, the subsystem of globular clusters, and interstellar hydrogen. It appears reasonable to use these data to redetermine structural parameters of the populations mentioned.

2. Nucleus

At the Basel IAU Symposium on Spiral Structure of the Galaxy we argued that different methods lead to different values for the mass of the nucleus of M31 (Einasto and Rümmel, 1970c). On the basis of photometric data (Redman and Shirley, 1937; Johnson, 1961; Kinman, 1965), and spectrophotometric data on the stellar content and mass-to-light ratio $\mathcal{M}/L = f = 16$ (Spinrad, 1966) we obtained (Einasto, 1969) for the mass of the nucleus the value $\mathcal{M} = 5 \times 10^7 \mathcal{M}_\odot$. On the other hand, the known mean radial velocity dispersion of stars in the nucleus $\sigma_r = 225$ km s^{-1} (Minkowski, 1962), mean radius $a_0 = 5$ pc (obtained from the photometric profile of the nucleus), and axial ratio of equidensity ellipsoids $\varepsilon = 0.8$ enable us to apply the tensor virial theorem. The resulting mass is $\mathcal{M} = 5 \times 10^8 \mathcal{M}_\odot$ (Einasto and Rümmel, 1970c).

The discrepancy in mass can be removed supposing, as Lynden-Bell (1969) does, that in the center of the galaxy a massive body exists. In this case the tensor virial theorem is to be modified. For the mean radial velocity dispersion of stars in the nucleus we have

$$\bar{\sigma}_r^2 = G a_0^{-1} \beta_r (\mathcal{M}_c + H_0 \mathcal{M}_0) \tag{1}$$

where G is the gravitational constant,

$$a_0^{-1} = \mathcal{M}_0^{-1} \int_0^\infty \mu(\alpha) \alpha^{-1} \, d\alpha \tag{2}$$

the harmonic mean radius of the stellar subsystem of the nucleus, $\mathcal{M}_c$, $\mathcal{M}_0$ are the masses of the central body and the stellar population of the nucleus respectively, and H_0, β_r dimensionless parameters. In formula (2), $\mu(\alpha)$ is the mass distribution function (mass of an ellipsoidal layer of equal density and unit equatorial thickness). The parameter H_0 depends only on the distribution of mass of the stellar population in the nucleus and is defined by the formula

$$H_0 = a_0 \mathcal{M}_0^{-2} \int_0^\infty M(\alpha) \mu(\alpha) \alpha^{-1} \, d\alpha, \tag{3}$$

where

$$M(\alpha) = \int_0^\alpha \mu(\alpha') \, d\alpha' \tag{4}$$

is the integral mass distribution function. The parameter β_r depends on the axial ratio of equidensity ellipsoids ε and on the inclination angle i of the equatorial plane of the system to the line of sight

$$\beta_r = \beta_R \cos^2 i + \beta_z \sin^2 i, \tag{5}$$

where

$$\beta_R = 1/2e^2 [(\text{arc sin } e/e) - \varepsilon], \tag{6}$$

$$\beta_z = \varepsilon^2/e^2 [(1/\varepsilon) - (\text{arc sin } e/e)] \quad \text{and} \quad e^2 = 1 - \varepsilon^2. \tag{7}$$

The values of H_0, β_R, β_z have been calculated for various models of stellar systems; the results will be published elsewhere. In particular for the nucleus of M31 we obtained from photometric observations $\varepsilon = 0.8$, and $i = 12°8$, which gives $\beta_r = 0.375$.

The space density of the stellar population can be fairly well represented by an exponential law, for which $H_0=0.312$. Adopting for the mean radius and the mass of the stellar component the values given above, we obtain for the central mass $\mathcal{M}_c=1.4\times10^8\mathcal{M}_\odot$.

After the Basel Symposium I requested Dr. H. Spinrad to determine the maximum value of mass-to-light ratio consistent with spectroscopic observations. Recently new data became available (Spinrad, 1970). According to a new model of stellar content the mass-to-light ratio increases to a value $f_V=45$, due to the necessity of adding faint M dwarfs. An upper limit to the number of red dwarfs is given by $V-K$, $V-L$ color observations (Sandage et al., 1969), which gives $f_V \leqslant 65$.

A re-examination of the photometric data mentioned above gives for the luminosity of the nucleus in visual light a value $L_V=1.42\times10^7 L_\odot$. This is four times higher than our previous estimate. The difference is due to the absorption and color corrections: $A_V=0.6$ in our Galaxy, $A_V=0.3$ in M31 (Arp, 1965) and $B-V=1.0$ (Sandage et al., 1969). The luminosity profile in the central part of M31 is shown in Figure 1. Using these data we obtain for the mass of the stellar component of the nucleus $\mathcal{M}_0=6.4\pm2.1\times10^8\mathcal{M}_\odot$. The mean error of the result is estimated allowing for uncertainties of L_V, and f_V.

On the other hand, applying a better model of the distribution of mass with $H_0=0.273$, we obtained from the virial theorem $\mathcal{M}_0=5.7\pm1.9\times10^8\mathcal{M}_\odot$. In this estimate we adopted $\mathcal{M}_c=0$; the sources of mean error are uncertainties in σ_r, and a_0.

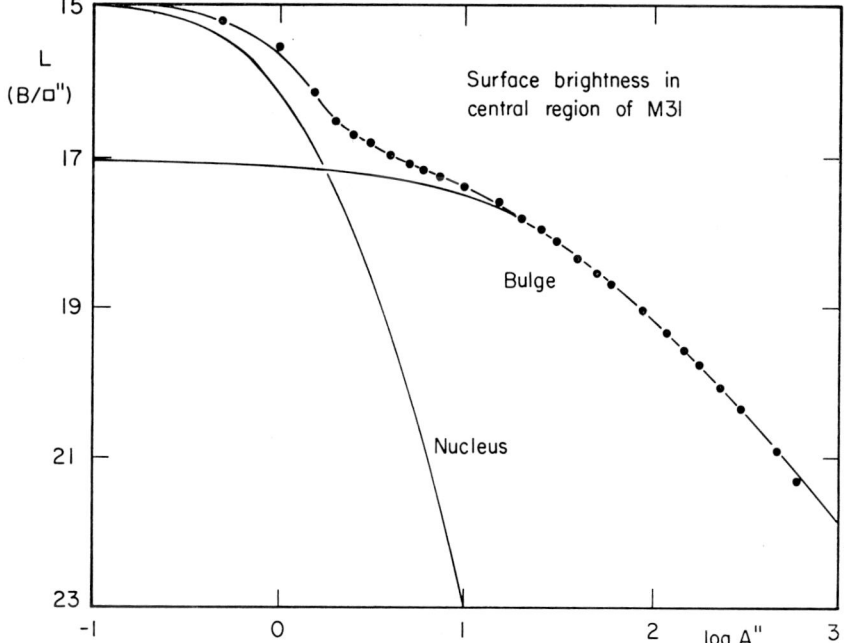

Fig. 1. Surface brightness of the nucleus and the bulge in the central region of M31. Dots represent observations, curves represent model profiles.

Good agreement between these two independent estimates shows that it is not necessary to suppose the existence in the center of M31 of a body of large mass. However, ultraviolet observations carried out from the OAO indicate the presence of a UV-source in the nucleus of M31. If this source has dimensions small compared with the mean radius of the stellar component of the nucleus, formula (1) may be applied to estimate an upper limit for the point mass, which may be attributed to the UV-source. Assuming $\mathcal{M}_0 \geqslant 4.0 \times 10^8 \mathcal{M}_\odot$, we find $\mathcal{M}_c \leqslant 0.5 \times 10^8 \mathcal{M}_\odot$.

Sandage et al. (1969) concluded from the U−B color variation along the major axis that the stellar content of the nucleus differs from that of the bulge. However, at present it is not clear whether the variation is due to the difference in stellar content or to the existence of a non-stellar UV-source in the center of M31. A difference in stellar content is not excluded because the nucleus is dynamically almost isolated and there is no appreciable exchange of stars between the nucleus and the bulge. To demonstrate this we have calculated, with the aid of a model of the mass distribution and the gravitational field of the nucleus, the apogalactic distances of stars moving through the center of the nucleus with various velocities (Table I). The majority

TABLE I

v	R_{apogal}
km s^{-1}	pc
225	1.8
450	8
1080	85

of the nucleus stars have velocities of some hundred km s^{-1} and do not move far off from the center. Only stars having large velocities exceeding the escape velocity with respect to the nucleus, 1080 km s^{-1}, go far away.

The mass density near the center of M31 according to our model is $2 \times 10^6\ \mathcal{M}_\odot\ \text{pc}^{-3}$, the angular velocity of circular motion 186 km s^{-1} pc^{-1}. The mean observed angular velocity is 12 km s^{-1} pc^{-1} (Lallemand et al., 1960). Using the tensor virial theorem and supposing a rigid-body rotation we find that the rotation energy is only 7.5% of the total kinetic energy of the nucleus. The binding energy of the nucleus (total negative energy per unit mass) is 7.5×10^4 km^2 s^{-2}.

3. The Subsystem of Globular Clusters

To reconstruct the past history of a galaxy it is important to know the structure and kinematics of populations of stars of various ages (cf. Einasto, 1970). As commonly accepted, the oldest stellar population belongs to the halo. Unfortunately the spatial density in the halo is very low and it is difficult to separate it by observation from bright disks and spirals. The study of elliptical galaxies shows that the transition from the halo to the bulge is a continuous one. Both populations have practically identical axial ratios of equidensity surfaces and can be described by the same density law of de Vaucouleurs' type. The mean radius of the spheroidal component of M31 (bulge and halo together) is (Einasto, 1969) $a_0 = 1$ kpc.

In the case of globular clusters we have the possibility of studying the overall spatial distribution of an old population without any significant interference by other populations.

Our sample of globular clusters has been collected on the basis of Veteŝnik's (1962) catalog. Photometric data on clusters were collected from various sources (Sharov, 1968). Probable open clusters are excluded from the general list; they lie in the V, B−V diagram to the left of the reddening line $A_V/E_{B-V} = 2.5$ (Van den Bergh, 1970), going through the point V = 18.0, B−V = 1.00. H_α-regions (Haro, 1950), objects without any photometric data, and very faint clusters (B > 19.0), were also excluded. The remaining sample was divided into two groups: bright clusters (B ⩽ 17.5) and faint clusters (17.5 < B ⩽ 19.0), which consist of 101 and 92 clusters respectively.

On the basis of measured (x, y) co-ordinates, published by Veteŝnik (1962), the galactocentric co-ordinates W, U along the major and minor axes of the galaxy, and the projected distance from the center $A = (W^2 + E^{-2}U^2)^{1/2}$ were calculated. E is the apparent axial ratio of equidensity ellipses. We found $E = 0.57$, which corresponds to the true axial ratio $\varepsilon = 0.54$.

The distribution of clusters in A is somewhat different in the two groups, which

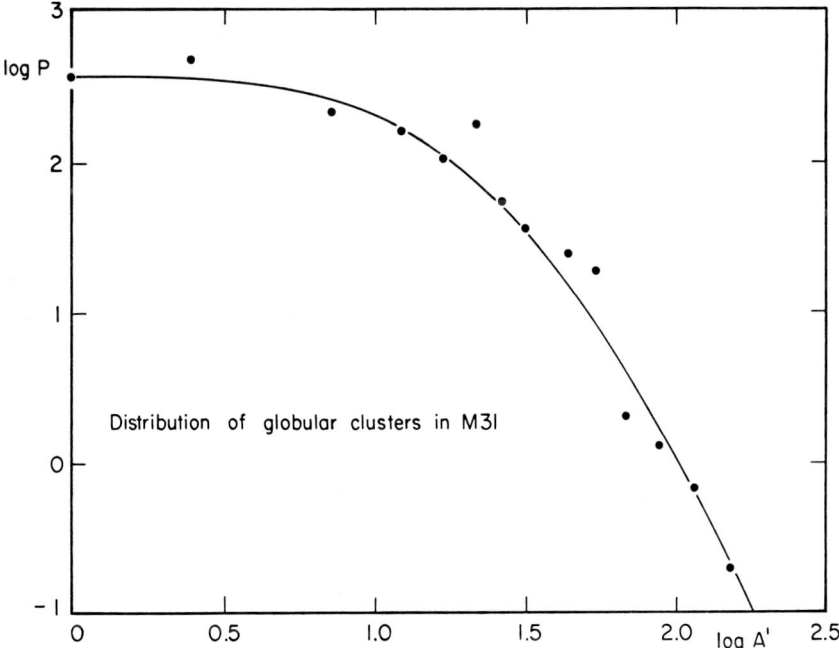

Fig. 2. Distribution of globular clusters in M31. As abscissae the logarithm of the major axes of equidensity ellipses (in minutes of arc) are plotted; as ordinates, the number of clusters per unit surface area (in arbitrary units). Dots represent observations; the solid curve represents the model.

can perhaps be explained by a selection effect in observations. In the central and outer halo region the relative number of faint clusters is smaller. After a slight correction of numbers of faint clusters both groups were united and a general distribution was found (Figure 2). The distribution can be fairly well represented by a modified exponential density law (Einasto, 1970) with mean radius $a_0 = 4.5$ kpc, $N = 4$, and $x_0 = 10.5$.

This result shows that the system of globular clusters has a much greater mean radius than the spheroidal component of M31. As we have no reason to believe that the spatial structure of the system of globular clusters differs significantly from that of the other old first generation stars, we come to the conclusion that the subsystem of old stars cannot be identified with the spheroidal component of the galaxy. In the central part of the galaxy star formation has probably taken place much longer than in the halo, giving rise to the formation of the bulge.

As to the kinematics of globular clusters, Van den Bergh (1970) has recently determined radial velocities for 44 bright globular clusters in M31. From his data the mean galactocentric radial velocity dispersion can be found. Adopting for the systemic velocity of the galaxy -300 km s^{-1} (Van den Bergh, 1970; Rubin and Ford, 1970) we obtained

$$\sigma_r = 122 \text{ km s}^{-1}$$

which is only 0.54 of the mean radial velocity dispersion in the nucleus (Minkowski, 1962).

Preliminary model calculations show that the velocity dispersion of the subsystem of globular clusters depends mainly on the total mass of the galaxy and is quite insensitive to the mass concentration towards the center.

The difference between the mean velocity dispersions of the nucleus and the subsystem of globular clusters has one important consequence. Velocity dispersions near the nucleus have been used to determine the total mass of a galaxy. For example, by this method Brandt and Roosen (1963) obtained for M87 $\mathcal{M} = 2.7 \times 10^{12} \mathcal{M}_\odot$, and $\mathcal{M}/L = 85$. However, a strong dependence of the dispersion on the distance R shows that the velocity dispersion near the center characterizes the mass and the size of the nucleus only. If the run of relative velocity dispersion in other galaxies is the same as in M31, the total mass of the galaxy and the respective mass-to-light ratio obtained from the virial theorem are approximately 4 times higher than the true ones.

4. Interstellar Hydrogen

The distribution and kinematics of interstellar hydrogen in M31 were analyzed on the basis of radio data in our report to the Basel Symposium (Einasto and Rümmel, 1970b). Recently new optical radial velocities of hydrogen clouds in M31 became available (Rubin and Ford, 1970). In the present paper the results of a preliminary study of these data are given.

In Figure 3 the velocities from Table I of the paper by Rubin and Ford are plotted as a function of R. If in one association several emission clouds have been observed, a weighted mean rotation velocity and distance have been used; points far from the major axis have been omitted. Hydrogen 21-cm rotation velocities according to our treatment (Einasto and Rümmel, 1970b) are also indicated in the figure. The solid line represents the rotational curve adopted by Rubin and Ford (model 5 in Table III of Rubin and Ford's paper).

New observations confirm earlier results (Babcock, 1939) that the rotation velocity has a sharp maximum near 0.5 kpc from the center and a deep minimum near 2 kpc. Identifying the rotation velocities with the circular ones, Rubin and Ford calculated a corresponding mass distribution model and found a negative projected density at about 2 kpc from the center.

TABLE II

Quantity	Unit	Nucleus	Bulge	Halo	Disc	Flat
ε		0.80	0.60	0.54	0.10	0.02
a_0	kpc	0.005	0.4	4.5	10	8
N		4	4	4	1	0.5
h		4.411	4.411	4.411	2.220	1.571
k		0.4165×10^{-4}	0.4165×10^{-4}	0.4165×10^{-4}	0.4245	1.128
L_V	$10^9 \, L_\odot$	0.014	4.5	15.3	12.5	3.1
$\mathscr{M}_\mathrm{I}$	$10^9 \, \mathscr{M}_\odot$	0.6	19	62	126	10
$\mathscr{M}_\mathrm{II}$	$10^9 \, \mathscr{M}_\odot$	0.6	12	1	184	20
f_I		42	4.2	4.1	10.1	3.2
f_II		42	2.7	0.07	14.7	6.5

However, it is difficult to adopt negative or very small positive projected densities, because photometric observations indicate no minimum in projected luminosity density in this region. A large negative gradient of circular velocity would imply instability of circular motion, also difficult to explain. Apparently low rotation velocity in this region is due to another cause.

Adopting a smooth distribution of mass and identifying maximum rotation velocities near $R=0.5$ kpc with the circular velocity, we can calculate parameters for a corresponding mass distribution model and find the run of circular velocity for the whole galaxy. The parameters of this model are given in Table II, the circular velocity in Figure 3 (Model II).

This model has no negative densities, but the mass-to-light ratios* of the bulge and the halo are improbably small, especially that of the halo. The model has also another essential deficiency. The observed velocity dispersion of stars in the central parts of M31 is comparable to the rotation velocity (see preceding section) and the corresponding term in the hydrodynamic equations cannot be neglected. This indicates that the circular velocity must be considerably greater than the rotational one.

---

* The luminosities given in Table II are converted into the V system adopting $B-V=1.0$ for the nucleus, the bulge and the halo, and $B-V=0.4$ for the disc and the flat component, with a correction for mean absorption $A_V=1.0$.

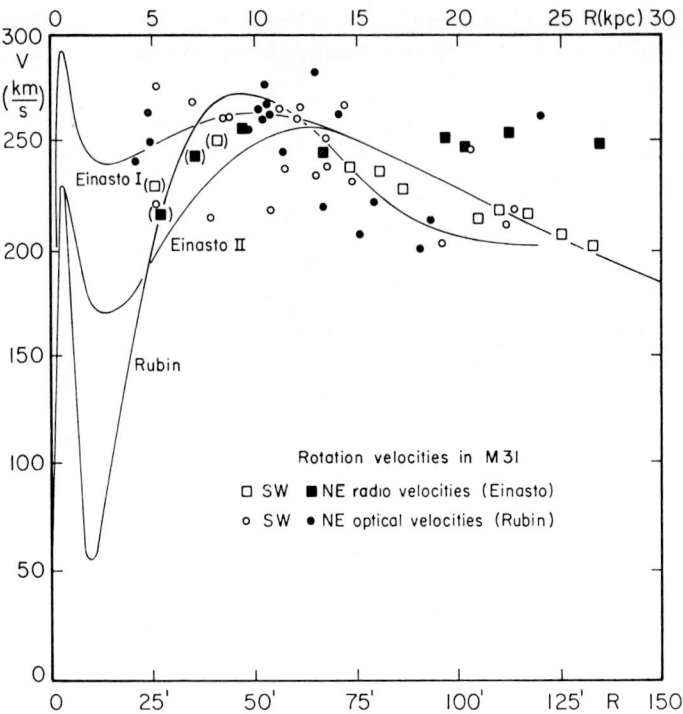

Fig. 3. Rotation of M31. Observed rotation velocities according to 21-cm radio data (Einasto and Rümmel, 1970b) and optical measurements (Rubin and Ford, 1970) are given with different signs. Circular velocity curves are shown by solid lines.

Both deficiencies are removed to a certain extent in another model whose parameters and circular velocity curve are also given in Table II and Figure 3 respectively (Model I). The mass-to-light ratios of the bulge and the halo seem to be still too low, compared with the mean mass-to-light ratio for the galaxy as a whole, $\mathcal{M}/L_V = 6.1$. However, if we assume a considerably greater circular velocity in the central region, the velocity dispersion becomes improbably large.

To solve the problem of mass distribution in the central region of M31, spectrophotometric data on the stellar content of the bulge are urgently needed.

Acknowledgements

I am indebted to Dr H. Spinrad, Dr S. van den Bergh, and Mrs Dr V. Rubin for sending me the results of their work prior to publication, to Dr G. Kuzmin for suggestions connected with the use of the virial theorem, and to Mrs L. Einasto and U. Rümmel for programming the calculations and for help in the observational data processing.

References

Arp, H.: 1965, *Astrophys. J.* **141**, 43.
Babcock, H. W.: 1939, *Lick Obs. Bull.* No. 498.

Bergh, S. van den: 1970, *Astrophys. J. Suppl.* No. 171.
Brandt, J. C. and Roosen, R. G.: 1969, *Astrophys. J. Letters* **156**, L59.
Einasto, J.: 1968, *Tartu Publ.* **36**, 357.
Einasto, J.: 1969, *Astrofizika* **5**, 137.
Einasto, J.: 1970, *Tartu Teated* No. 26, 1.
Einasto, J. and Rümmel, U.: 1970a, *Astrofizika* **6**, 241.
Einasto, J. and Rümmel, U.: 1970b, in W. Becker and G. Contopoulos (eds.), 'The Spiral Structure of our Galaxy', *IAU Symp.* **38**, 42.
Einasto, J. and Rümmel, U.: 1970c, in W. Becker and G. Contopoulos (eds.), 'The Spiral Structure of Our Galaxy', *IAU Symp.* **38**, 51.
Haro, G.: 1950, *Astron. J.* **55**, 66.
Johnson, H. M.: 1961, *Astrophys. J.* **133**, 303.
Kinman, T. D.: 1965, *Astrophys. J.* **142**, 1376.
Lallemand, A., Duchesne, M., and Walker, M. F.: 1960, *Publ. Astron. Soc. Pacific* **72**, 76.
Lynden-Bell, D.: 1969, *Nature* **223**, 690.
Minkowski, R.: 1962, in G. G. McVittie (ed.), *Problems of Extra-Galactic Research*, Macmillan Co., N.Y., p. 112.
Redman, R. O. and Shirley, E. G.: 1937, *Monthly Notices Roy. Astron. Soc.* **97**, 416.
Rubin, V. C. and Ford, W. K., Jr.: 1970, *Astrophys. J.* **159**, 379.
Sandage, A. R., Becklin, E. E., and Neugebauer, G.: 1969, *Astrophys. J.* **157**, 55.
Sharov, A. S.: 1968, *Astron. Zh.* **45**, 146.
Spinrad, H.: 1966, *Publ. Astron. Soc. Pacific* **78**, 367.
Spinrad, H.: 1970 (private communication).
Vetešnik, M.: 1962, *Bull. Astron. Inst. Czech.* **13**, 180.

THE POPULATION I CONTENT OF THE ELLIPTICAL COMPANIONS OF M31

P. W. HODGE

University of Washington, Seattle, Wash., U.S.A.

(To be published elsewhere)

Abstract. Twenty years ago Walter Baade pointed out the remarkable fact that two of the four elliptical galaxy companions of M31 (namely NGC 185 and 205) contained apparent Population I in their central areas, evidenced by the presence of dust clouds and a dozen or so OB stars. I wish to report here on some observations of these stars and absorption regions, made in an attempt to understand the origin of this anomalous material and its relevance, if any, to the pattern of galaxy development.

1. Methods of Observation

Direct photoelectric measures with small diaphragms were made every few seconds along the NS and EW axes of the galaxies, all in the U, B, V system. Schmidt plates taken with the 48-inch Palomar instrument were used to produce isophotes over a wide range of distances from the galaxy centers. For the OB stars, Arp's sequence in M31 was used with astrophotometer measures to establish local standards near NGC 205, and then Baade's 200-in. plates were used to transfer magnitudes therefrom to the embedded OB stars.

2. Color Distribution

The color distributions of NGC 185 and 205 are quite similar. Both show a blue excess at the center that decreases outward to a limit, and the color then remains constant further outward. The data are:

NGC 205: $B-V$ (outer) $= +0.72$
          $B-V$ (nucleus) $= +0.52$
          difference   0.20 mag.

NGC 185: $B-V$ (outer) $= +1.15$
          $B-V$ (nucleus) $= +0.95$
          difference   0.20 mag.

The color anomaly in NGC 205 is noticeably offset from the center in the same sense as the OB stars.

3. The OB Stars

The best of Baade's 200-in. telescope plates show as many as 61 OB stars in NGC 205 and nearly as many in NGC 185. In NGC 205 the centroid of these stars is offset from the nucleus by 30 pc west and 33 pc north. For NGC 185 they are nearly centered. The sizes of the Population I stellar arrays are 335 pc for NGC 205 and approximately 270 pc for NGC 185. The stars in NGC 205 are arranged in an irregular elliptical grouping with its major axis lined up with that of the galaxy's old stars. The brightest star in NGC 205 has $M_B = -5.2$, implying an age of 4×10^6 yr, if at the main sequence turn-off. Masses for the Population I components for the two galaxies are 5×10^5 $\mathcal{M}_\odot$ for NGC 205, based on the available luminosity function assumed to extend parallel to the van Rhijn function to $M_V = +3.5$, and 2×10^5 $\mathcal{M}_\odot$ for NGC 185, based on the total color and a similar assumption about the luminosity function.

4. The Dust Lanes

NGC 205 has two large dust lanes and more than a dozen small, faint ones. For the largest the measured color excess is $\Delta(B-V) = +0.13$ implying total absorption of $\Delta V = 0.4$ mag., using the usual assumptions. For the two absorption regions in NGC 185 the total absorption measures average 0.3 and 0.15, and total masses are estimated to be 200 and 25 $\mathcal{M}_\odot$, respectively.

Discussion

Heidmann: Have you any comment on the fact that in NGC 205 the dust is on one side, and the O and B stars are on the other side?

Hodge: The dust clouds are clustered in an area southwest of the NGC 205 nucleus at a distance almost identical to that of the OB star centroid, in the opposite direction. It is conceivable that this causes obscuration of the fainter stars and that the asymmetry in the OB star distribution is only apparent, the result of the real asymmetry in dust. Alternatively, the southwest part of the Population I material, if it is truly symmetrical in mass, may be relatively less condensed into stars there, in which case perhaps neutral hydrogen should be searched for near the dust regions. The total dust mass cannot be more than a percent or so of the computed mass contributed by the Population I stars.

ELECTRONOGRAPHIC PHOTOMETRY OF STARS IN THE GLOBULAR CLUSTERS OF THE MAGELLANIC CLOUDS

M. F. WALKER

Lick Observatory, Calif., U.S.A.

Abstract. Due to the linearity, high quantum efficiency, and high storage capacity of electronographic image intensifiers, their application to astronomical photometry constitutes an advance equal in importance to those that resulted from the introduction of the photographic plate and the photomultiplier. With existing electronographic image tubes and telescopes of 60-in. aperture stellar photometry is possible to about magnitude 23. Consequently a major advance in the study of the stellar contents of galaxies in the Local Group may now be made.

During 1968–69, the spectracon image intensifier was used on the 60-in. reflector of the Cerro Tololo Observatory for B, V observations of 14 globular clusters in the Magellanic Clouds. Details of the observing program and discussions of the observations so far reduced will appear shortly in papers in the *Astrophysical Journal* and in *Sky and Telescope*.

GAS IN THE NUCLEUS AND DISK OF M31

VERA C. RUBIN* and W. KENT FORD, Jr.*

Dept. of Terrestrial Magnetism, Carnegie Institution of Washington

Abstract. Image tube spectra of emission regions in M31 have been studied to determine relative line strengths and abundances. Lines of H, He, [NII], [SII], [OI], [OII], [OIII] are observed. The analysis indicates: (1) that there are ions whose line strengths, relative to Hα, are a function of the position of the emission region in M31; (2) abundance differences or different excitation mechanisms exist in regions separated by only a few hundred parsecs.

1. Introduction

For several years we have been obtaining image tube spectra of emission regions in M31. The spectra are centered at Hα, and include the Balmer lines Hβ, Hγ, Hδ; [SII] $\lambda\lambda$ 6717, 6731; [NII] $\lambda\lambda$ 6548, 6583; [OIII] $\lambda\lambda$ 5007, 4959; [OII] λ 3727; [OI] λ 6300; HeI λ 5876. On a few well exposed spectra, lines of [AIV] λ 7238, [AIII] λ 7136, HeI λ 6678, [NII] λ 5755, and [SII] λ 4071 are also seen. From the observed line strengths, it is now possible to answer the following questions. (1) Are there ions whose line strengths, relative to hydrogen, are a function of the position of the emission region in the galaxy? The answer is yes. (2) Is it possible to explain the observed line ratios by a single HII region chemical abundance, with only changes in T_e, N_e, and ionization from region to region? The answer is no: abundance differences or different excitation mechanisms exist in regions separated by only a few hundred parsecs. We now indicate how these results are obtained.

2. Relative Line Strengths

Line intensities, relative to $I(\mathrm{H}\alpha)$, have been determined for about 50 regions; for half of these we have two or more plates. Details of the photometry and a catalog of line intensities will be published elsewhere. All intensities are corrected for extinction and spectral response. From the duplicate values, we estimate that the mean error of the line ratio for a strong or moderate line is less than 20%; for a weak line the error may be as great as 30%. Because the ratios $I([\mathrm{OIII}])/I([\mathrm{OII}])/I([\mathrm{OI}])$ cannot be reconciled with collisional excitation mechanisms, we must assume that the excitation is radiative (Peimbert, 1968). Then the $I(\mathrm{H}\alpha)/I(\mathrm{H}\beta)$ ratio is predicted to be 2.84.

From the observed $I(\mathrm{H}\alpha)/I(\mathrm{H}\beta)$ ratio for each region, the logarithmic reddening correction, c, has been computed, adopting the reddening function tabulated by Seaton (1960). Values of c range from 0.1 to 2.0. Near the nucleus, there is a large spread in values, with $\bar{c}=1.0$. Near $R=24$ kpc, the spread has decreased, and $\bar{c}=0.5$.

---

* Visiting Astronomer, 1966, 1967, 1968, 1969 Lowell Observatory, and Visiting Astronomer 1967, 1968, 1969 Kitt Peak National Observatory, operated by the Association of Universities for Research in Astronomy, Inc. under contract with the National Science Foundation.

Except for the general decrease in reddening with increasing distance from the nucleus, there is no marked correlation of reddening with position of the spiral arms in M31. For the stars in M31, a value of $c=0.1$ is typical; hence, most of the reddening observed here arises in the emission regions. For emission regions in our Galaxy, Orion, M8, and M17, values of c (Peimbert and Costero, 1969) cover the range 0.38 to 1.7. All line strengths discussed below are corrected for reddening, although the ratios are generally insensitive to this correction because the wavelength region considered is usually small.

The intensity ratio $I([N{\sc ii}]\,\lambda\,6583)/I(H\alpha)$ is greater than unity only for emission regions near the nucleus and decreases to <0.2 for R near 22 kpc. Values of this ratio

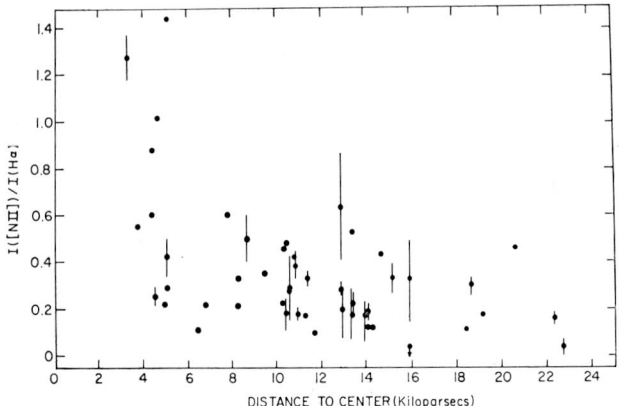

Fig. 1. Intensity ratio $I([N{\sc ii}]\,\lambda\,6583)/I(H\alpha)$ for emission regions in M31, as a function of distance from the center. Error bars indicate the total range of values from two or more plates.

are plotted in Figure 1. From the integrated spectra of galaxies, it is known that $I([N{\sc ii}])>I(H\alpha)$ in the nuclei of some galaxies. This effect is seen here for individual emission regions as well.

The ratio $I([S{\sc ii}]\,\lambda\,6717+\lambda\,6731)/I(H\alpha)$ exhibits an equally striking variation with R. For emission regions near the nucleus, this ratio is as large as 2, but for $R>5$ kpc it is generally near 0.1. In all cases with $I([N{\sc ii}])>I(H\alpha)$, then $I([S{\sc ii}])>I(H\alpha)$ also. Even for regions with $I([S{\sc ii}])$ and $I([N{\sc ii}])<I(H\alpha)$, there is a correlation of $I([N{\sc ii}])$ with $I([S{\sc ii}])$.

The $[O{\sc iii}]/H\beta$ intensity ratios indicate Aller excitation classes from 2 to 4, with the regions of higher excitation generally at greater distances from the nucleus. The principal exceptions to this relation are some regions close to the nuclear bulge ($R=4$ or 5 kpc), which exhibit high $[O{\sc iii}]/H\beta$ ratios.

The observed values of the helium to hydrogen ratio, $I(He\,\lambda\,5876)/I(H\alpha)$, show a large scatter at all R, with values from 0.01 to almost 0.1. For 21 regions, no He $\lambda\,5876$ line is seen, indicating an upper limit of 0.03 to this ratio. There is no indication of an increase of the helium/hydrogen ratio with decreasing R.

Perhaps more significant than these variations in the relative line strengths with distance from the nucleus is the fact that the 50 spectra can be placed into 8 classes, with similar strengths of [OI], [OII], [OIII], [NII], and [SII], relative to Hα, within each class. In what follows we shall discuss two of these spectral groups.

We show in Table I the relative line strengths for 10 emission regions, identified by Baade and Arp (1964) as the emission regions closest to the nucleus of M31. These spectra fit into two very different groups on the basis of relative line strengths.

TABLE I

Relative line ratios for HII regions near nucleus of M31.

Region Arp No.	R (kpc)	$\frac{I([SII])}{I(H\alpha)}$	$\frac{I([NII])}{I(H\alpha)}$	$\frac{I([OIII])}{I(H\beta)}$	$\frac{I([OII])}{I(H\alpha)}$	c
55	3.2	1.34	1.69	3.1	$\geqslant 1$	0.40
415	3.7	1.74	0.73	–	–	–
521	4.4	0.76	1.17	2.4	> 1	0.66
23	4.6	1.00	1.35	1.5	$\gg 1$	0.74
416	5.1	1.68	1.92	4.4	$= 1$	0.78
521a	4.4	< 0.12	0.80	–	$\ll 1$	–
519	4.5	0.14	0.32	0.5	–	1.2
74	4.9	0.05	0.29	0.5	< 1	1.1
75	5.0	0.06	0.59	0.2	< 1	1.1
423	5.1	0.29	0.56	0.2	< 1	1.1

[SII]: $\lambda 6717 + \lambda 6731$
[NII]: $\lambda 6548 + \lambda 6583$
[OIII]: $\lambda 5007$
[OII]: $\lambda 3727$

For the strong line group, the intensities of the forbidden sulphur doublet and the forbidden nitrogen doublet are greater than the Hα intensity; [OIII], [OII] and [OI] are also strong. It is unlikely that this is an ionization effect; where [OIII] is strong, [SII] is generally weak, for its ionization potential is low, so most [SII] would be ionized to [SIII]. In the second group, (the normal line group) the sulphur intensity is only one-tenth that of Hα; all other ions are relatively weaker also.

On Baade's 100-in. plates, there are no significant differences between the appearances of these 10 emission regions, nor are they separated spatially into two groups. Regions 521 (strong line) and 521a (normal line) are located within a hundred parsecs of each other, and spectra of both objects are contained on a single plate. Their radial velocities are in no way peculiar, although the velocity dispersion near $R=4$ kpc is the largest observed. (Rubin and Ford, 1970.) Curiously, the reddening coefficient, c, is significantly different in the two types of regions; for the strong line group, $\bar{c}=0.64$, while for the normal line group, $\bar{c}=1.2$. This difference must indicate either different properties of the interstellar medium at the positions of the two types of regions, or different excitation mechanisms, producing an Hα/Hβ ratio different from that predicted by radiative excitation.

We show in Figure 2 the line strengths, relative to Hα, for all 8 spectral groups. For each group, the mean distance to the nucleus, $\bar{R}$, and the number of emission regions in the group, n, are indicated. For $R \geqslant 5$ kpc, all lines are weaker than Hα, [N II] generally decreases with R, and [S II] is generally near 0.1, except for a group of 10 regions centered near 12 kpc, in which [S II]/Hα is near 0.3. For comparison, corresponding line strengths for Orion, M8, M17 (Peimbert and Costero, 1969) are included. About half of the emission regions in M31, $R \gtrsim 8$ kpc, are 'Orion-like', while the remaining regions have [N II] and [S II] lines which are about twice as intense, relative to Hα, as the corresponding lines in Orion.

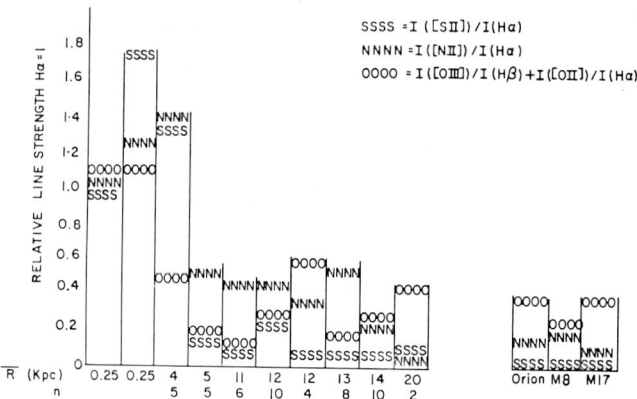

Fig. 2. Line strengths, relative to Hα, for emission regions in M31. Regions have been grouped by spectral properties. The mean distance to the center $\bar{R}$, and the number of regions in each group, n, are shown. For $R = 0.25$ kpc, line intensities come from spectra taken across the nuclear bulge.

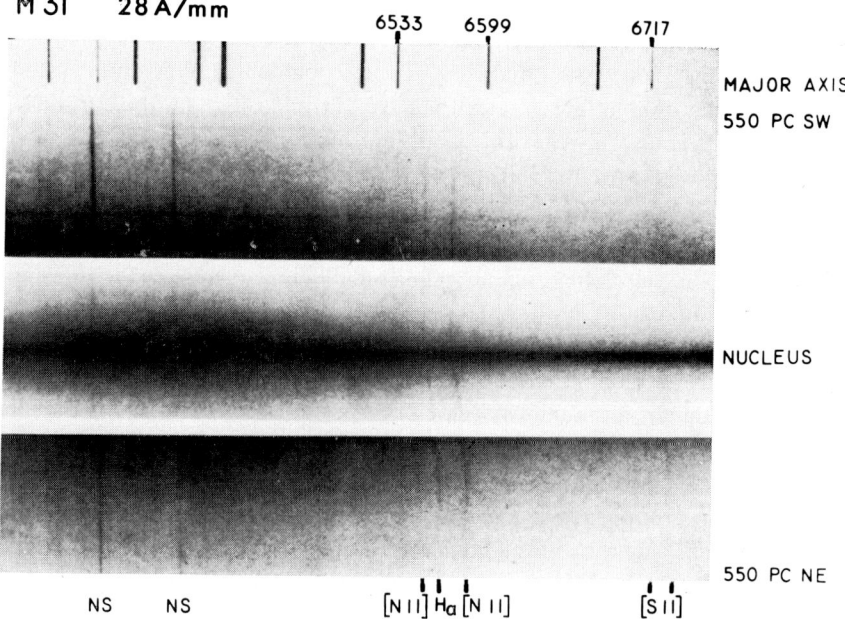

Fig. 3. Image tube spectra across the nuclear bulge in M31, dispersion 28 Å/mm, showing emission lines of [N II] λ 6548, Hα, [N II] λ 6583, and [S II] $\lambda\lambda$ 6717 and 6731. Slit along major axis. Exposure times 5^h43^m, 2^h, 5^h46^m, with 84-in. Kitt Peak reflector.

For $R<3$ kpc, no emission regions are identified in M31. However, high dispersion spectra across the nuclear bulge show emission lines of [N II] λ 6548, Hα, [N II] λ 6583, [S II] $\lambda\lambda$ 6717 and 6731, and [O III] λ 5007. The intensity ratio $I([\text{N II}]\,\lambda\,6583)/I(\text{H}\alpha)$ is generally about 2, although the ratio changes to 1 in some regions. Abrupt changes in the line ratios occur in dust lanes across the nuclear bulge. We show in Figure 3 three spectra taken along the major axis of M31, showing the emission lines in the red spectral region. Relative line intensities, measured at $R=0.25$ kpc SW and 0.25 kpc NE are plotted in the first two columns of Figure 2. Just as in the strong line group of emission regions near the nucleus, [S II], [N II], [O III], and [O II] are strong in the nuclear regions.

3. Relative Abundances

We will discuss only very generally the transformation from relative line strengths to relative abundances. If we assume radiative excitation, then the absence of the [N II] λ 5755 line implies $T_e < 11\,000°$ in general, while the ratio of the [S II] doublet implies $N_e \leqslant 10^4$, except in the nucleus where high densities exist. The ratio of total chemical abundances of oxygen to hydrogen, $N(\text{O})/N(\text{H})$, is then given by $(N(\text{O}^+) + N(\text{O}^{++}))/N(\text{H}^+)$, where the ratios of the ions are obtained from expressions given by Peimbert and Costero (1969). The procedure is similar for $N(\text{S})/N(\text{H})$. For N^{++}, which is not observed, and S^{++}, which is not observed, we use ratios of ions of oxygen with similar ionization potentials.

For the strong line group of emission regions near the nucleus of M31, no choice of T_e can reproduce the solar abundance for O, N, and S. If $N(\text{O})/N(\text{H})$ is chosen to be solar, then $T_e \sim 8000°$, nitrogen is overabundant by a factor of about 2, and sulphur is overabundant by a factor of about 25. Even for a high T_e, sulphur is still overabundant, and oxygen then becomes underabundant.

For the normal line group, nitrogen and oxygen have the solar abundance for $T_e = 7000°$, and sulphur is overabundant by a factor of 2, values which are like those for emission regions in our Galaxy (Peimbert and Costero, 1969). The remaining emission regions in M31 have relative abundances which do not vary by more than factors of 2 or 3 from those in galactic emission regions. Details for all regions will be published elsewhere.

The mean helium abundance, for 15 regions with measurable He I λ 5876 lines, is $N(\text{He})/N(\text{H}\alpha) = 0.12$. However, if we include 21 additional regions with low values for the upper limit to this line intensity, then $N(\text{He})/N(\text{H}\alpha) < 0.08$. The He II λ 4686 line is never observed: this implies $N(\text{He}^{++})/N(\text{H}^+) < 5 \times 10^{-4}$.

4. Conclusions

If the observed line ratios in M31 are transformed to relative chemical abundances, using conventional radiative excitation theory, then the abundance of sulphur become abnormally high in the nuclear bulge and in some emission regions near the nucleus. The remaining emission regions have abundances which differ only by small factors

(2–3) from galactic emission regions. Hence it appears that chemical abundance differences must exist in regions separated by only hundreds of parsecs, or excitation mechanisms must differ over the same distances. From the present observations, it is not possible to choose between these two possibilities, although no simple mechanism that can enhance the sulphur lines is apparent. The low Hα/Hβ ratio observed for the strong line regions makes it unlikely that the excitation is collisional, unless T_e is very high ($\sim 40000°$). At this T_e, the intense sulphur line would not be predicted, so models with various regions and various T_e would have to be invoked.

In our Galaxy, emission regions with abnormally strong sulphur lines are unknown, although observations are scarce, and do not extend to regions near the nucleus. Only the Crab nebula (Minkowski, 1942; Woltjer, 1958) and supernova remnants (Parker, 1964) have [S II] lines stronger than Hα. In M31, a single supernova remnant would be difficult to observe. If many supernovae recently enriched the interstellar medium in some regions near the nucleus of M31, the helium abundance would be expected to be higher than is observed. It is possible that the mechanism enhancing the sulphur lines in supernova remnants is also operating near the nucleus of M31. For the galactic supernova remnants, it is generally concluded that the [S II] originates in a zone of lower temperature than do the lines of higher ionization.

Among our collection of spectra of nuclei of external galaxies, only for two, M51 and NGC 1052, is the intensity of the sulphur doublet ($\lambda 6717 + \lambda 6731$) greater than the H$\alpha$ intensity. For both these nuclei also, $I([N II]) > I(H\alpha)$. These spectra are reproduced in Figure 4 and values of the line strengths, corrected for extinction,

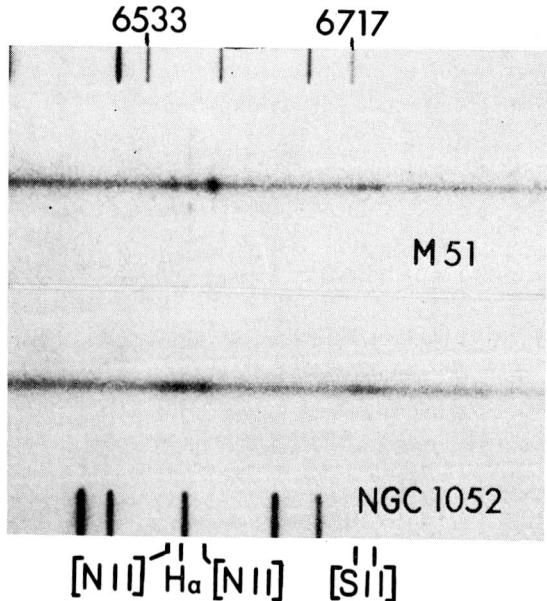

Fig. 4. Image tube spectra of M51 and NGC 1052, dispersion 66 Å/mm. Exposure times 38^m, 48^m at the Perkins 72-in. telescope. For M51, emission extends 75″, slit EW. Spectrum of NGC 1052 (larger redshift) has been displaced so emission features are in line with those in M51.

TABLE II

Emission line strengths for nuclei of M51, NGC 1052, and M81; $H\alpha = 1.0$

Nucleus	[N II] λ 6583	[S II] λ 6717	[S II] λ 6731	[S II] λ 6717 + λ 6731
M51	3.8	0.95	0.98	1.9
NGC 1052	1.1	0.82	0.68	1.5
M81	1.6	0.14	0.15	0.3

reddening and spectral response, are listed in Table II. In M81, the forbidden nitrogen lines are stronger than Hα, but $I([S II])/I(H\alpha) = 0.3$. For each galaxy, we have only a single calibrated plate, so the values may be revised later. In no other galaxies have we observed [S II] lines as great as one-tenth of the Hα intensity. More observations and studies of the [S II] doublet are desirable, because an understanding of its peculiar strength in the nuclei of some galaxies and some emission regions would add to our small store of knowledge concerning galactic nuclei.

Acknowledgements

We wish to thank Dr John Hall, Director of the Lowell Observatory, and Dr N. U. Mayall, Director of the Kitt Peak National Observatory, for making telescope time available.

References

Baade, W.: 1963, *Evolution of Stars and Galaxies*, Harvard University Press, Cambridge, p. 59.
Baade, W. and Arp, H. C.: 1964, *Astrophys. J.* **139**, 1027.
Minkowski, R.: 1942, *Astrophys. J.* **96**, 199.
Parker, R. A. R.: 1964, *Astrophys. J.* **139**, 493.
Peimbert, M.: 1968, *Astrophys. J.* **154**, 33.
Peimbert, M. and Costero, R.: 1969, *Bol. Obs. Tonantzintla Tacubaya* **5**, 3.
Rubin, V. C. and Ford, W. K., Jr.: 1970, *Astrophys. J.* **159**, 379.
Seaton, M. J.: 1960, *Rep. Prog. Phys.* **23**, 313.
Woltjer, L.: 1958, *Bull. Astron. Inst. Neth.* **14**, 39.

Discussion

King: Will the velocity fields be measured and published?

Mrs. Rubin: Velocities of the H II regions are published. (Rubin and Ford: 1970, *Astrophys. J.* **159**, 379.) Velocities of the excited gas in 16 position angles across the nuclear bulge have been measured but are not yet understood.

DISTRIBUTION OF DUST AND
HII REGIONS IN SPIRAL GALAXIES

BEVERLY T. LYNDS

Steward Observatory, Tucson, Ariz., U.S.A.

Abstract. The bright HII regions of a galaxy are always found near or in regions of high obscuration. It is also generally true that the dust lanes of a galaxy better define a spiral pattern than do HII regions. There appears to be a correlation between the size of the central region and the number of HII regions in a galaxy.

This is the second progress report on the study of the distribution of dark nebulae in galaxies. The ultimate goal of this investigation is to try to answer the questions "Which constituents of a spiral galaxy best define the basic pattern of its structure; and do the different constituents delineate the same design?". In the present study, I have concentrated on a comparison of the distribution of interstellar dust, as detected by its occultation of luminous regions of a galaxy, with the positions of the bright HII regions, as detected on Hα interference filter photographs of the spirals. The observational data of the paper consist of direct photographs, both in blue light and in the light of the Balmer alpha line, of 33 galaxies. For twelve of these objects, I used the Hα material obtained by Sandage at the 200-in.; the remaining 21 galaxies were photographed with the University of Arizona's 90-in. telescope equipped with a Carnegie image tube. The blue photographs used were gathered from the Hale Observatories' files, from Mrs Burbidge's collection, and from the 90-in. observations. The conclusions presented in this report are based on a sample of 3 S0, 2Sa, 11 Sb, and 16 Sc galaxies.

The general conclusions presented in my first progress report (Lynds, 1970) which was based on studies of 16 Sc galaxies (eight of which had Hα photographs available), are confirmed and extended to earlier type spirals. The bright HII regions of a galaxy are always found either next to or embedded in regions of high obscuration. It is also generally true that the dust lanes of a galaxy better define a spiral pattern than do the HII regions.

Figure 1 is an example of the technique now used in studying the 90-in. data. A series of interference filters, each with almost the same transmission curve of half-width of about 75 Å and centered at the wavelengths of 6563, 6607, 6650, and 6782, are used. Each galaxy is photographed through the appropriate filter so that the Balmer alpha line is recorded and then the galaxy is also photographed in the red through a filter which does not transmit this emission line. The two top photographs of the figure show NGC 5194 (M51) as photographed through the 6563 filter (left, which transmits the emission line) and the 6650 (right, which does not transmit the line). The lower left-hand photograph is the same galaxy photographed through a broad-band blue filter. The lower right-hand picture is a scale drawing of the location of the darkest dust lanes of the galaxy. The drawing itself was made from the superb

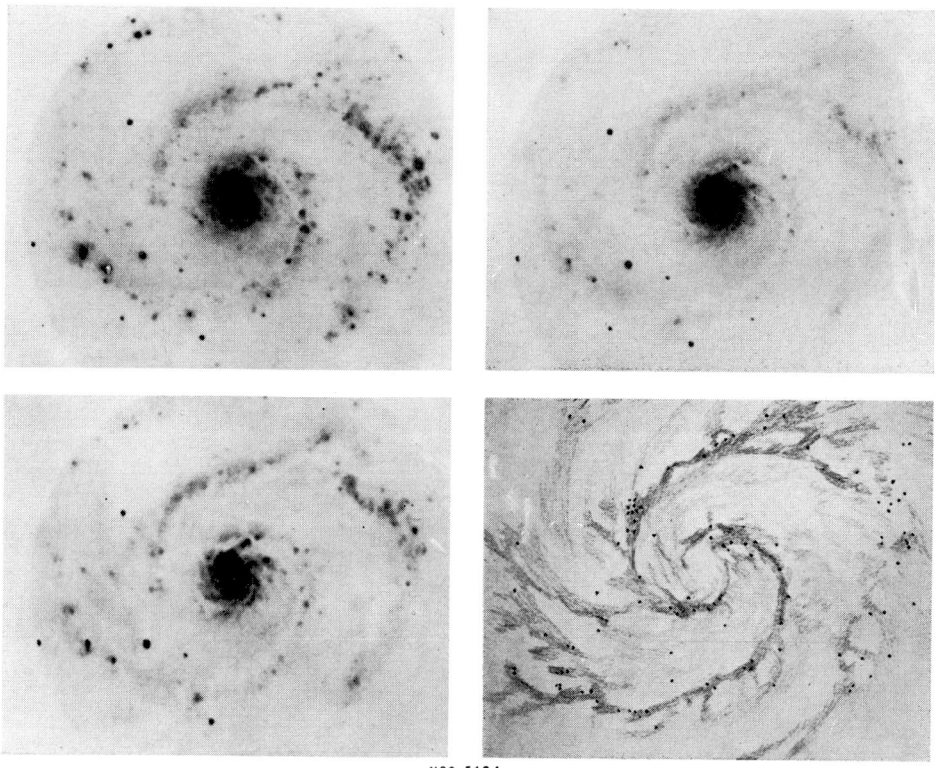

NGC 5194

Fig. 1. NGC 5194 (M51) photographed (top left) through a 6563 Å filter which transmits the Balmer alpha emission line and (top right) through a 6650 Å filter which does not transmit this line, and through a broad-band blue filter (lower left). The lower right-hand picture is a scale drawing of the location of the darkest dust lanes of the galaxy.

blue photograph taken by Humason with the 200-in. A careful comparison of the two top photographs enables one to identify the H II regions of NGC 5194, and these regions are marked as black dots in the drawing. NGC 5195 would lie just to the right of the photographs, and one could easily imagine that there is some evidence for a disturbance on the right-hand side of the drawing. The characteristics of the distribution of dust in an Sc galaxy are well illustrated in this case. The H II regions are invariably found in obvious dark lanes. Very often, they are found on the edge of the primary dust lane which lies on the inside edge of a luminous arm, but they are also found in feathers of dust which cross the lanes and in the more diffuse outer wisps of dark nebulosity. Frequently, H II regions occur at a branching of a dark lane.

The search for bright H II regions in spirals of earlier type is not so fruitful. NGC 4736, seen in Figure 2, is an Sb galaxy rather rich in H II regions (as seen on the Hα photograph on the right of Figure 2); but it is quite evident that the complex spiral structure of this galaxy is defined by the dust in the regions interior to the zone of bright H II regions as well as in the outermost parts of the galaxy (as seen on the blue photograph

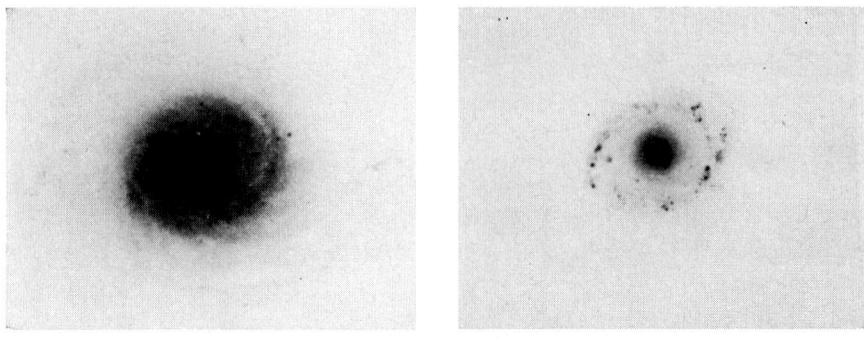

Fig. 2. NGC 4736, an Sb galaxy rich in H II regions photographed in the blue (left) and in Hα (right).

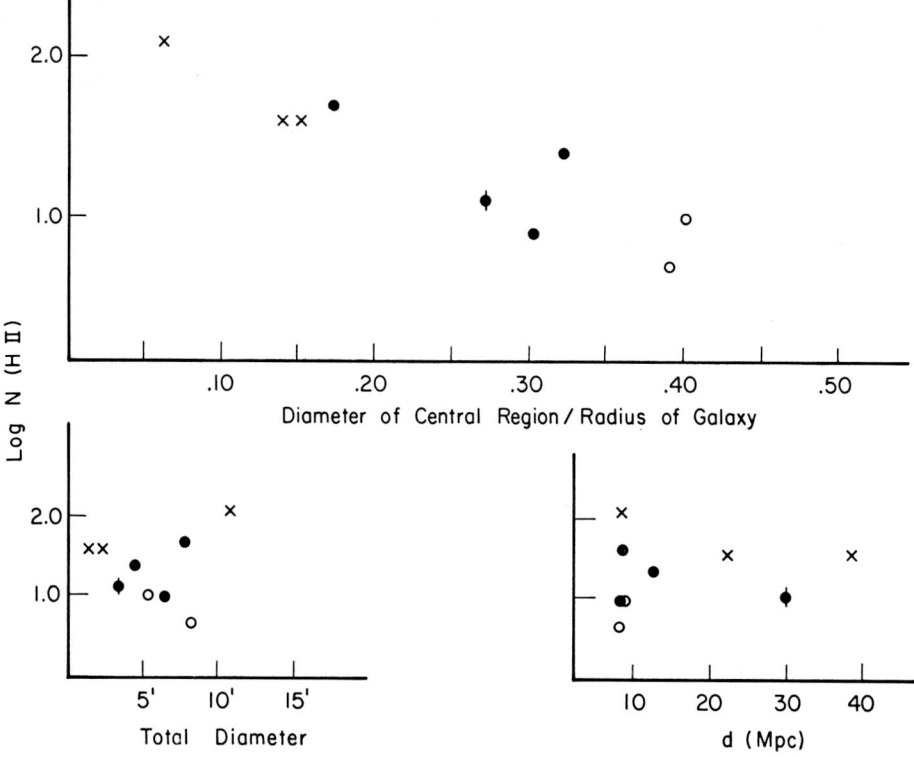

Fig. 3. (upper diagram) The size of the central region (in units of the total radius of the galaxy) vs the log of the number of H II regions detected on a standardized photograph taken with the 90-in. telescope. The log of the number of H II regions detected vs total diameter (lower left) and vs linear diameter of the central region (lower right).

on the left of Figure 2). It is possible to divide each of the Sa and Sb galaxies into three distinct regions: a central zone in which no H II regions exist (the exception to this is a group of galaxies in which very intense H II regions are found in the nucleus itself), an intermediate belt in which the brightest of the H II regions are located, and an extended low-luminosity region. In all regions, spiraling dust patterns are detected.

Figure 3 shows that there appears to be a correlation between the size of the central region and the number of H II regions in a galaxy. In the upper diagram, I have plotted the size of the central region (in units of the total radius of the galaxy) vs the log of the number of H II regions detected on a standardized photograph taken with the 90-in. telescope. These nine galaxies represent the best data available; they consist of three Sc galaxies (crosses), three Sb (filled circles) and one SBb (filled circle with line), and two Sa galaxies (open circles).

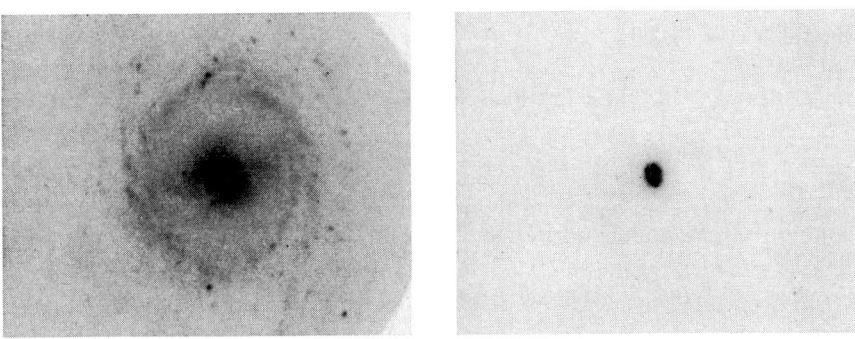

Fig. 4. NGC 3351 photographed in the blue (left) and in Hα (right).

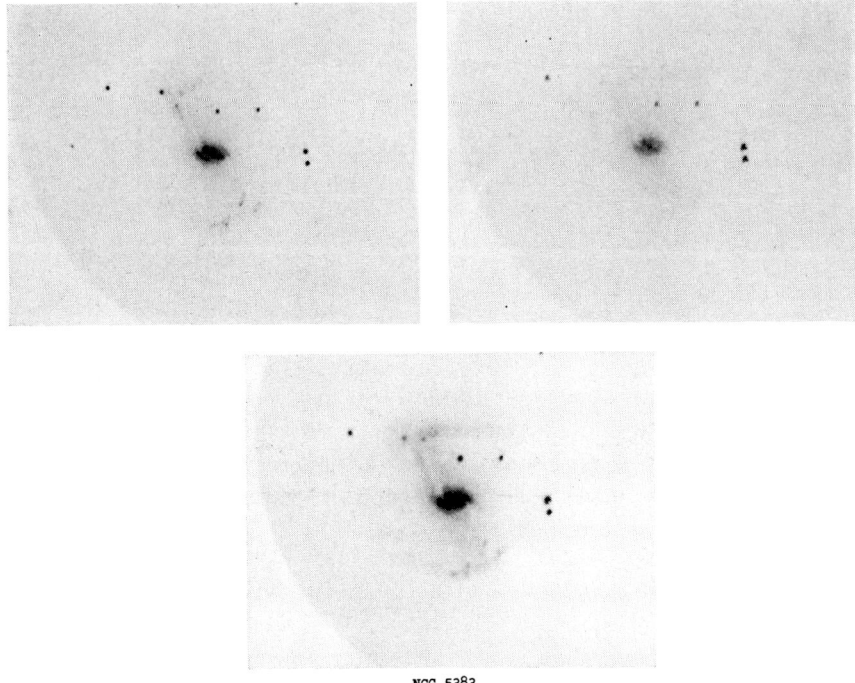

Fig. 5. NGC 5383 photographed in the light of Hα (top left), redward of Hα (top right) and in blue light (lower photograph).

The fact that the Sc, Sb, and Sa galaxies are arranged in the diagram in the sense of increasing diameter of the nuclear region is a reflection of the manner in which Hubble set up his morphological classification. The figure itself illustrates that the smaller the central amorphous region, the more H II regions are found exterior to it. The two lower diagrams demonstrate that there appears to be no selection effect in the H II counts as far as the angular size of the galaxy and its distance are concerned.

Six of the galaxies studied have several very bright H II regions in their nuclei. One example, NGC 3351, is shown in Figure 4; the six knots seen in the Hα photograph (right) are all very much brighter than the center of the galaxy. The strong central dust lanes wind in to these regions as will be seen on a sketch of this galaxy. Another example of a galaxy in which these bright central emission regions are found is NGC 5383, shown in Figure 5. The two top photographs are in the light of Hα (left) and redward of Hα (right); the lower one is a blue photograph. Note the difference in relative brightnesses of these emission patches as photographed in the emission line and in the continuum nearby.

Figure 6 summarizes the dust and H II characteristics of Sb galaxies by means of sketches of two typical systems, NGC 3351 and NGC 4579 – a central region, possibly

NGC 4579   NGC 3351

Fig. 6. Sketches of the dust and H II characteristics of NGC 4579 and 3351.

containing intense H II regions in the very nucleus, has its spiral features often solely defined by strong dust lanes; a zone next occurs in which H II regions are found together with the most luminous arms which are often sandwiched between heavy dust lanes; and finally, there is an outer region of relatively low luminosity, fragmentary dust lanes, and occasional H II regions.

Acknowledgements

Grateful acknowledgement is made to Dr Allan Sandage and to Dr Horace Babcock for making the Hale Observatories' material available for this investigation. The author is also indebted to Mrs Burbidge for her material and for helpful discussions.

Reference

Lynds, Beverly T.: 1970, in W. Becker and G. Contopoulos (eds.), 'The Spiral Structure of our Galaxy', *IAU Symp.* **38**, 36.

Discussion

Salpeter: Is the dust density higher in the inner regions, devoid of H II, than further out in the galaxy?

Mrs Lynds: The central dust lanes are certainly more *regular* and appear more dense, but that may be because the outer ones are diffuse and fragmentary.

King: With what 'arms' do you compare the distribution of H II and dust?

Mrs Lynds: The bright blue arms were what I was referring to.

Mrs Rubin: Can you say anything about the presence of H II regions near the nuclear regions of galaxies? Are the regions absent, or are they just not visible in the light from the nuclear bulge?

Mrs Lynds: The bright H II regions are exterior to this zone. We see only occasionally H II regions in the nucleus.

Ozernoy: Dr Pikelner explains both the dark lanes and 'feathers' in Sc galaxies, which were observed by Mrs Lynds, as the result of the compression of interstellar gas by shock waves in the framework of Lin's theory of spiral arms. The rarefied gas forms a front, but clouds penetrate into the arm without great deceleration. The degree of compression in the shock is determined by equality of internal pressure (mainly magnetic) and pressure of the stream. According to observations, the density of the material in the dark lanes is 10 cm^{-3} or more. From this Pikelner predicts that such high compression is possible, if magnetic fields in Sc galaxies are 0.9 μG or less and random velocities 2 km s^{-1} or less, i.e. field and velocities in Sc are considerably less than those in Sb and, in particular, in our Galaxy.

SPIRAL ARM PATCHES IN Sc AND SBc GALAXIES

I. PRONIK and K. CHUVAEV

Crimean Astrophysical Observatory, Nauchny, Crimea, U.S.S.R.

Multicolor observations of galaxies are being carried out at the prime focus of the 2.6-m Schajn telescope using an image intensifier and 6–9 color filters. The effective wavelengths of the bandpasses used are approximately 3600, 3730, 4400, 4680, 5090, 5280, 6090, 6600 and 7400 Å. The filters for 3730, 5090 and 6600 Å are centered on emission lines.

Observations of the Sc galaxies NGC 628, 4254, 5194 and SBc galaxies NGC 925, 1073, 3359, 4088 and 7741 were made in 1965–1969.

The energy distributions for the central region and for the bright patches in the spiral arms of every galaxy have been determined. For absolute calibration extrafocal star images were used. The background light of the nearest spiral arm has been subtracted.

Most of the patches show Hα-emission (Pronik and Chuvaev, 1967, 1969, 1970, 1971a, b; Pronik, 1972).

Color indices in Tifft's system have been calculated and two-color diagrams (1–3), (3–4) for the central regions and the spiral arm patches are shown in Figures 1–4. From these figures one can see that:

(1) The color indices of spiral arm patches form a sequence.

(2) This sequence is a continuation of the blue end of the sequence for the central regions of Sc galaxies (Tifft, 1961, 1963, 1969).

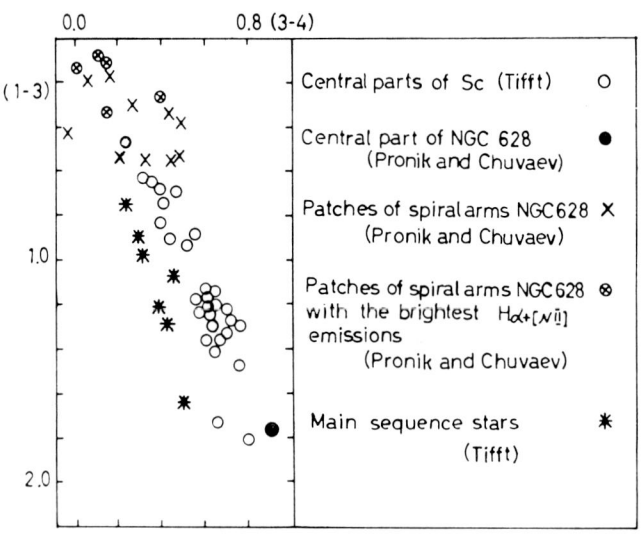

Fig. 1.

(3) The position of this sequence compared with that for the main sequence stars (Figure 1) shows that most patches contain not only blue stars but yellow ones as well.

(4) Very few of the bluest patches have colors like that of bluest stars.

(5) The reddest patches of all galaxies under consideration have colors like the bluest central regions of Sc galaxies.

(6) Galaxies having the bluest central regions have the bluest patches too.

(7) The reddening of the (1–3) colors of most of the patches in NGC 4254 (Figure 3) may be caused by interstellar dust absorption in the spiral arms of this galaxy.

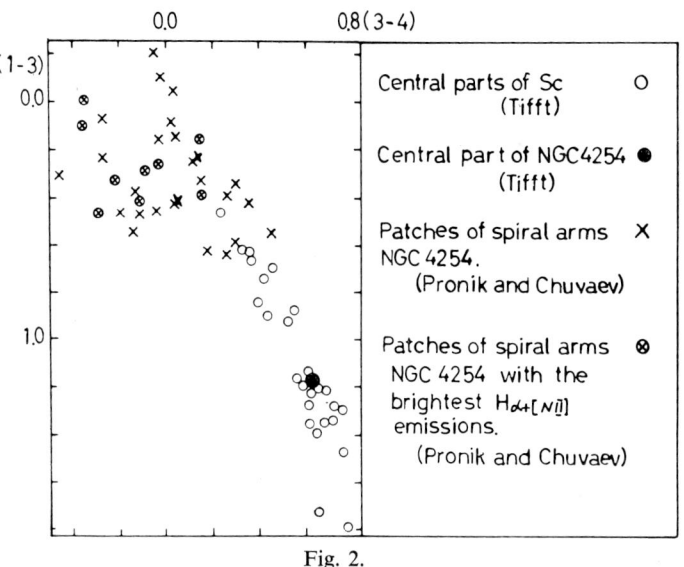

Fig. 2.

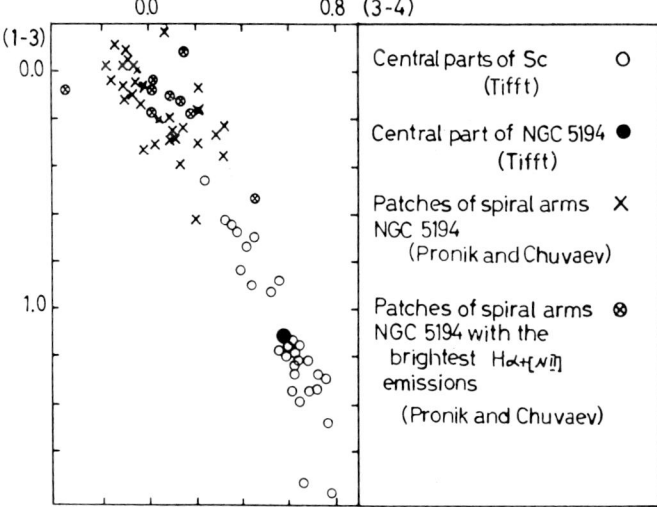

Fig. 3.

(8) The central regions and spiral arm patches of five SBc galaxies studied (Figure 4) are bluer than those of the three Sc galaxies investigated.

Thus we can see that spiral arm patches or associations are not so young as the youngest stars, their ages are determined by the yellow stars present in the structures. Moreover one may suspect that the color-color sequences for spiral arm patches have an evolutionary nature. They are evolving into those for groups of stars having a spectral energy distribution like that of the central regions.

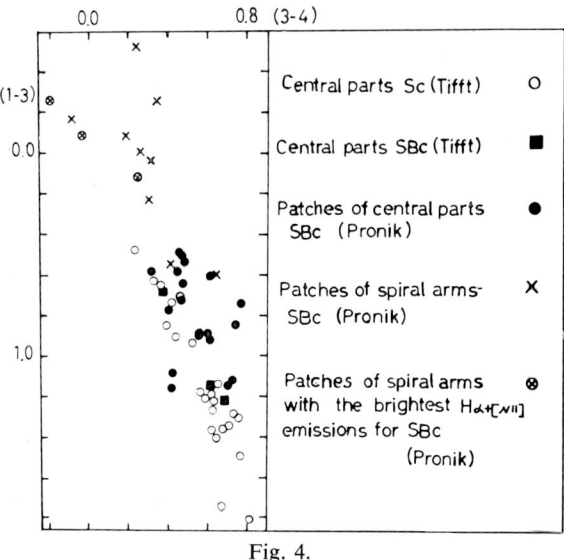

Fig. 4.
Figs. 1–4. Two-color diagrams for Sc and SBc galaxies.

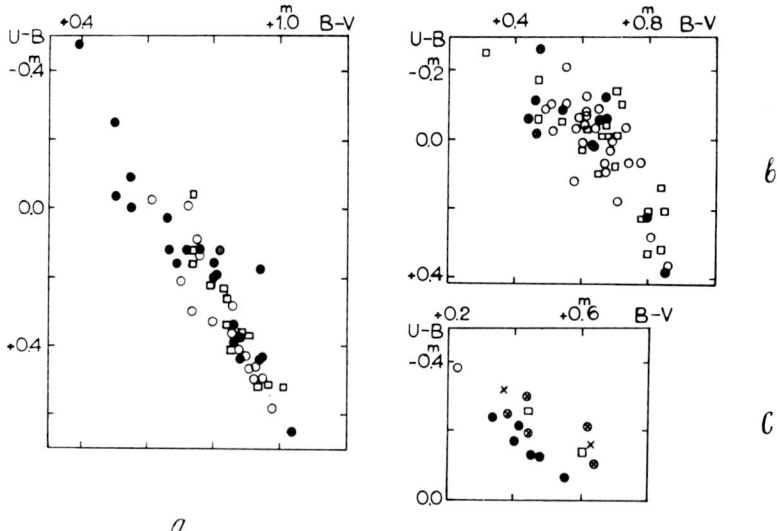

Fig. 5. Two colour diagram (U−B), (B−V) for different morphological types (according to de Vaucouleurs): (a) 1–3 types, (b) 4–6 types, (c) 7–10 types. ○ = A galaxies, □ = AB galaxies, ● = B galaxies.

In connection with point 8 consider Figure 5. This shows three two-color diagrams in the UBV system for S galaxies as a whole according to de Vaucouleurs and de Vaucouleurs (1964). From this figure one can see that SB galaxies are statistically bluer than the ordinary ones of the same morphological type. Among the latest types barred spirals predominate. It would be interesting to know with what kind of galaxy structures this effect is connected (bars, blue patches or both).

Acknowledgement

We are indebted to Mrs A. Bruns for detailed numerical work and for drawing the figures.

References

de Vaucouleurs, G. and de Vaucouleurs, A.: 1964, Reference Catalogue of Bright Galaxies, University of Texas Press, Austin.
Pronik, I. and Chuvaev, K.: 1967, *Izv. Krymsk. Astrofiz. Obs.* **38**, 219.
Pronik, I. and Chuvaev, K.: 1969, *Izv. Krymsk. Astrofiz. Obs.* **40**, 96.
Pronik, I. and Chuvaev, K.: 1970, in W. Becker and G. Contopoulos (eds.) 'The Spiral Structure of our Galaxy', *IAU Symp.* **38**, 83.
Pronik, I. and Chuvaev, K.: 1971a, *Izv. Krymsk. Astrofiz. Obs.* **43**, in press.
Pronik, I. and Chuvaev, K.: 1971b, *Izv. Krymsk. Astrofiz. Obs.* **44**, in press.
Pronik, I.: 1972, *Izv. Krymsk. Astrofiz. Obs.* **45**, in press.
Tifft, W.: 1961, *Astron. J.* **66**, 390.
Tifft, W.: 1963, *Astron. J.* **68**, 301.
Tifft, W.: 1969, *Astron. J.* **74**, 354.

OBSERVATIONS OF H_{II} REGIONS IN Sc GALAXIES

LEONARD SEARLE

Hale Observatories, Pasadena, Calif., U.S.A.

(To be published elsewhere)

Abstract. A spectral survey of H II regions in the galaxies M33, M51, M101, NGC 2403 and NGC 1232 has been carried out. Absolute fluxes have been measured photoelectrically. The energy distribution of the embedded O-association and the intensities of the emission lines have been obtained. The emission spectra can be classified in a one-parameter sequence which is the same for all the galaxies studied. The spectral type is independent of size, shape, surface brightness and density of the H II region, but does depend on the location of the region in its galaxy.

In the galaxies surveyed there is a close correlation between the distance of an H II region from the center of the galaxy and the appearance of its spectrum. The line ratios [O III]/Hβ, Hα/[N II], and [O II]/[N II] all increase by large factors as one passes from regions in the inner spiral arms to those in the outermost arms. The average value of these ratios in a galaxy increases on going from early Sc galaxies to late Sc galaxies. The results show that the N/O abundance ratio (and probably also the O/H abundance ratio) is lowest in irregular galaxies and in the outer parts of late Sc galaxies. It increases towards the center of the spirals and it is highest in the inner spiral arms of the early Sc galaxies. In contrast, the He/H abundance ratio is constant across galactic disks and along the morphological sequence.

Discussion

Terzian: The electron densities of the H II regions which you mentioned are considerably smaller than those which we find in the H II regions of our own Galaxy. Do you have an explanation for this?

Searle: The H II regions which I have observed are less dense but much larger and more massive than the H II regions in the solar vicinity. They are more comparable with the Struve-Elvey emission regions than they are with Orion, for example.

THE NEUTRAL HYDROGEN DISTRIBUTION IN SPIRAL AND IRREGULAR GALAXIES

R. D. DAVIES

University of Manchester, Nuffield Radio Astronomy Laboratories, Jodrell Bank, Cheshire, Great Britain.

Abstract. An investigation of several nearby and irregular galaxies shows significant changes in the neutral hydrogen distribution with morphological type. M31, an Sb, has a central hole while M33, an Sc, has relatively a much smaller hole or no hole at all and NGC 6822, an irregular galaxy, is centrally concentrated. The H II regions in all these morphological types are found in the regions of highest neutral hydrogen density.

Studies of nearby spiral and irregular galaxies have been undertaken at Jodrell Bank with sufficient angular and velocity resolution to make a comparison between the neutral hydrogen distributions in the two types of galaxy. Observations have been made with the 250 ft radiotelescope illuminated to give a halfpower beamwidth of approximately $18' \times 14'$. The frequency resolution in the various surveys was 8.6 km s^{-1} or less. The spiral galaxies investigated were M31 (Sb), M33 (Sc) and M101 (Sc) and the irregular (Type I) galaxies were NGC 6822, IC 10 and IC 1613. The Sc galaxies, NGC 2403 and IC 342, have also been observed but the reduction of the data is not yet complete. The main discussion which follows will be directed towards M31, M33 and NGC 6822 where the angular resolution is best relative to the size of the galaxy.

1. M31

The integrated neutral hydrogen distribution in M31 has a central region of low surface density extending to about 5 kpc from the centre (Roberts, 1967; Davies and Gottesman, 1970). Outside this lie ridges of high surface density that can be traced out to 30 kpc from the centre. These ridges which are not fully resolved in the present observations are probably the counterparts of the neutral hydrogen 'spiral' arms in the Milky Way. The brightest ridges are associated with the spiral arms in M31 designated S4, N4, S5 and N5. A considerable amount of neutral hydrogen is found to be associated with the arms S6, N6, S7 and N7 which are fainter optically. Even beyond the optically identifiable arms there is significant neutral hydrogen emission.

A particular study has been made of the distribution of H II regions catalogued by Baade and Arp (1964) in relation to the distribution of neutral hydrogen. Firstly, on the scale of spiral arms these are found to lie in the regions of greatest neutral hydrogen density. Then, on a smaller scale (~ 2 kpc), within spiral arms the individual contour maps at fixed frequencies show concentrations which in most cases are correlated in position with groups of H II regions. This correlation is illustrated in the accompanying film which shows the sequence of neutral hydrogen maps at adjacent frequencies; these are superimposed on a plot of the distribution of H II regions.

The original maps of M31 and also M33 are published by Gottesman and de Jager (1970). These data can be used to throw some light on the problem of the rate of star formation as a function of gas (neutral hydrogen) density. I am taking the surface density of H II regions in any given area as an index of the present rate of star formation in that area. The surface density of H II regions was compared with the neutral hydrogen surface density in 126 adjacent areas in a strip 15' wide along the major axis of M31 and the results are summarized in Figure 1. If a thickness of 200 pc is assumed

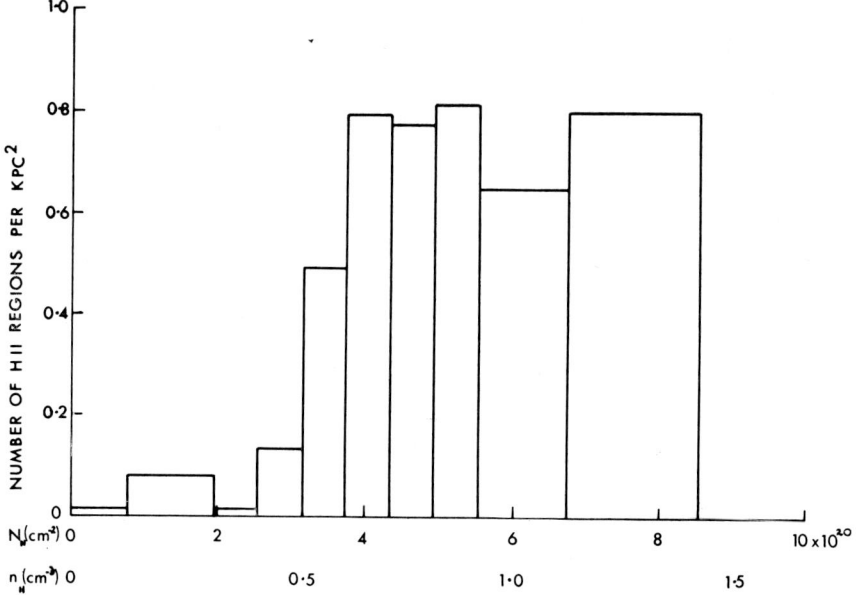

Fig. 1. The correlation between the surface density of H II regions and neutral hydrogen in M31. N_H is the surface density of neutral hydrogen projected onto the plane of M31 measured in atoms cm^{-2}. n_H is the volume density calculated for an assumed thickness of 200 pc. Each block of the histogram represents an approximately equal area of M31.

for the neutral hydrogen layer the mean volume density n_H, is as indicated on the abscissa. The main feature of Figure 1 is the apparent threshold in mean density at 0.5 cm^{-3} required before star formation begins. It was also found that the lack of H II regions at low neutral hydrogen densities was characteristic of both the central regions and of the far outer regions. Remembering that these neutral hydrogen data were obtained with a beam which is ~15' in diameter, the threshold effect is remarkably clear; observations with higher angular resolution may well show an even more abrupt effect. Since the densities given here are the mean value as seen with this angular resolution we can expect that the local density at the H II region where the stars are actually forming may be rather higher.

2. M33

The neutral hydrogen maps of M33 at different velocities have been added to give the surface density distribution projected onto the plane of the sky. This is shown in Figure 2 along with a schematic indication of the stellar and H II region distributions taken from the Sky Atlas prints and from Carranza *et al.* (1968). It is evident from Figure 2 that an angular resolution of $10' \times 13'$ (to which the original observations were deconvolved) is insufficient to resolve any central hole in the neutral hydrogen distribution of M33. Indeed the hole, if it exists, is smaller when scaled for the size of the galaxy than the hole in M31. Optically there is a distinct difference between

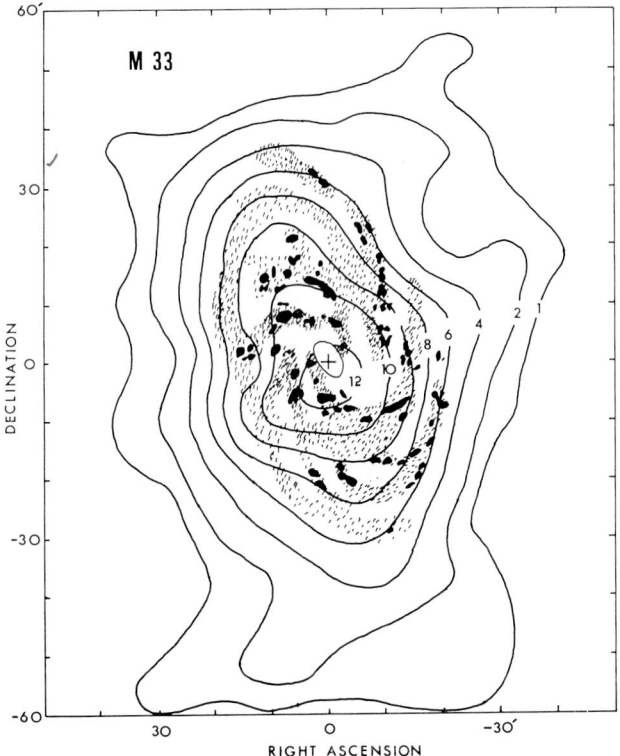

Fig. 2. The distribution of neutral hydrogen in M33 compared with the optical features. Hatched areas represent the principal stellar arms and the filled areas are the main H II regions. The contour levels are proportional to the surface density of neutral hydrogen.

the two galaxies in relation to the Population I distribution in the central regions. M33 has H II regions scattered through the spiral arms and extending into the central regions in contrast to M31 which is devoid of H II regions in its centre. The neutral hydrogen distribution in M33 appears therefore to follow the same general trend as the H II regions.

M33 has one unusual feature which will now be discussed. At velocities near the turnover velocity at either end of the rotation curve the neutral hydrogen is displaced in the clockwise sense from the major axis. This displacement can be seen even in the outer contour levels of the integrated neutral hydrogen distribution shown in Figure 2. By contrast the inner contours follow the optical shape very closely. The distortion in M33 is more clearly seen in the contour maps at fixed frequency. Two at either end of the spread in rotation velocity are illustrated in Figure 3 and show that the main concentration lies on the major axis but substantial wings of emission lie displaced clockwise from the main concentration in areas where there are no optical

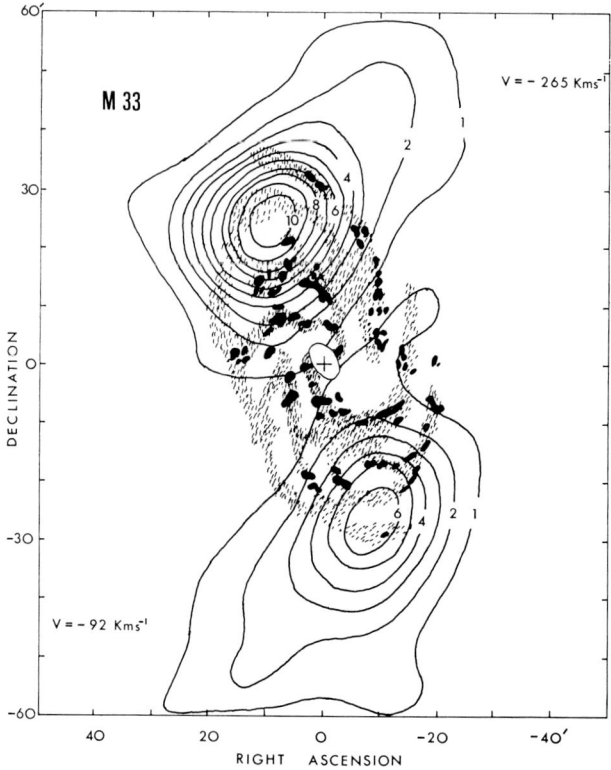

Fig. 3. Contour maps at fixed frequencies demonstrating the distortion in the neutral hydrogen distribution of M33. The northern contours are for a velocity of -265 km s^{-1} and the southern contours are for -92 km s^{-1}. Each map plots the measured signal in a band 8.6 km s^{-1} wide.

features. At present the origin of these distortions in the neutral hydrogen is conjectural. They may represent an exaggerated form of the tilt in the plane seen in the Milky Way and proposed to explain certain features of the optical and neutral hydrogen structure of M31 (see, for example, Davies and Gottesman, 1970). This tilt is often believed to originate in gravitational tides amplified by a resonance effect in the companion galaxy. The most likely disrupting galaxy for M33 would seem to be M31 although a direct gravitational tide is insufficient to produce the observed effect. Further amplification is required as is the case for the Milky Way.

3. NGC 6822

NGC 6822 is an Irr I galaxy with optical dimensions 20′ × 20′ as given by Holmberg (1958). Its neutral hydrogen emission covers a much larger area as may be seen from Figure 4 and can be traced over an elliptical region at least 59′ × 28′ (after removal of the beam smearing effect) with a major axis at position angle 118.5°±2.5°. The bright central optical region is elongated at an angle of about 70° to the main neutral hydrogen concentration and is suggestive of a bar rather like the main body of the Large Magellanic Cloud. In fact the faint stellar extensions from the main bar of NGC 6822 could be embryo spiral arms condensing out of the main mass of neutral

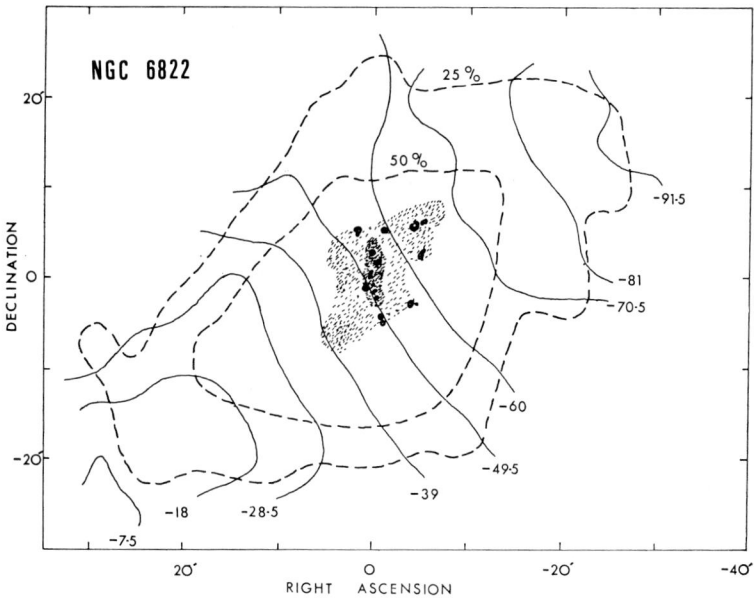

Fig. 4. The distribution of neutral hydrogen in NGC 6822 compared with optical features. The dashed contours represent neutral hydrogen surface density levels at 25 and 50% of the maximum while the full lines are lines of constant median velocity measured in km s^{-1} relative to the Sun. Hatched areas show the main distribution of stars and the filled areas are H II regions.

hydrogen. Several H II regions in NGC 6822 have been catalogued by Hodge (1969) and these, as shown in Figure 4, are confined to the main optical regions of the galaxy.

Evidently star formation is only proceeding at a significant rate in the central regions of the NGC 6822 protogalaxy. The surface density of neutral hydrogen in this central region of star formation is $\sim 2 \times 10^{21}$ cm^{-2} which, for an assumed line of sight thickness of 200 pc, corresponds to a density of ~ 1.5 cm^{-3}. If the true axial ratio of irregular galaxies is as found by Hodge and Hitchcock (1966) this density becomes ~ 0.3 cm^{-3}. At densities of half this value or less no star formation seems to have occurred.

4. Large Magellanic Cloud

The Large Magellanic Cloud will be mentioned here because its neutral hydrogen structure is intermediate between that of M33 and NGC 6822. It is an irregular galaxy in which H II regions and older stars are spread throughout the neutral hydrogen distribution (see, for example, McGee and Milton, 1966). Because of its closeness detailed comparisons can be made between the neutral hydrogen and stellar distributions which show the presence of H II regions only in the densest knots of neutral hydrogen. This situation is similar to that found in M31, the only other galaxy studied with sufficiently high resolution to distinguish the concentrations of neutral hydrogen.

5. Conclusion

An investigation of the neutral hydrogen distribution relative to the H II and stellar distributions in M31, M33, the LMC and NGC 6822 shows a progression from an H I and H II distribution with a central hole, through one with a more uniform distribution over the galaxy to one in which the H I and H II have a strong central distribution. Moreover the H II regions are found principally within the regions of higher neutral hydrogen density. It appears that averaged neutral hydrogen volume densities of 0.5 to 2 cm^{-3} are required for star formation.

References

Baade, W. and Arp, H.: 1964, *Astrophys. J.* **139**, 1027.
Carranza, G., Courtès, G., Georgelin, Y., Monnet, G., and Pourcelot, A.: 1968, *Ann. Astrophys.* **31**, 63.
Davies, R. D. and Gottesman, S. T.: 1970, *Monthly Notices Roy. Astron. Soc.* **149**, 237.
Gottesman, S. T., and Davies, R. D.: 1970, *Monthly Notices Roy. Astron. Soc.* **149**, 263.
Gottesman, S. T. and de Jager, G.: 1970, *Mem. Roy. Astron. Soc.* **74**, 67.
Hodge, P. W. and Hitchcock, J. L.: 1966, *Publ. Astron. Soc. Pacific* **78**, 79.
Hodge, P. W.: 1969, *Astrophys. J. Suppl.* **18**, 73.
Holmberg, E.: 1958, *Medd. Lunds Astron. Obs., Ser.* II, No. 136.
McGee, R. X. and Milton, J. A.: 1966, *Australian J. Phys.* **19**, 343.
Roberts, M. S.: 1967, in H. van Woerden (ed.), 'Radio Astronomy and the Galactic System', *IAU Symp.* **31**, 189.

Discussion

Mrs Rubin: From your observations, have you determined a rotation curve for M31? Does it agree with the optical rotation curve? Is there velocity asymmetry between the NE and SW sides of the galaxy?

Davies: Yes, we have a rotation curve for M31 which agrees within the accuracy of both sets of data with the optical data from various sources. There is a small and apparently significant difference between the northern and southern major axes. We do not find asymmetry of more than 20 km s^{-1} between the preceding and following sides of the major axes. These results are discussed by Gottesman and myself in the reference given in my paper.

Arp: Are the occasional protuberances in the hydrogen profiles in M31, particularly along the minor axis, real? If you consider them real, what interpretation do you give to them?

Davies: These effects show only in the outer contour in each fixed velocity contour map which is only twice the rms noise level. Consequently, only small weight should be assigned to them in the

M31 maps. The distortions discussed in the text for M33 are certainly real; I have already suggested an interpretation for them.

Rees: The apparent sharp cut-off in the star formation rate when n_H falls below ~ 0.5 cm^{-3} may perhaps be taken as evidence for the 'two phase' model of the interstellar medium discussed by, for example, Field *et al.* In this picture, when the density is below a critical density (whose value depends on the heating rate) all the matter is in the hot phase and no cool clouds can exist.

de Vaucouleurs: We have radial velocities for about 15 H II region in M101 from spectrum scans of N_1, N_2, at McDonald Observatory which may help in the analysis of the 21-cm data. What is the position angle of the line of nodes from the 21-cm velocities?

Rogstad: 30° to 40°–45°, depending on the interpretation of the 21-cm velocity map.

Weliachew: I would like to say a few words about a paper by M. Guélin and myself that is to be published in the next issue of *Astron. Astrophys.*, about the neutral hydrogen in M101. The main features which have come out of this study are (1) the asymmetry in hydrogen content between northern and southern halves as already noted by other observers, and (2) large-scale agreement between projected hydrogen density maxima and main spiral features.

ABSORPTION BY NEUTRAL HYDROGEN IN THE IRREGULAR GALAXY M82

M. GUÉLIN

Observatoire de Meudon, 92-Meudon, France

and

L. WELIACHEW

California Institute of Technology, Pasadena, Calif. U.S.A.

(To be published in *Astron. Astrophys.*)

Abstract. A neutral hydrogen survey of the irregular galaxy M82 has been carried out with the transit radio telescope at Nançay, France. The resolving power was 4′ in right ascension and 34′ in declination. The velocity resolution was 59 km s^{-1}.

Drift scans covering 58′ in right ascension were taken across the center of the galaxy where the radio source is located; fourteen scans were averaged.

Line profiles were derived every 2′ in right ascension. The profiles at $\pm 4'$ from the radio source are similar within the measurement errors. In particular, they do not show any rotation effect within ± 15 km s^{-1}. They were averaged in order to provide an estimate of the expected emission profile at the radio source position.

Subtraction of this average from the line profile measured in front of the radio source yielded significant negative temperatures at all velocities from 180 to 360 km s^{-1} and no positive temperatures at other velocities.

These negative temperatures were assigned to absorption of the radiation from the radio source by neutral hydrogen in M82.

Absorption is running from 3 to 6% in depth. The average velocity of the absorption profile is lying between the central emission velocity and the optically-determined velocity which are known to show a large disagreement.

The width of the absorption profile shows a velocity gradient of 200 km s^{-1} across the 35″ × 20″ radio source. Such a large velocity spread across this angular extent is only shown by the excited gas showing emission lines at optical wavelengths (Burbidge *et al.*, 1964).

Since then, the absorption in M82 at the neutral hydrogen wavelength has been confirmed by measurements done with the Owens Valley Radio interferometer at 2 spacings where any emission is completely resolved (interfringes were 3′ and 1′.5).

In addition to the conclusions of the single dish observations done at Nançay, the interferometer data have shown that the steeper velocity gradient occurs along the major axis of the galaxy which coincides with the major axis of the radio source.

Reference

Burbidge, E. M., Burbidge, G. R., and Rubin, V. C.: 1964, *Astrophys. J.* **140**, 942.

THERMAL RADIO EMISSION FROM NORMAL GALAXIES

YERVANT TERZIAN

Cornell University, Ithaca, N.Y., U.S.A.

Abstract. Thermal radiation from normal spiral galaxies may be detectable at centimeter and millimeter wavelengths. Predictions have been made assuming free-free radiation from H II regions at $T_e = 7000$ K, and a range of mean electron densities and radii.

1. Introduction

The observed radio spectra of normal galaxies are characteristic of the type of spectra found for non-thermal sources. However, thermal emission due to free-free transitions in ionized hydrogen regions inside normal galaxies also contributes to the radio radiation. Observations and theory suggest that the non-thermal radiation from normal galaxies is more intense than the thermal radiation by several orders of magnitude at radio frequencies below 5 GHz. At higher frequencies the contribution of the thermal radiation becomes important and perhaps it may even be the dominant source of energy at frequencies higher than 30 GHz.

This paper is a preliminary study in estimating the contribution of the thermal radiation from normal galaxies. Recently Hodge (1969) has completed the work 'An Atlas and Catalog of H II Regions in Galaxies'. This work is a search for Hα-emission in ninety galaxies which led to the discovery of H II regions in sixty of them. Here we use Hodge's work for nine galaxies to estimate their thermal radio emission from the catalogued H II regions. In this study we also include the Andromeda galaxy (M31) using the H II data from Baade and Arp (1964), and M33 using the data by Sandage (1962), Courtès and Cruvellier (1965), and Carranza *et al.* (1968).

The predicted thermal radio spectra are then compared with the available radio observations. These observations are primarily from Mathewson and Rome (1963), Heeschen and Wade (1964), De Jong (1966, 1967), Lang and Terzian (1969), Kuril'chick *et al.* (1970), and Whiteoak (1970). The radio observations for M31 are from Kraus *et al.* (1966), and Howard and Maran (1965).

2. Thermal Emission

The available data on emission nebulae in normal galaxies put some restrictions on the methods one can use to compute the thermal radio emission. The most reliable technique for computing the thermal radio emission is to use the recombination theory together with observed absolute fluxes of Balmer lines integrated over an entire galaxy. Terzian (1965, 1968) has shown that the ratio of the fluxes at a radio frequency v and at Hβ for an optically thin thermal region is given as

$$\frac{S_v}{S_{H\beta}} = \frac{N_i}{N_p} \frac{T_e}{\langle b_4 \rangle} \frac{1.664 \times 10^{-19}}{e^{0.986 \times 10^4/T_e}} \ln\left(49.5 \frac{T_e^{3/2}}{v}\right) \tag{1}$$

where N_i/N_p is the ratio of the number of ions to protons and indicates the relative abundance of helium, which can be estimated to be 10 to 15%. T_e is the electron temperature which is of the order of 10^4 K, $\langle b_4 \rangle$ describes the departure from thermodynamic equilibrium for the Hβ transition and it is equal to 0.20 (Burgess, 1958), and v is the radio frequency. Clearly the dependence on frequency for an optically thin thermal source is very small. If we let $v = 5$ GHz, we find that

$$\frac{S_v}{S_{H\beta}} = 3.2 \times 10^{-14}. \tag{2}$$

Hence the thermal radio spectrum of a source can be computed if one has a knowledge of the Hβ flux. Similarly a knowledge of the fluxes of other Balmer or Paschen lines may allow us to predict the thermal radio spectra (Pipher and Terzian, 1969).

Unfortunately the above method cannot be used for the present study because absolute Balmer line fluxes are not available for H II regions in normal galaxies. One should emphasize that such data will be extremely useful in computing the free-free emission from other galaxies.

In this preliminary study we compute the thermal radio spectra making use of the free-free emission theory together with the total number (and apparent area) of the emission regions in each galaxy measured from the work of Hodge. Since the measurement of emission region areas is a very uncertain quantity, we also compute thermal spectra by assuming certain mean sizes to the emitting regions.

The flux density of a radio source is normally given as

$$S_v = \frac{2kv^2}{c^2} \int_\Omega T_b \, d\Omega \tag{3}$$

where k is Boltzmann's constant, c is the velocity of light, T_b is the brightness temperature, and Ω is the solid angle of the source. If one assumes constant brightness for a source, then letting $\Omega = A/D^2$ where A is the projected area of the source and D is its distance, we can then write

$$S_v = \frac{2kv^2}{c^2} \frac{T_b A}{D^2}. \tag{4}$$

The brightness temperature is obtained from the solution of the equation of radiative transfer as $T_b = T_e (1 - e^{-\tau_v})$, where τ_v is the optical depth. Since the thermal radio emission will only be important at the very high frequencies where generally $\tau_v \ll 1$, we can approximate $T_b \approx T_e \tau_v$, and the last expression becomes

$$S_v = \frac{2kv^2}{c^2} \frac{A T_e \tau_v}{D^2}. \tag{5}$$

The optical depth for thermal emission can be derived from the free-free absorption coefficient (Oster 1961, Terzian 1968) as

$$\tau_v = \frac{9.78 \times 10^{-3}}{v^2 T_e^{3/2}} \ln\left(4.95 \times 10^7 \frac{T_e^{3/2}}{v}\right) \int N_e^2 \, ds \tag{6}$$

where N_e is the electron density in cm^{-3} and v is the frequency in units of Hz. One can assume that the H II regions inside a galaxy are spherical and have uniform electron densities, so that $\int N_e^2 \, ds = N_e^2 s$ where s is the diameter of a single emission region in a galaxy. If A is the total projected area of emission regions in a galaxy and N is the total number of such regions, then $A/N = \pi (s/2)^2$, and $s = 2(A/\pi N)^{1/2}$, so that

$$N_e^2 s = 2 N_e^2 \left(\frac{A}{\pi N} \right)^{1/2}. \tag{7}$$

Using Equations (6) and (7) with (5) we finally have

$$S_v = \frac{3.91 \times 10^{-2} k}{c^2 (\pi)^{1/2}} \frac{A^{3/2}}{D^2 (N)^{1/2}} \frac{N_e^2}{(T_e)^{1/2}} \ln \left(4.95 \times 10^7 \frac{T_e^{3/2}}{v} \right). \tag{8}$$

The last expression can now be used to compute the thermal radio spectra of normal galaxies. The distances for the galaxies are taken from Roberts (1969) and van den Bergh (1968). The values for N are those given by Hodge (1969), Baade and Arp (1964) for M31, Sandage (1962) and Courtès and Cruvellier (1965) for M33.

TABLE I

Number of H II regions in galaxies and distances

Galaxy	Type	Distance (Mpc)	Number of H II regions
M31	Sb	0.69	688
M33	Sc	0.69	369, 101
NGC 628	Sc	7.8	193
NGC 2403	Sc	3.2	109
NGC 2903	Sc	7.0	74
NGC 3368	Sa	7.3	3
NGC 3627	Sb	7.3	25
NGC 3628	Sb	7.0	18
NGC 4449	Ir I	3.3	81
NGC 5457	Sc	3.5	189
NGC 6946	Sc	4.1	39

Table I gives the distances to the galaxies and the number of H II regions in each of them. In the case of M33 Sandage has reported 369 emission nebulae and Courtès and Cruvellier 101. The latter work includes primarily the H II regions which were photographically resolved. In order to make use of Equation (8) one has to assume values for T_e and N_e. From extensive research on emission nebulae in our own galaxy we find that $T_e \approx 7000$ K with a variation of a few thousand degrees. Mean values for N_e are more difficult to estimate; however, we know that in extreme cases N_e may have values of ~ 1 cm^{-3} to $\gtrsim 10^6$ cm^{-3} (as in compact H II regions). For the average regions, however, the range in N_e can be taken to be 10 cm^{-3} to 100 cm^{-3}.

In practice it is extremely difficult to measure reliable areas for most of the emission regions in other galaxies because of the limited resolution of the optical photographs. Nevertheless, we have tried to measure the total area of H II regions given by Hodge for several galaxies and have applied the results in Equation (8). The results were very unrealistic. In particular the mean sizes of the emission regions were found to be several hundred parsecs in size as compared to 10–80 parsec for the sizes of normal H II regions in our own galaxy. However, one can argue that at distances of 3 to 7 Mpc we detect only the giant H II regions.

For the reasons given above we have eliminated the parameter A from Equation (8) and assumed several mean radii for the individual H II regions in other galaxies. Thus, thermal radio spectra were computed by using $A^{3/2} = r^3 (\pi N)^{3/2}$ in Equation (8), where r is the mean radius of an H II region.

3. Results

A series of computations of thermal spectra was performed with wide ranges in r, N_e and T_e. As can be seen from the previous section, the dependence of T_e on the spectrum when $\tau_\nu \ll 1$ is very small, hence, we have taken $T_e = 7000$ K for the results reported below.

Table II and Figure 1 show the main results of the computations. In Table II we compare the observed radio emission at $\lambda 3$ cm and $\lambda 3$ mm wavelengths (these are extrapolated fluxes; references for radio observations are given in Section 1) with the theoretically predicted pure thermal radiation. The computed flux densities due to thermal radiation in Table II are those at $\lambda 3$ cm. The thermal spectra are almost independent of wavelength when $\tau \ll 1$ and the fluxes at $\lambda 3$ cm are essentially the same

TABLE II

Flux densities at $\lambda 3$ cm, $\lambda 3$ mm, extrapolated from the observed spectra, compared with predicted thermal fluxes

Galaxy	3 cm (obs.)	3 mm (obs.)	Flux Density (10^{-26} W m^{-2} Hz^{-1})[a]		
			Thermal (1)	Thermal (2)	Thermal (3)
M31	1.50	0.11	0.32	60.5	505.0
M33[b]	0.31	0.03	a) 0.169	32.5	271.3
			b) 0.046	8.9	74.3
NGC 628	0.025	0.003	0.0007	0.135	1.13
NGC 2403	0.15	0.07	0.0023	0.448	3.75
NGC 2903	0.07	0.009	0.0003	0.063	0.53
NGC 3368	0.002	0.00002	0.00001	0.002	0.02
NGC 3627	0.14	0.03	0.0001	0.020	0.17
NGC 3628	0.11	0.02	0.00008	0.015	0.13
NGC 4449	0.05	0.004	0.0016	0.314	2.62
NGC 5457	0.06	0.007	0.0036	0.691	5.77
NGC 6946	0.30	0.05	0.0005	0.97	0.81

[a] Computed thermal fluxes are at $\lambda 3$ cm. The thermal spectra are almost flat at this wavelength region so that the $\lambda 3$ mm fluxes are essentially the same as those at $\lambda 3$ cm.
[b] a) 369 H II regions, b) 101 H II regions, (see text for observed flux).
Thermal (1) $\langle N_e \rangle = 20$ cm^{-3}, $\langle r \rangle = 10$ pc.
Thermal (2) $\langle N_e \rangle = 70$ cm^{-3}, $\langle r \rangle = 25$ pc.
Thermal (3) $\langle N_e \rangle = 100$ cm^{-3}, $\langle r \rangle = 40$ pc.

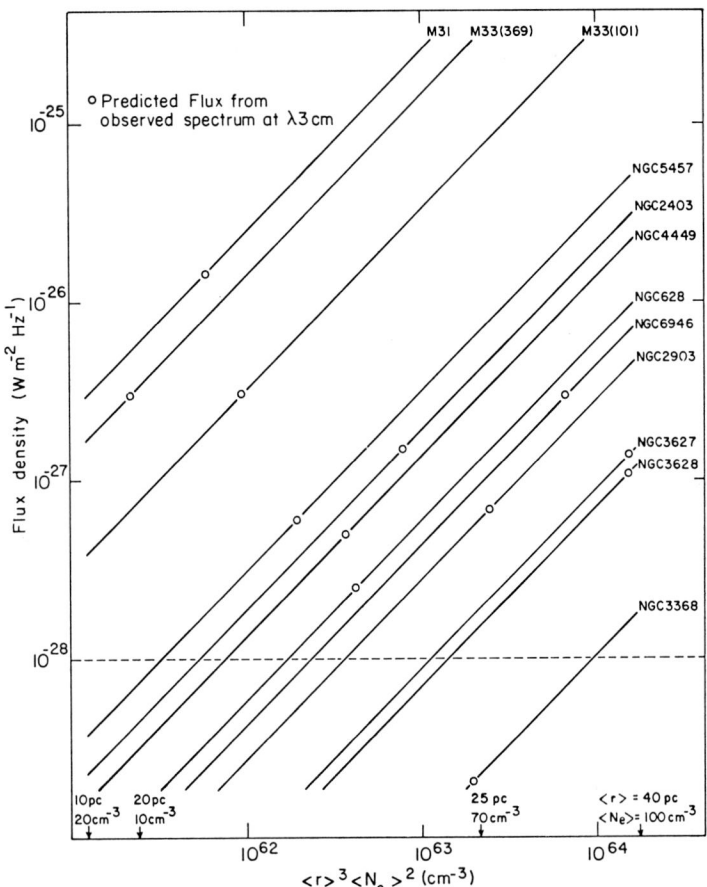

Fig. 1. Predicted flux densities due to thermal radiation for a number of galaxies as a function of $\langle r \rangle^3 \langle N_e \rangle^2$. $\langle r \rangle$ is the mean radius of the H II regions and $\langle N_e \rangle$ is the mean electron density. The dashed line indicates the lower limit of detectability for a radio source with present-day instrumentation. The circles on each curve show the predicted flux for the corresponding galaxies at $\lambda 3$ cm extrapolated from their observed spectra.

as those at $\lambda 3$ mm. Three cases of computed thermal spectra are given in Table II with regard to mean radii and mean densities of the H II regions; (1) $\langle N_e \rangle = 20$ cm^{-3}, $\langle r \rangle = 10$ pc; (2) $\langle N_e \rangle = 70$ cm^{-3}, $\langle r \rangle = 25$ pc; (3) $\langle N_e \rangle = 100$ cm^{-3}, $\langle r \rangle = 40$ pc.

The data on $\langle N_e \rangle$ and $\langle r \rangle$ for our own galaxy can be obtained from Murdin and Sharpless (1968). These authors have compiled more than seventy emission regions and their data indicate that $\langle N_e \rangle \sim 70$ cm^{-3} and $\langle r \rangle \sim 25$ pc. One can clearly note that if the mean population of H II regions in other spiral galaxies is similar to our own, then case (2) in Table II shows that the radio radiation of the thermal regions is in general of the same order as the radio non-thermal radiation at $\lambda \lesssim 3$ cm. One also notes that a comparison of the predicted thermal radiation with the observations may put a limit on the mean values of r and N_e (actually on the product $\langle r \rangle^3 \langle N_e \rangle^2$).

Figure 1 shows the relationship between flux density due to thermal radiation (at $\lambda 3$ cm, $T_e = 7000$ K) and the parameter $\langle r \rangle^3 \langle N_e \rangle^2$ for the H II regions in eleven galaxies. The dashed line represents the low flux density limit for a detectable radio source with the present-day available instrumentation.

The arrows show four examples of combinations of $\langle r \rangle$ and $\langle N_e \rangle$. The open circles indicate the flux densities from the observed spectra of the galaxies (extrapolated to $\lambda 3$ cm). As an example, if $\langle r \rangle = 20$ pc and $\langle N_e \rangle = 10$ cm^{-3} for M33, where Sandage reports 369 H II regions, then we see that the flux due to the H II regions is of the same order as the flux due to the non-thermal radiation ~ 0.3 f.u. (Recent observations of M33 by Terzian and Pankonin (1971, in preparation) show a flux at $\lambda 3$ cm of ~ 4 f.u.)

One should emphasize that these computations include only the free-free radiation from normal H II regions. No attempt has been made to include the so called 'background' thermal component, which appears in our own galaxy, and which is probably due to ionization of the neutral hydrogen by low energy cosmic rays.

As a conclusion, we have shown that thermal radiation from normal spiral galaxies may be important at centimeter wavelengths, and certainly very important at millimeter wavelengths. More accurate predictions can be made if one obtains integrated absolute fluxes of Balmer or Paschen lines of normal spiral galaxies.

Acknowledgements

The author wishes to acknowledge the computations done by Mr V. Pankonin on thermal spectra using the projected areas of H II regions from Hodge's catalogue. This work was supported by the Arecibo Observatory. The Arecibo Observatory is operated by Cornell University under contract to the National Science Foundation and with partial support from the Advanced Research Projects Agency.

References

Baade, W. and Arp, H.: 1964, *Astrophys. J.* **139**, 1027.
Bergh, van den, S.: 1968, *Commun. David Dunlap Obs.* **195**, 1.
Burgess, A.: 1958, *Monthly Notices Roy. Astron. Soc.* **118**, 477.
Carranza, G., Courtès, G., Georgelin, Y., Monnet, G., and Pourcelot, A.: 1968, *Ann. Astrophys.* **31**, 63.
Courtès, G. and Cruvellier, P.: 1965, *Ann. Astrophys.* **28**, 683.
De Jong, M. L.: 1966, *Astrophys. J.* **144**, 553.
De Jong, M. L.: 1967, *Astrophys. J.* **150**, 1.
Heeschen, D. S. and Wade, C. M.: 1964, *Astron. J.* **69**, 277.
Hodge, P. W.: 1969, *An Atlas and Catalog of H II Regions in Galaxies*, University of Washington, Washington, D.C.
Howard, W. E., III and Maran, S. P.: 1965, *Astrophys. J. Suppl. Ser.* **10**, 1.
Kraus, J. D., Dixon, R. S., and Fisher, R. O.: 1966, *Astrophys. J.* **144**, 559.
Kuril'chik, V. N., Andrievskii, A. E., Ivanov, V. N., and Spangenberg, E. E.: 1970, *Soviet Astron.* **13**, 881.
Lang, K. R. and Terzian, Y.: 1969, *Astrophys. Letters* **3**, 29.
Mathewson, D. S. and Rome, J. M.: 1963, *Australian J. Phys.* **16**, 360.

Murdin, P. and Sharpless, S.: 1968, in Y. Terzian (ed.), *Interstellar Ionized Hydrogen*, W. A. Benjamin Press, New York, p. 289.
Oster, L.: 1961, *Astron. J.* **134**, 1010.
Pipher, J. L. and Terzian, Y.: 1969, *Astrophys. J.* **155**, 475.
Roberts, M. S.: 1969, *Astron. J.* **74**, 859.
Sandage, A. R.: 1962, in G. C. McVittie (ed.), 'Problems of Extra-Galactic Research', *IAU Symp.* **15**, 359.
Terzian, Y.: 1965, *Astrophys. J.* **142**, 135.
Terzian, Y.: 1968, in D. E. Osterbrock, and C. R. O'Dell (eds.) 'Planetary Nebulae', *IAU Symp.* **34**, 87.
Whiteoak, J. B.: 1970, *Astrophys. Letters* **5**, 29.

RADIO-EMISSION FROM SUPERNOVAE REMNANTS IN DISTANT GALAXIES

P. NOTNI, H. OLEAK, and G.-M. RICHTER

Zentralinstitut für Astrophysik, Potsdam, G.D.R.

Abstract. It is suggested that some 5C2 radio sources previously identified with normal spiral galaxies, are associated with supernova remnants in these galaxies. This hypothesis is tested by the relative radio and optical positions and by luminosity estimates. It could be further tested by possible coincidences of Ohio radio sources with known supernovae, and also by predicted decreases in radio emission.

Some sources of the 5C2 radio survey were identified by G. G. Pooley and S. Kenderdine (1968) with normal spiral galaxies. Exact position measurements on plates of the Schmidt telescope of the Karl Schwarzschild Observatory, Tautenburg, show that the radio sources prefer the region of the spiral arms (Figure 1). In Table I columns 4, 5 and 6 give the observed offsets of the radio positions from the center of each galaxy and the probability (in per cent) that this distance is the result of observational error. The distance modulus (column 8) for all but one (5C 2.108) is obtained by adopting a mean absolute magnitude of -18^m for the galaxies. For 5C 2.108 Pooley's identification ('blue star') proved to be a supernova of March 1953 in a faint $19^m\!.5$ spiral or

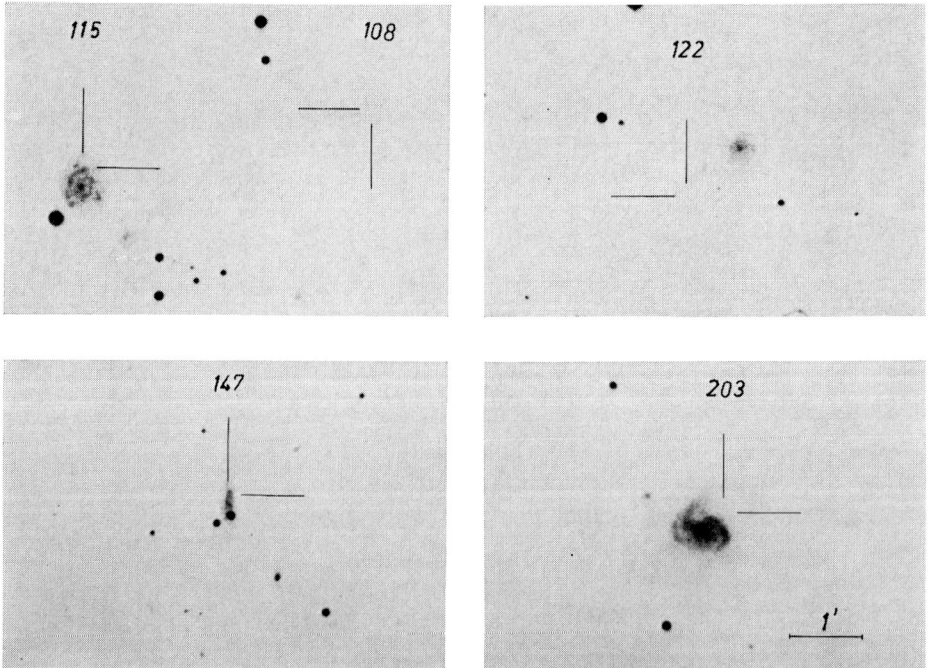

Fig. 1. Radio source positions relative to some spiral galaxies.

TABLE I
Candidates for supernova remnants

5C 2	No. VV[a]	Type	$\Delta\alpha$	$\Delta\delta$	p	m_B	$m-M$	lg S_{408}	lg P_{408}	Notes
115	8-20-76	SA $(r)b$	$+1''$	$-15''$	4	14.4	32.4	-27.85	20.2	Blue knot near radio position
122	8-20-79	SA c	$-44''$	$+38''$	≪1	15.2	33.2	-28.05	20.2	A blue stellar object, 19^m, $\Delta\alpha=-26'', \Delta\delta=+5''$, $p\ll 1$, is closer to the radio position
147	8-20-86	SB	$+0$	$+6''$	42	15.1	33.1	-27.77	20.5	
203	8-21-8	SB b	$+32''$	$-14''$	≪1	11.6	29.6	-26.84	20.0	NGC 3583
108	–	S?, Ir?	$+5''$	$+8''$	18	19.5	–	-27.80	20.8–21.6	
–	–	–	$+6''$	$+4''$	25	17	34–36	–	–	Supernova in 5C 2.108

[a] Designation according to Vorontsov-Velyaminov and Krasnogorskaya (1962)

irregular galaxy. Its detection is the reason for including this galaxy in Table I. The supernova has been used to estimate the distance of the parent galaxy which thereby proved to be obviously a system of low luminosity.

The estimated radio luminosity (P_{408} in W Hz^{-1} ster^{-1}) of these galaxies (column 10) is of the same order as expected for normal galaxies. Since the radiation obviously does not originate in the nucleus or in the halo as a whole and no optical peculiarities (such as, e.g. jets in weak elliptical radio galaxies) except for knots in the spiral arms are visible, only two possibilities remain for the source of radio emission: H II-regions and supernova remnants. H II regions can be excluded because of their much lower radio luminosity ($<10^{18}$) whereas supernova remnants might possess at some earlier stage of evolution the desired high luminosity of $P \approx 10^{20}$ or 10^{21} W Hz^{-1} ster^{-1}. It is

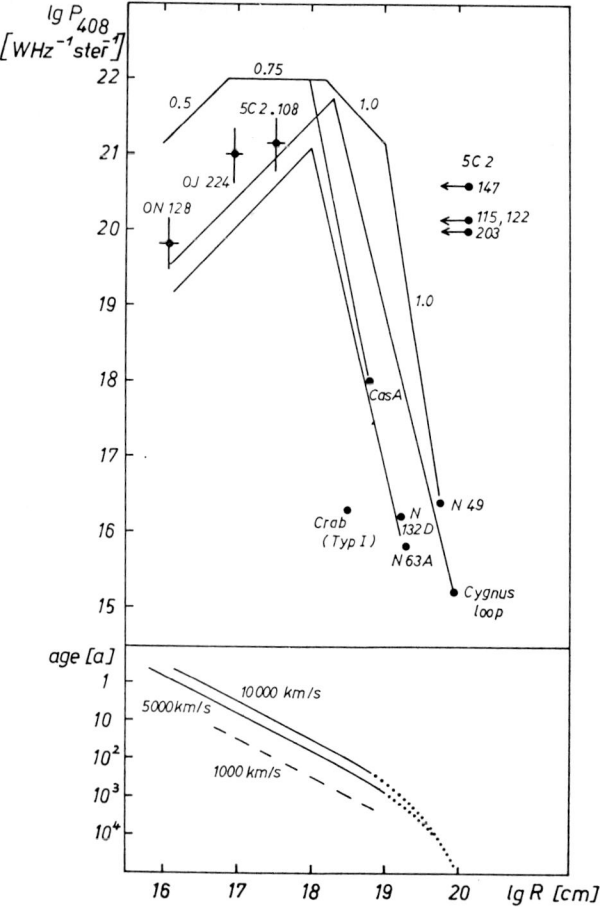

Fig. 2. Extrapolated dependence of radio flux at 408 MHz on source radius for supernovae of Type II. $P \sim R^{5-2\gamma}$ until the moment at which the source becomes optically thin at decimeter wavelengths, $P \sim R^{-2\gamma}$ afterwards, where $\gamma = 2\alpha + 1$ and α = spectral index. For N 49 the assumed variation of α is marked on the curve. The observed spectral indices are 1.0 (N 49), 0.75 (Cas A), 0.5 (Cygnus loop, N 63 A, N 132D), 0.3 (Crab). The lower part gives the ages for various assumed expansion velocities.

possible, see e.g. I. S. Shklovsky (1966, § 8), to extrapolate the observed fluxes of the galactic supernovae backwards to the moment at which the source was optically thick. For spectral indices $\alpha < 0.75$ this is the stage of maximum radio luminosity reached about 10 to 100 yr after the outburst. However, the theory must be modified for sources which would reach a flux exceeding about 10^{22} W Hz^{-1} ster^{-1}. The total energy-content of the relativistic particles being of the order of 10^{49} erg (Shklovsky 1966, § 6), such a strong radiation of about 10^{22} W Hz^{-1} ster^{-1} implies a maximum lifetime of a few tens of years. In the simple theory the energy loss by radiation is assumed to be negligible, the spectrum remaining constant. If, however, the energy losses amount to a significant proportion of the total energy the spectrum will steepen. Therefore, if all supernovae start with approximately the same low spectral index, e.g. $\alpha = 0.5$, the stronger ones, which attain the greatest maximum luminosity should leave the maximum with greater spectral index. Some examples of this behavior are shown in Figure 2, starting from the observed nearby supernova remnants. It can be seen that the radio-maxima of supernovae of Type II probably lie between 10^{20} and 10^{22} W Hz^{-1} ster^{-1}.

The supernova observed at the position of 5C 2.108 should have had at the time of the radio survey (1966) a radio luminosity of about 10^{21} W Hz^{-1} ster^{-1} lying near its maximum radio brightness. The radio luminosity estimated from the distance modulus 34 to 36 mag (see Table I) is of the same order of magnitude and supports the identification. The other candidates of Table I might also be supernova remnants presumably lying on the descending branch, aged about 10^2 to 10^3 yr.

Stimulated by the above results we looked for radio sources near the positions of known extragalactic supernovae. Among 77 supernovae in regions covered by the Ohio survey (Scheer et al., 1967; Dixon et al., 1968; Fitch et al., 1969) nine coincide with sources on the Ohio maps, two of which are included in the Ohio source list. Only about three coincidences are expected to result by chance. To be conservative we include only the two sources contained in the final Ohio list (Table II). Because of the comparatively low positional accuracy of the Ohio survey, we cannot exclude the possibility in these cases that we are observing the 'normal' radiation of the whole galaxy. As can be seen in Figure 2, ON 128 would lie on the ascending branch and therefore its radio flux should increase during the next decennium by about an order

TABLE II
Suggested identifications of Ohio sources with supernova remnants

No.	NGC	Type	m_B	SN	p'	$m-M$	lg P_{408}	References
ON 128	4254	SA(s)c	10.5	1967 h	2%	28.5	19.8	Zwicky: 1968, *Publ. Astron. Soc. Pacific* **80**, 462.
OJ 224	Anon.	SB c	14.5	1962 f	4%	32.5	21.0	Zwicky et al.: 1963, *Publ. Astron. Soc. Pacific* **75**, 236.

p' = Probability of chance agreement of positions.

of magnitude. The apparent diameters of all sources are expected to be very small; about 10^{-4} seconds of arc for 5C 2.108, ON 128 and OJ 224.

We expect a strong supernova of Type II to possess a radio luminosity of more than 10^{20} W Hz^{-1} ster^{-1} for about 100 yr. This time interval being nearly the rate of occurrence of supernovae in an average spiral, a high percentage of them should emit a radio flux at this level from a 'point' source. This is in accordance with the findings in the 5C 2 survey: three out of six spirals of magnitude 14–15.5 within the central section of the survey (diameter 2°) are included in our Table I.

Note added in proof. Because of a subsequently detected systematic error in the radio positions, the observed offsets from the centers of the galaxies are changed by small amounts (*Astron. Nachr.*, in press). The conclusions of the paper remain unaltered. – From Table II the source ON 128 must be deleted; it has been found to be an extended source (O. de la Beaujardière *et al.*, *Ann. Astrophys.* **31**, 387, 1968).

References

Dixon, R. S. and Kraus, J. D.: 1968, *Astron. J.* **73**, 381.
Fitch, L. T., Dixon, R. S., and Kraus, J. D.: 1969, *Astron. J.* **74**, 612.
Pooley, G. G. and Kenderdine, S.: 1968, *Monthly Notices Roy. Astron. Soc.* **139**, 529.
Scheer, D. J. and Kraus, J. D.: 1967, *Astron. J.* **72**, 536.
Shklovsky, I. S.: 1966, in *Supernovae*, Moscow.
Vorontsov-Velyaminov, B. A. and Krasnogorskaya, A. A.: 1962, Morphological Catalogue of Galaxies, Part I, *Trudy gos. Astron. Inst. Sternberga*, **32**.

MASS-LUMINOSITY RATIOS AND SIZES OF GIANT ELLIPTICAL GALAXIES

IVAN R. KING

Berkeley Astronomy Dept.

and

RUDOLPH MINKOWSKI

Radio Astronomy Laboratory, University of California, Berkeley, Calif. U.S.A.

Abstract. Observations of velocity dispersion and of light distribution at the centers of 22 galaxies, mostly ellipticals give M/L values of 15 to 40 for giants of all luminosities with smaller values for intrinsically faint galaxies. Core radii within which surface brightness drops by a factor of two are typically 50–100 pc with a range from 2 to several hundred parsecs. Elliptical galaxies fit the modified isothermal models used for clusters, while later types have a flatter density gradient. Complete results will be published elsewhere.

Spectroscopic determination of the stellar velocity dispersion at the centers of giant galaxies makes it possible to determine central mass-luminosity ratios without recourse to the unverified assumptions implied by direct use of the virial theorem. (Those assumptions are that stars of all luminosities and in all locations have the same velocity dispersion and that the mass is distributed the same as the light.) The present technique involves a self-contained analysis of a small central region; essentially it compares an rms velocity with a scale height to find the force of gravity, which leads to a central mass density. In addition to the velocity dispersions the new data consist of photometry of the central peak of brightness. The remaining uncertainties come from the correction for atmospheric seeing and of course the uncertain Milky Way absorption and cosmic distance scale. (The results below assume $H=100$ km s^{-1} Mpc^{-1} and $A_{B,\text{pole}} = 0\overset{m}{.}3$.)

Values of the central M/L have been derived for 18 elliptical galaxies, 2 S0's, and 2 Sb's. The results lead to the following generalizations. (1) There is no clear difference in M/L between the centers of ellipticals and the centers of S0's and Sb's. (2) Systems with integrated absolute magnitudes fainter than -19 have a distinctly lower central M/L. (3) Among the brighter systems there is no noticeable correlation of central M/L with absolute magnitude. Values from 15 to 40 for giant systems probably represent a real range. (4) The M/L values at the centers of giant galaxies are in good agreement with the results derived by Spinrad through photoelectric analysis of the spectra of a few systems.

The photometry was aimed at high resolution in the centers of the galaxies but was also extended far out into the envelopes. Its conclusions are as follows. (1) The central brightness peaks are small, with the surface brightness dropping by a factor of two from its central value typically in a 'core radius' of 50 to 100 pc (1 to 2 s of arc at 10 Mpc). (2) Core radii range from 2 pc for M31 and M32 to many hundreds of parsec for some of the Virgo giants. (3) Except for a scale factor, elliptical galaxies have remarkably similar profiles. They fit the modified isothermal models previously developed for star clusters; the envelopes are also very close to the laws suggested by Hubble and by de Vaucouleurs, but neither of those laws fits the cores. (4) Up to the point where the present study loses the elliptical galaxies in the sky noise, the galaxies show no clear limits. (5) A small amount of data on the later-type galaxies confirms their well-known flatter density gradient.

This project also includes studies of ellipticity profiles and rotations of several of the galaxies. These will be reported on at a later date.

Discussion

Woltjer: Could you give an estimate of the upper limit to the mass of a massive body at the center of your systems?

King: You can put a moderately massive object in the center, but you *cannot* use the high stellar velocity dispersion as an argument for its existence. My dynamical argument says that the velocity dispersion must be due to gravitating material that has the same extended distribution as the stars. An additional massive object is limited to a size that will not appreciably affect the gravitational field within one core radius, because the stars would then show an unmistakably different luminosity profile. I think that the practical upper limit to a central object is perhaps a tenth of the core mass.

Tifft: Do you have star/light distributions for galaxies with sharp centers, more sharp that is than NGC 4472 which is relatively diffuse?

King: Yes, I have some cases with sharper centers, where the central profile is nevertheless resolved.

ON THE MASS-TO-LIGHT RATIOS FOR DOUBLE GALAXIES

R. J. DICKENS

Royal Greenwich Observatory, Herstmonceux, Great Britain

and

J. V. PEACH

Dept. of Astrophysics, University of Oxford, Great Britain

Abstract. Statistical mass-to-light ratios are found for spiral and irregular galaxies, and E and S0 galaxies, using 57 multiple systems including 43 pairs. The results are consistent with those of the earlier study by Page. Preliminary conclusions point out the great uncertainties in such mass-to-light ratios due to inaccurate magnitudes and relative radial velocities. The derived masses are relatively insensitive to the choice of the distribution function of projected separations.

The critical importance of galaxy masses and mass-to-light ratios for studies of their stellar content and for the resolution of such problems as the stability of groups and clusters of galaxies makes any observational evidence bearing on the masses of galaxies, and especially E and S0 galaxies particularly valuable. There is little information available on the masses of ellipticals from rotation curves, and only slightly more from the observationally difficult measurements of internal velocity dispersion. The use of the double galaxy method appears competitive with these, and we have reinvestigated the method in the light of two advances which suggest that such a reinvestigation is now opportune. These are, firstly, that since the work of Page (1952, 1966) and Holmberg (1954) there are now more velocities available for double and multiple systems, and secondly, that since the publication of Zwicky's *Catalogue of Galaxies and Clusters of Galaxies* it is possible to redetermine the frequency function of space separations of physical pairs, which is needed in the reduction of the radial velocity differences and angular separations to give masses.

1. Outline of the Method

Assuming circular orbits and point masses we can write an average over all pairs

$$\overline{rv^2} = 8\pi^2 \bar{M} \,\overline{\cos^3 x}\, \overline{\cos^2 y}$$

where r is the projected linear separation, v the differential radial velocity, $2M$ the mass of the pair, and x and y are the angles between the radius vector and the celestial sphere and between the direction of orbital motion and a plane containing the radius vector and the Sun respectively. For a random distribution of x and y, $\overline{\cos^3 x} = 3\pi/16$ and $\overline{\cos^2 y} = \frac{1}{2}$. While the distribution of y is certainly random, that of x depends on the actual distribution of true space separations between physical pairs $F(R)$. Now it can be shown that

$$f(r) = \int_r^{R_m} \frac{rF(R)\,dR}{R\sqrt{R^2 - r^2}},$$

where $f(r)$ is the number distribution of physical pairs as a function of projected separation and R_m is the maximum separation of physical pairs. $f(r)$ can be determined from counts of galaxies around selected central galaxies chosen at random with an appropriate correction for optical companions. Then we can write

$$\overline{v^2} = 4\pi^2 \bar{M} g(r, R_m)$$

where $g(r, R_m)$ contains integrals involving $F(R)$. A regression of $\overline{v^2}$ on $g(r, R_m)$ leads to $\bar{M}$.

2. Determination of $F(R)$

The distribution of physical pairs around galaxies with known velocities and with $m_{pg} \leqslant 13.0$ was determined as follows. The positions of all galaxies with $m_{pg} \leqslant 15.5$ in circles of 2° radius centred on these galaxies were taken from Zwicky's Catalogue and their distances from the central galaxies calculated assuming that they had the same modulus, the modulus being computed from the velocity with $H_0 = 100$ km s^{-1} Mpc^{-1}. The number of galaxies as a function of true separation is determined by combining data from all (~ 300) fields with appropriate correction for optical companions, assuming the distribution of these on the celestial sphere is random. The excess over a random distribution gives the distribution of projected separations of physical pairs, $f(r)$ from which $F(R)$ is obtained by deprojection. Hence $g(r, R_m)$ can be calculated for application to the pair velocity data.

3. Radial Velocities and Angular Separations

All galaxies with published radial velocities were investigated for possible physical companions; both pure pairs and multiple systems were isolated, the multiple systems consisting of more than two galaxies which could be treated as dynamical pairs. It proved possible to assemble reliable velocities for about 60% more systems than were considered in the treatment of double galaxy masses by Page (1966). Weights were assigned to the velocity differences on a uniform system. Angular separations were measured from the Sky Survey prints. The photographic magnitudes were taken from the Reference Catalogue of Bright Galaxies (de Vaucouleurs and de Vaucouleurs, 1964) and corrected for a galactic absorption of 0.24 mag at the poles.

4. Results and Discussion

Preliminary results of least squares solutions for various groups of the data are given in Table I where they are compared with the results of Page (1966). Masses and M/L

TABLE I

Masses and mass-to-light ratios for multiple systems ($H_0 = 100$ km s^{-1}Mpc^{-1})

		Pure pairs only		Pure + multiple systems	
		Page (1966)	This investigation	Page (1966)	This investigation
SP, IRR	$10^{-10} M/M_\odot$	1.6 ± 2.3	2.3 ± 4.1	4.0 ± 4.2	5.9 ± 7.6
	M/L (solar units)	1.4 ± 1.8	3.0 ± 5.4	3.2 ± 4.2	7.0 ± 9.0
	number of systems	10	28	16	36
E, S0	$10^{-10} M/M_\odot$	64 ± 38	44 ± 20	66 ± 29	45 ± 14
	M/L (solar units)	92 ± 92	60 ± 27	98 ± 68	42 ± 13
	number of systems	13	15	18	21

ratios for S and Irr systems are essentially unchanged. The large M/L ratio for E galaxies in pairs should be noted as it seems difficult to reconcile with the smaller (~ 50) values obtained from other dynamical methods and with the larger (~ 800) values obtained from the application of the virial theorem to clusters of galaxies. The discrepancies in the M/L ratios can partly be ascribed to a systematic difference of about 0.2 mag between the magnitudes of the Reference Catalogue and those used by Page.

A detailed account of this investigation will be published elsewhere. We should like to point out the following preliminary conclusions. Firstly it is not generally realised how great are the uncertainties in M/L ratios determined by this type of statistical analysis due to the generally inaccurate magnitudes available. These are such as to render the total luminosities and hence M/L ratios uncertain by about a factor two and M/L is probably systematically too large. Secondly, velocity differences based on individual velocity determinations are of such low weight as compared with those based on simultaneous observation of both objects that their influence on the solutions is small and leads to larger errors. Thirdly, the derived masses are relatively insensitive to the choice of $F(R)$.

References

Holmberg, E.: 1954, *Medd. Lunds Astron. Obs. Ser. I*, **186**, 1.
Page, T. L.: 1952, *Astrophys. J.* **116**, 63.
Page, T. L.: 1966, in *Proc. Fifth Berkeley Symposium on Mathematical Statistics and Probability*, Univ. of California Press, Berkeley, **3**, 31.
Vaucouleurs, G. de and Vaucouleurs, A. de: 1964, in *Reference Catalogue of Bright Galaxies*, Univ. of Texas Press, Austin.

Discussion

Lewis: In NGC 253-7 (my observations) and 3226-7 (Rubin and Ford) it is seen that the position angle of the major axis of the less massive galaxy is very near to the position vector joining the two galaxies. This seems to suggest that the less massive galaxy is approximately in its own plane of symmetry. If this were generally true, it suggests that the distribution of position vectors between the two galaxies and the orbits on the plane of the sky are related. This may mean that some of the quantities you assume to be independent and randomly distributed are not in fact so.

Dickens: While this effect *may* be generally present, one cannot base an analysis on two cases only. If these distributions prove likely to be related by further observations, such correlations could be incorporated into the theory. However, the large errors already inherent in the method would not appear to justify consideration of such effects, at least for the present.

King: (1) Does your method involve the assumption that the orbits are circular? (2) Did you derive a mean M/L directly, or did you divide a mean M by a mean L?

Dickens: (1) Yes. However, Page has shown that even the assumption of purely radial orbits should not change the mean masses by more than a factor of two.

(2) The results shown used a mean L for each group to obtain a mean M/L. However, we intend finally to solve separately the regression curve for M/L.

DO GALAXIES EVOLVE ALONG THE $R-\omega$ SEQUENCE?

WILLIAM C. SASLAW

Berkeley Astronomy Dept., University of California, and Institute of Theoretical Astronomy,
University of Cambridge, and Center for Advanced Studies, University of Virginia

Abstract. The stellar contents of Sb and Sc galaxies do not appear to be related to their rotation properties. This lack of correlation suggests that the position of a galaxy on the $R-\omega$ sequence is determined to a large extent by the environment in which the galaxy formed, rather than by subsequent evolution.

Suppose we are given some physical property of a galaxy, such as its rotation curve. It may then be asked how quickly this property changes with time. If the change is very rapid, then the property can be used to set up an evolutionary sequence for an ensemble of galaxies. If the change is sufficiently slow, then the property is the remnant of initial conditions when the galaxy formed. We consider the internal rotation of Sb and Sc spiral galaxies from this point of view. For further discussion see Saslaw (1970, 1971a, 1971b).

Nineteen rotation curves for fairly normal Sb and Sc galaxies have been measured, mainly by the Burbidges and their collaborators. These rotation curves generally have a spatial resolution of about 6 or 7 arc sec, which is sufficient to determine average dynamical properties of the galaxy. *To a first approximation*, the curves have an inner region of solid body rotation and an outer region of constant rotational velocity. We can estimate the linear extent, R, and angular velocity, ω, of the central region of each galaxy. These show an approximate correlation of the form $R \sim \omega^{-1}$. Possibly the Sb's have a different correlation from the Sc's, but the present observational uncertainties make this difficult to determine. An alternative statement of the $R \sim \omega^{-1}$ correlation is that the rotational velocity at which the curve turns over is generally 200 ± 50 km s^{-1}, even though values of both R and ω can vary by more than an order of magnitude.

The position of a spiral galaxy on the $R-\omega$ sequence gives us a direct measure of its dynamical properties. This measure is purely observational and independent of any model for the mass distribution of the galaxy. Of course, when such a model is available we can use the rotation curve to derive the mass $M(R)$ interior to R. A very crude illustrative model, supposing the mass $M(R)$ to be collected at the center, would give $M \sim R^3 \omega^2$. To obtain $R \sim \omega^{-1}$ would require $M(R) \sim R$. Thus the $R-\omega$ sequence can also be viewed as a mass sequence. However, this would introduce new uncertainties.

We ask whether galaxies evolve along the $R-\omega$ sequence, or whether this sequence is a relic of initial conditions. To search for observational hints, we ask more specifically whether there is a relation between a galaxy's location on the $R-\omega$ sequence and the stellar content of its nucleus. If among many galaxies there were a monotonic relation between the dynamical properties and the stellar content, for example, the smaller the solid body region the older its stars, this would suggest that the galaxies began from similar initial conditions, but that R decreases with time. On the other hand, if there is a random relation between dynamical properties and stellar content, it would suggest that initial conditions still dominate the observed properties of the dynamics, the stellar content, or both. This second situation, in fact, appears to be the case.

To begin examining this question, Spinrad and I have scanned selected blue wavelengths of nine galaxies at the Lick 120-in. prime focus. The galaxies are selected to have bright, fairly compact nuclei, and the wavelengths are chosen to give line indices which are sensitive to the stellar content (Tables I and II).

TABLE I

Rest wavelength (Å)	Feature	Line index
4040	Continuum	–
4100	Hδ	(4040)/(4100)
4200	CN λ4215	(4040 + 4500)/2(4200)
4300	CH λ4313	(4040 + 4500)/2(4300)
4340	Hγ	(4040 + 4500)/2(4340)
4500	Continuum	–
	Blueness	(4040)/(4500)

The notation (λ) denotes the intensity at wavelength λ.

TABLE II

NGC	Type	Observed systemic velocity km s^{-1}	Distance Mpc	R_{turnover} arc sec	R_{turnover} kpc	ω km s^{-1} kpc^{-1}
157	Sc	1710	24	50	5.8	30
224	Sb	−300	0.67	2100	6.5	45
1084	Sc	1465	19.2	22	2.0	65
1832	Sb	1935	25.3	7	0.7	170
2903	Sbc	600	7.9	70	2.1	100
3521	Sb	815	8.5	38	1.6	150
5055	Sb	560	10.3	30	1.5	150
5248	Sc	1190	15.4	10	0.75	200
6181	Sc	2350	32.9	6	0.9	210

Differences of line indices among galaxies are caused mainly by differences in stellar population. To make absolute quantitative models of the stellar content of these nuclei we need much more data. However, here we are concerned with the trend of relative differences among nuclei, and this can be estimated qualitatively directly from the line indices. Figures 1 and 2 show the results. Error bars show the standard deviations for several scans of a given line index.

DO GALAXIES EVOLVE ALONG THE R–ω SEQUENCE

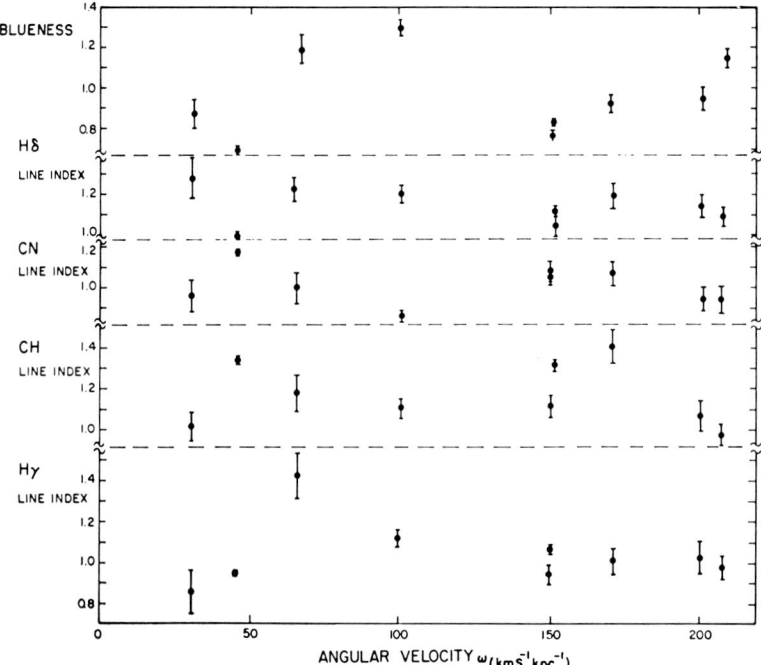

Fig. 1. Angular velocity vs spectral line indices for spiral galaxies.

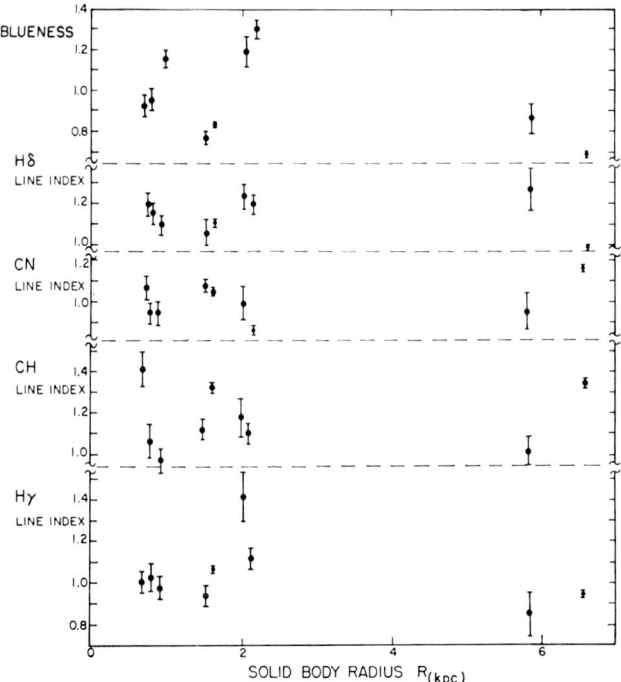

Fig. 2. Solid body radius vs spectral line indices for spiral galaxies.

The various indices appear to vary quite randomly with R and with ω. This is especially true of the blueness index, which turns out to be a useful rough measure of stellar content for these galaxies.

The lack of correlation between dynamical properties and stellar content suggests that different degrees of evolution, starting from the same initial conditions, cannot explain what we observe, even for galaxies of similar type. It also suggests that the position of a galaxy on the $R-\omega$ sequence (and perhaps the stellar content as well) is governed to a large extent by the environment in which the galaxy formed.

There is another, indirect, line of evidence which also tends to support this conclusion. Turbulent and dusty galaxies do not appear to cluster in any particular region of the $R-\omega$ sequence. If these qualities indicate either young or old galaxies, but not both, this result also argues against a predominantly evolutionary interpretation. Therefore, in rotation curves we may find new clues to the origin of the galaxies.

References

Saslaw, W. C.: 1970, *Astrophys. J.* **160**, 11.
Saslaw, W. C.: 1971a, *Astrophys. J.* **163**, 249.
Saslaw, W. C.: 1971b, *Monthly Notices Roy. Astron. Soc.* **152**, 341.

Discussion

Ozernoy: A simple argument against any evolution along the Hubble sequence is that, according to Holmberg's data, the mean mass of a galaxy increases from irregulars to ellipticals. Apparently the same is valid for the spiral subsequence.

CLASSIFICATION OF COMPACT OBJECTS: QSS, QSOs, N-TYPE AND COMPACT GALAXIES, SEYFERT AND GALACTIC NUCLEI

W. W. MORGAN

Yerkes Observatory, Williams Bay, Wis., U.S.A.

Abstract. Some methods currently in use for the classification of the optical forms of the 'compact' galaxies and quasi-stellar objects are reviewed. It is shown that the category 'Seyfert Galaxy' is basically a spectroscopic (rather than a form) classification.

An optical form-classification is described which is, in principle, identical with published classification criteria for QSO, N-type, and compact objects. The importance of maintaining rigid form-standards is emphasized.

1. Introduction

The forms of compact extragalactic objects differ from ordinary spirals and ellipticals; and it is not possible to interpolate satisfactorily the compact objects in the Hubble or Yerkes sequences of standard types. The optical forms observed for galaxies having eruptive nuclei have their own peculiar shapes, which require a special morphology.

This has led to definitions of certain categories. For example, Sandage (1967) has given a short definition and distinction between the QSS and the N-type galaxies:

"These systems, called N-galaxies, have some resemblance in optical appearance to quasi-stellar sources (QSS) in that most of the light comes from the pointlike nucleus. However, QSS and N-galaxies differ; all N-galaxies show an outer nebulous envelope on plates taken with the 48-in. Schmidt, while all QSS are indistinguishable from stars..."

In the same paper, Sandage showed that the N-galaxies and QSS are cleanly separated in the UBV two-color plot – and that the radio-quiet QSO occupy the area defined by the QSS.

Later, C. R. Lynds (1968) published a discussion of the forms and spectra of certain galaxies classified as N-type. He showed that the object Braccesi 234, while completely stellar in appearance, falls in the N-type domain of the two-color plot. This object has a low redshift, and sharp emission hydrogen and forbidden lines. He also states that there are galaxies classified as N-type that show no emission lines, but that do contain H and K, and the G-band in absorption. The high level of the critical approach used by Lynds on both form and spectral characteristics sets a standard for similar investigations of the future.

The basic problem, then, is to examine and describe the forms and spectra of the compact objects (QSO, N, compact, Seyfert) with the greatest individual precision – while at the same time constructing the most useful groupings of the objects in some morphological scheme.

2. Comments on the Classification of QSO, N-Type and Seyfert Galaxies

There seems to be general agreement on the definition of quasi-stellar objects (G. R. Burbidge, 1970): "QSOs are defined as a class of objects which have a star-like appearance on direct plates and have very large redshifts in their spectra". However, two of the three first discoveries (Matthews and Sandage, 1963) were noted as having "exceedingly faint wisps of nebulosity" associated with the star-like images. The surveys of Sandage and Véron (1965) and Braccesi et al. (1968) furnished spectrophotometric criteria for candidates for non-radio QSOs.

The N-galaxies have been defined (Morgan, 1958) as "systems having small brilliant nuclei superposed on a considerably fainter background"; later, N-type radio-galaxies were defined by Matthews et al. (1964) as "galaxies having brilliant star-like nuclei containing most of the luminosity of the system. A faint nebulous envelope of small visible extent is observed".

As has been noted by a number of investigators, the above-mentioned definitions follow rather closely the earlier comments by Seyfert on the forms of the nuclear region of the galaxies in his classical paper (Seyfert, 1943): "...their most consistent characteristic being an exceedingly luminous stellar or semi-stellar nucleus which contains a relatively large percentage of the total light of the system".

These similar definitions complicate the classification problem; and if each is literally applicable there would seem to be no reason for the introduction of the N-category. However, re-examination of the optical form of the twelve original Seyfert galaxies indicates that Seyfert's optical description quoted above does not hold for some of the galaxies in his paper. Burbidge et al. (1963) showed that four of the twelve galaxies listed by Seyfert (NGC 2782, 3077, 4258, and 6814) do not satisfy their reformulation of Seyfert's criteria:

"(i) it has a small bright nucleus.

(ii) the spectrum of the nucleus contains emission features which are not normally seen in the spectra of galaxies. Some of these should be lines indicating higher excitation than that required to excite [O III] and [Ne III]...

(iii) the emission features; or at least the hydrogen emission lines, must be of great width, which, if interpreted as Doppler motions, corresponds to velocities in the range ± 500 to ± 4250 km s^{-1}, according to Seyfert".

Recent observations by Nolan Walborn of the forms of seven of the eight remaining galaxies in Seyfert's list with the new 41-in. reflector at the Yerkes Observatory indicate that two others (NGC 1275 and 3516) do not accord with Seyfert's optical criterion: also, the small nucleus of NGC 1068 can hardly be said to contain "a relatively large percentage of the total light of the system". From the optical forms of the eight currently accepted members of Seyfert's group, only five (NGC 3227, 4051, 4151, 5548, and 7469) satisfy all optical conditions for membership in the group (the data for NGC 3227 were taken from Rubin and Ford (1968)).

We now have two questions which require comment: Does the 'N type' classification differ from the 'Seyfert' classification? And: What is the nature of the Seyfert classification?

The N-type classification is based on a certain appearance of the optical image. The description of the class is formulated for images containing a relatively small number of picture elements, for the sake of use over large ranges in distance without systematic error. When the number of picture elements is increased (through proximity or the use of a larger telescope) it is seen that many N-galaxies are spiral in nature; the class "N ', therefore, includes a subclass 'NS'. Under these terms, one (NGC 5548) of the eight Seyfert galaxies is an outstanding example of the N-type, while two (NGC 1275 and 3516) are different – and could not properly be described as of N-type. We therefore have the following situation: The form of one of the original Seyfert galaxies can be considered as the prototype for class N, as defined above; certain other of the Seyfert galaxies could not properly be classified as N, while the forms of still others are intermediate in appearance. The 'Seyfert' category does not define a particular form-class; it *does* define, however, a rather particular combination of spectroscopic characteristics. If we oversimplify somewhat the latter, we can describe a spectroscopic category which has some characteristics of a 'natural group': strong, relatively narrow N_1 and N_2 emissions combined with very broad emission Balmer lines.

If we apply the term 'Seyfert' as a *spectroscopic* description, then we avoid the fundamental ambiguity between categories 'Seyfert', 'N', and 'quasi-stellar object'. The 'Seyfert' classification can be used unambiguously as a spectroscopic class; the types 'QSO', 'N', and 'C' (to be discussed below) then furnish an unambiguous group of form-categories. The 'Seyfert spectrum' category (as suggested recently by Arp (1970)), then also becomes unambiguous in use and can be applied as information to many C-type, N-type, and QSO objects – in the sense of additional, rather than competing data. When considered as a spectral, rather than a form, classification, the Seyfert types as illustrated in the spectral tracings in Seyfert's paper are found to be present in members of all three basic form categories (QSO, N, Compact). In addition, the varied spectroscopic characteristics described by Lynds (and later by Sargent and others), furnish material of unique value for astrophysical investigation and for possible relationships between luminosity on one hand, and form and spectrum on the other.

3. The Form-Classification of the Zwicky, Markarian and Haro Objects

The most careful consideration of the nature of objects classified as 'compact' by Zwicky has been carried out by Sargent (1970b). The definitions given below are based on his conclusions. Zwicky (1964) has defined a compact galaxy in these terms:

"We call compact galaxies those which can just be distinguished from stars on plates taken with the Palomar 48-in. telescope and which have diameters of $2''-5''$..."

"We shall call a galaxy *moderately compact* if its image, on photographs taken with the 18-in. Palomar Schmidt telescope, can just barely be distinguished from stars of the same apparent brightness. These systems have diameters of about $5''-10''$..."

Sargent summarizes the results of his own observations of a number of the galaxies in Zwicky's lists as follows:

"An inspection of the *Palomar Sky Survey* prints reveals that while, in accord with the definition above, the majority of the objects are small, a considerable minority are extended, often with bright knots, jets, distorted spiral arms, bridges to other nearby galaxies, rings, or some such unusual appendage. These, presumably, are the eruptive and post-eruptive galaxies and the galaxies with compact parts... The feature that is common to the objects in Zwicky's lists is that all have a region of high surface brightness."

From the above it can be seen that a simple, precise form-classification for such a varied collection of remarkable objects is impossible. In his beautiful discussion Sargent has recognized this, and has proceeded to give careful, individual descriptions of the forms and spectra of the objects observed. Such a procedure is a necessity in laying the groundwork for a later form-classification, and a detailed spectral classification as well.

At the present time, the basic spectroscopic approach used by Sargent for separation into astrophysical categories is certainly the most fruitful general procedure. The forms of a certain number of the Zwicky objects can be classified on the form system described below, but they constitute a minority of the objects observed by Sargent. Warning should be given against an overclassification of the Zwicky optical forms by relaxing or straining the criteria; such a procedure would further complicate, rather than clear, the present situation.

The seventy galaxies listed by Markarian possess abnormally strong ultraviolet continua, as observed by objective prism camera with a dispersion of ~ 1800 Å mm^{-1} near Hγ. Most of these galaxies are compact, with sharp, star-like nuclei. The selection, then, is from a spectroscopic criterion, and differs from that of Zwicky.

E. Khachikian has summarized the results of joint investigations of forty of these objects (Khachikian, 1968; Weedman and Khachikian, 1968, 1969). The range in spectroscopic characteristics is great, represented by the five groups:

(1) Narrow emission and absorption lines;
(2) Narrow, strong emission lines only;
(3) Strong and diffuse emission lines; N_1, N_2 much stronger than the hydrogen lines;
(4) No strong emission lines;
(5) Very broad hydrogen emission lines, typical of Seyfert galaxies.

Sargent notes that the Zwicky emission-line compact objects resemble the Markarian galaxies and the Haro (1956) galaxies.

In another paper, Sargent (1970a) discusses the nature of the Markarian galaxies in terms of the latter's classification and his own new spectrograms. The Markarian classification of the sharpness and strength of the ultraviolet continuum has resulted in the discovery of four new Seyfert-type spectra (Markarian 50, 69, 9, 10). The first two of these are described by Sargent as "structureless, blue, compact objects". The last two are classified below. From the thirty galaxies observed by Sargent the conclusion is drawn that the Markarian objects comprise a non-homogeneous group.

The Haro galaxies were selected on the basis of excesses in U, of U, B, V images on Tonantzintla Schmidt plates. An investigation by DuPuy (1968a, 1968b) showed that on the Palomar Sky Survey prints almost all of the Haro objects show "... structural distortions or highly irregular forms". From this it can be concluded that classification of their optical forms will be affected by the same difficulties as in the cases of the Zwicky and Markarian objects; and in all three catalogues the spectroscopic approach seems to be the most promising at present.

4. A Compilation of Criteria for Form-Classification: A Catalogue of Form-Types

We make use of three basic form classes: Q, N, C: with two additional subclasses: N− and NS:

Q: Objects having a star-like appearance on direct plates, with large redshifts.

N: Galaxies having small, brilliant nuclei containing a considerable fraction of the total luminosity, superposed on a considerably fainter main body.

C: Small, high surface-brightness galaxies which are slightly resolved on medium and large-scale photographs. (Adapted from definition given by Zwicky to Arp (private communication, 1970)).

To this must be added (to minimize selection effects with distance), the following: "structureless, blue, compact objects" (Sargent, 1970a).

N−: Less pronounced N-galaxies. (This is included to minimize selection effects with resolution or distance.)

NS: N-galaxies having well-developed spiral arms. (This is included to minimize selection effects with resolution or distance.)

Some standards for form classes are given below:

TABLE I

Object	Type	Source of illustration
NGC 5548	N	Walborn, Yerkes 41-in.
3C 48	N	Sandage and Miller (1966)
III Zw 2	N	Arp (1968)
3C 120	N	Arp (1968)
Markarian 9	N	Khachikian (1968)
Zw 0039.5 + 4003	N	Zwicky et al. (1970)
NGC 7469	N	Walborn, Yerkes 41-in., low resolution
B 264[a]	N	Arp (1970)
Markarian 10	NS	Khachikian (1968)
NGC 7469	NS	Walborn, Yerkes 41-in., high resolution; Burbidge et al. (1963)
NGC 4051	N−	Morgan (1958)
NGC 3516	C	Walborn, Yerkes 41-in.
Ton 256	C	Arp (1970)
B 264[b]	C	Arp (1970)

[a] From 200-in. print
[b] From 48-in. red print

5. Conclusion

One of the most remarkable characteristics of the Zwicky and Markarian objects is their great variety of appearance on large-scale photographs, as shown by Sargent. This characteristic makes form-classification especially difficult; but the fact of the existence of this variety is a discovery of great importance. It is to be hoped that the remainder of these objects can be observed, both spectroscopically and by direct photography.

The relationship of the N-type galaxies to the quasi-stellar objects is of crucial importance. A recent paper by Komberg and Ozernoy (1970) defines an epoch in this field and suggests many fruitful subjects for future research.

A remarkable and finely balanced general summary of the field of nuclei of galaxies has just been published by G. R. Burbidge (1970). This accurately delineates the state of the field in early 1970, and raises many fundamental questions for further investigation.

Acknowledgements

I am indebted for discussions of the classification of Seyfert and compact galaxies to Drs V. A. Ambartsumian, W. L. W. Sargent, and J. B. Oke; Drs Halton Arp, E. M. Burbidge, Sargent, and J. N. Bahcall sent preprints of work in this field. I am especially indebted to Dr G. R. Burbidge for his preprint (Burbidge, 1970). Photographic prints of B 264 and Ton 256 from Dr Arp made possible the classification of these objects. The plates of NGC 3516, 4051, 4151, 5548 and 7469 were taken by Mr. Nolan Walborn. Mr. J. W. Tapscott carried out necessary photographic work with distinction. The investigation was supported by a grant from the National Science Foundation.

References

Arp, H. C.: 1968, *Astrophys. J.* **152**, 1101.
Arp, H. C.: 1970, *Astrophys. J.*, in press.
Braccesi, A., Lynds, C. R., and Sandage, A. R.: 1968, *Astrophys. J. Letters* **152**, L105.
Burbidge, G. R.: 1970, *Ann. Rev. Astron. Astrophys.* **8**, 369.
Burbidge, G. R., Burbidge, E. M., and Prendergast, K. H.: 1963, *Astrophys. J.* **137**, 1022.
Haro, G.: 1956, *Bol. Obs. Tonantzintla Tacubaya* **2**, 8.
Khachikian, E.: 1968, *Astron. J.* **73**, 891.
Komberg, B. V. and Ozernoy, L. M.: 1970, *Astrophys. Space Sci.* **7**, 31.
Lynds, C. R.: 1968, *Astron. J.* **73**, 88.
Matthews, T. A., Morgan, W. W., and Schmidt, M.: 1964, *Astrophys. J.* **140**, 35.
Matthews, T. A. and Sandage, A. R.: 1963, *Astrophys. J.* **138**, 30.
Morgan, W. W.: 1958, *Publ. Astron. Soc. Pacific* **70**, 364.
du Puy, D.: 1968a, *Astron. J.* **73**, 882.
du Puy, D.: 1968b, *Publ. Astron. Soc. Pacific* **80**, 29.
Rubin, V. C. and Ford, W. K.: 1968, *Astrophys. J.* **154**, 431.
Sandage, A. R.: 1967, *Astrophys. J. Letters* **150**, L9.
Sandage, A. R. and Miller, W. C.: 1966, *Astrophys. J.* **144**, 1238.
Sandage, A. R. and Véron, P.: 1965, *Astrophys. J. Letters* **142**, L412.
Sargent, W. L. W.: 1970a, *Astrophys. J.* **159**, 765.
Sargent, W. L. W.: 1970b, *Astrophys. J.* **160**, 405.

Seyfert, C.: 1943, *Astrophys. J.* **97**, 28.
Weedman, D. W. and Khachikian, E.: 1968, *Astrofizika* **4**, 587.
Weedman, D. W. and Khachikian, E.: 1969, *Astrofizika* **5**, 113.
Zwicky, F.: 1964, *Astrophys. J.* **140**, 1467.
Zwicky, F., Oke, J. B., Neugebauer, G., Sargent, W. L. W., and Fairall, A. P.: 1970, *Publ. Astron. Soc. Pacific* **82**, 93.

Discussion

Miley: Radio quasars can be divided into two main classes according to their radio properties – those such as 3C 273 with an active component radio nucleus coincident with the optical QSO, and those such as 3C 47 with similar radio components located symmetrically about the optical QSO. These latter quasars tend to have steeper radio spectra which do not vary with time. It will be very interesting to discover whether the optical type can tell us anything about these two radio classes.

Tifft: NGC 4303 is a *blue* nuclear system photometrically. This concurs nicely with your note on *masking* of the blue spectrum by a blue continuum source. NGC 4501 is a *red* nuclear system photometrically. This concurs with an enriched dwarf system. There is obviously a fine case for smooth continuity between 'normal' and 'abnormal' nuclei. For example Seyfert – photometrically abnormal nuclei – normal.

Arp: The value of a precise and completely empirical classification on the basis of form is that we can now ask what is the transition between the star-like QSOs on the one hand and compact and N-galaxies on the other hand. One kind of transition could take place by an increasingly larger and brighter disk appearing around a stellar nucleus. Another kind of transition could take place if QSOs and compact objects had an increasing number of jets emerging from the nucleus until they formed the disk of an N-galaxy. Possible examples of this last kind of transition would be 3C 48 and 3C 120.

Wlérick: Just a remark concerning the galaxy NGC 4051. I am a bit shy to make it after the publication of the wonderful photograph of NGC 4151 taken with the Stratoscope II. Still I would like to report that on photographs taken with the electron camera at the Haute Provence Observatory in March 1968, the 'nucleus' of NGC 4051 appears elongated, in the V and B colors. Although the 'seeing' was medium, between 1" and 2", all the plates are consistent with a model in which there is one true 'nucleus', nearly stellar, and two relatively strong 'jets' that on the photographs are embedded in the apparent elliptical nucleus; the two-dimensional nebulosity starts beyond this nearly one-dimensional structure.

Dent: In this classification scheme, how would quasi-stellar objects with jets be classified? Would they be N-type, since the jets are, after all, a nebulosity? I'm thinking of 3C 273 and 3C 279 as examples.

Morgan: I would not classify 3C 273 as an N-type galaxy from the isolated jet which has been observed. If later observations should reveal a nebulous envelope, the classification should be re-examined.

Wray: It appears to me that the general problem of direct observation of these types of objects would be a rewarding area of work for someone possessing an electronographic camera. The large dynamic range would provide an improved capacity to record both the relatively bright nucleus and the fainter amorphous outer regions in a single exposure, while the relatively low noise would permit maximum information retention for the nearly stellar objects and would also improve the possibility of achieving at least a limited deconvolution of the point image spread caused by the atmospheric degradation and image brightness.

Khachikian: What do you think the lower limit to the width of a galaxy's H-lines is, to call it a Seyfert type?

Morgan: This is a matter for later decision by the spectroscopists. The extreme spectra are easy to classify, but there will also be borderline cases.

ASTROPHYSICAL STATISTICS OF 745 COMPACT GALAXIES NEAR THE GALACTIC NORTH POLE

L. RICHTER, N. B. RICHTER, and P. SCHNELLER

Karl Schwarzschild Observatory, Tautenburg, G.D.R.

Abstract. A brief description is given of methods used to find 745 compact galaxies in a 36 sq deg-field around M3. These constitute 24–40% of all galaxies in this area, brighter than 18.5 magnitude in B. Significant clustering found is related to Zwicky clusters of compact galaxies. Photographic 3-color photometry has been carried out. There seems to be no physical difference between cluster and field objects. The catalogue is to be published elsewhere, together with finding charts and detailed discussion of the photometry and statistics.

In a field of 36 sq deg round M3 a search for spherical concentrated (compact) galaxies was made on plates taken with the 134/200/400 cm Tautenburg Schmidt camera.

1. Method of Selection

The selection was made independently by two observers (Richter and Richter, 1968), using V or R-plates which were taken at times of good definition (seeing of about 1″). The two observers used different pairs of plates in the stereocomparator. The results were then compared and those objects selected by both observers were listed in the final catalogue. These comprise about 90% of the whole number. Those objects, which were selected only by one of the observers were once more discussed and inspected on further plates in order either to exclude them or to include them in the final catalogue. So we think the material may be very homogeneous.

2. Definition of Compact Galaxies: Morphological Criteria

In order to learn what Zwicky means when selecting compact galaxies we inspected such Zwicky objects on Tautenburg plates. After this process we selected the objects in 3 categories according to the following principles:

(1) Very concentrated: starlike, with a slightly diffuse border only distinguishable by a well-trained observer and very careful comparison with stars of the same brightness in the neighborhood;

(2) Concentrated: nearly starlike, but showing clearly a diffuse border;

(3) Medium concentrated: central part very concentrated like a star. Round this a halo of width not larger than the diameter of the central starlike part.

From the 745 objects selected and inspected for concentration 13% belonged to category 1, 49% to category 2 and 38% to category 3.

In general selection was confined to spherical concentrated objects, to avoid difficulties with photometry by iris photometer.

3. Limiting Magnitude of Selection

The excellent quality of the plates, the large scale ($1'' = 0.020$ mm) and good resolving power of the plates and light-gathering power of the large telescope allowed us to distinguish compact galaxies with certainty down to the 18.5 mag in B, in some cases even to the 19th mag.

4. Results

A. DISTRIBUTION IN THE TEST FIELD

All 745 objects were mapped according to their positions in the test field. There are obvious apparent accumulations, of which the reality has been investigated by statistical methods used by Meurers (1957, 1968) and his co-workers in Vienna for the investigation of star aggregates.

The results are shown in Figure 1. The squares marked with Roman numbers contain aggregates of compact galaxies which are with the highest degree of probability not random.

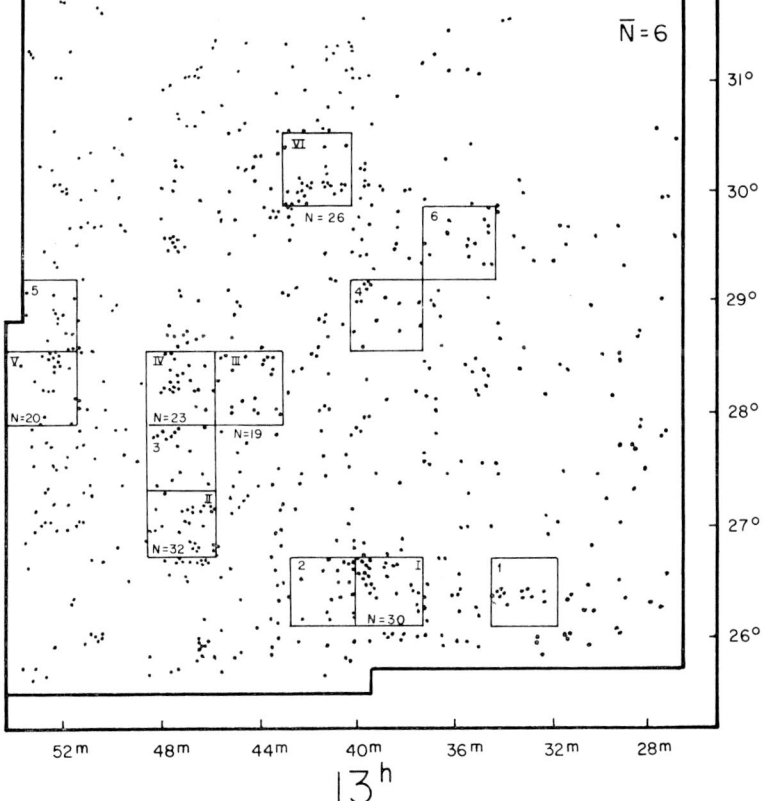

Fig. 1. Positions of 745 compact galaxies in the field round M 3. The squares enclose aggregates which are with highest probability not random according to statistical tests.

The squares with arabic numbers contain aggregates which are with great probability also not random.

The next step in the statistical investigation was to compare these aggregates with the positions of clusters of galaxies in Zwicky and Herzog's Catalogue.

In Figure 2 the borders of these clusters are marked on our test field. Clearly all except one our aggregates of compact galaxies are related to normal clusters of galaxies in the Zwicky catalogue.

Seventy-five percent of our objects belong to such clusters, and the remainder are situated in the field between them.

For the clusters X, Y, Z of Figure 2 we determined the proportion of spherical compact galaxies to the total number of galaxies counted in the test field down to the limiting magnitude of selection.

About 40% of galaxies of all types, visible on Tautenburg plates down to 18.5 mag in B, are compact galaxies according to our selection criteria 1, 2 and 3. If we assume that the objects of category 3 are normal E0 galaxies and should not be counted as compact galaxies there remains still 24%. This seems to be a very high percentage.

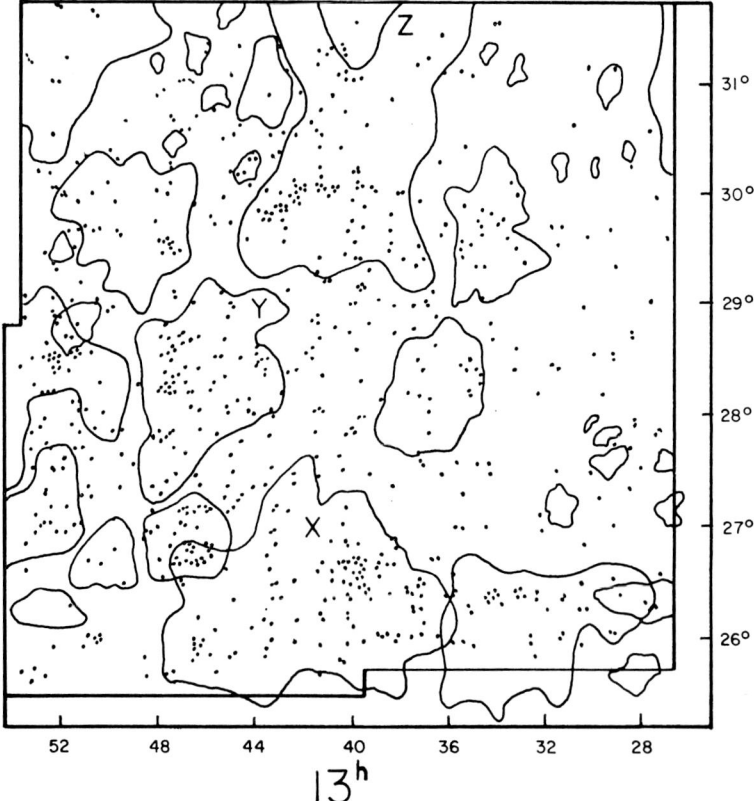

Fig. 2. Positions of 745 compact galaxies and the borders of Zwicky's clusters of galaxies in the test field.

B. PHOTOMETRIC STATISTICS

In a former publication (Richter and Richter, 1968) the authors have shown that we must expect systematic errors from integrated photometry of compact galaxies by iris photometry. There is a strong dependence of the brightness of those objects on the limiting magnitude of the plate, especially in system V. The fainter this limiting magnitude the brighter is the measured galaxy. In systems U and B this effect is still small, but in V the systematic error averages more than 0^m.32 for a difference of 1 mag in the limiting magnitudes of two different plates. There is also a dependence on intrinsic magnitude of the galaxies. The reasons for these effects are known and will be discussed in the final extensive publication.

For the purposes of a photometric catalogue of these objects and for preliminary astrophysical statistics we define quasi-brightnesses (maximal brightnesses) in the different color systems for the plates with the faintest photometric limiting magnitudes.

TABLE I

Brightness (V)	All galaxies	Cluster galaxies	Field galaxies
$m < 14.00$	23	16	7
$14.00 < m < 14.99$	37	25	12
$15.00 < m < 15.99$	127	87	40
$16.00 < m < 16.99$	228	163	65
$17.00 < m < 17.99$	270	191	79
$18.00 > m$	60	39	21
	745	521	224

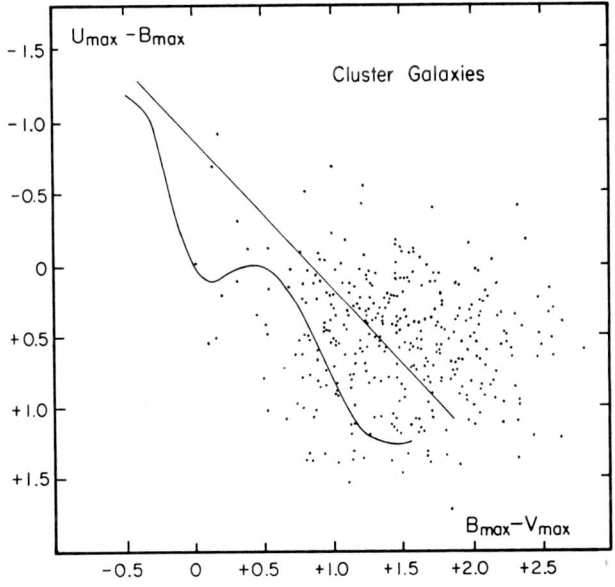

Fig. 3. Two-color diagram of compact cluster galaxies.

These are the following: U: $20^m\!.0$; B: $20^m\!.4$; V: $20^m\!.1$.

In Table I are shown the numbers of compact galaxies for different intervals of magnitude for the whole test field.

Special statistics for the large clusters in the field will be published in the final work together with many other statistical investigations. This publication will also include the whole catalogue and finding charts for the 745 compact galaxies.

With some restrictions it is interesting to construct from the quasi-magnitudes of the catalogue two-color diagrams for cluster and field galaxies separately (Figure 3

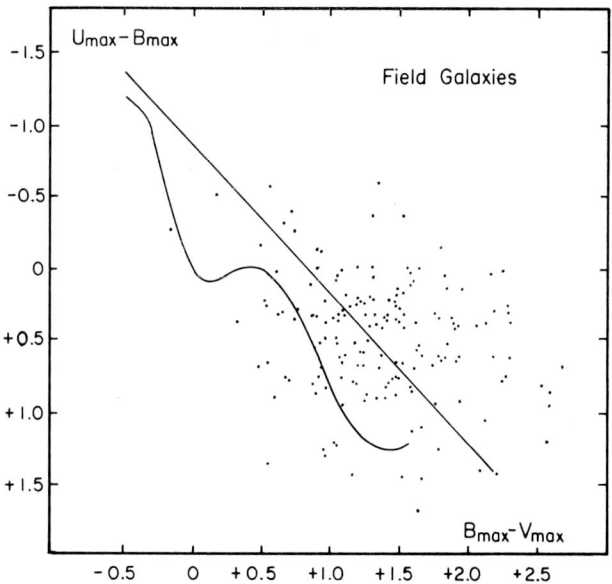

Fig. 4. Two-color diagram of compact field galaxies.

and Figure 4). Of course these diagrams are not representative of the real intrinsic positions of the single objects in the diagrams. The reason we have discussed before. But something can be derived in any case. There seems to be no physical difference between objects in the clusters and in the field around them. And there exists surely a number of blue compact galaxies. All these will be listed in a special investigation to put them at the disposal of spectroscopists.

References

Meurers, J.: 1957, *Veröff. Univ. Sternw. Bonn* No. 45.
Meurers, J.: 1968, *Ann. Univ. Sternw. Wien* No. 5.
Richter, L. and Richter, N.: 1968, *Mitt. Karl-Schwarzschild Obs.* No. 39.

OPTICAL SPECTRA OF COMPACT OBJECTS

E. M. BURBIDGE

University of California, San Diego and Institute of Theoretical Astronomy, Cambridge, England

Abstract. On the basis of their optical spectra, compact objects are divided into 3 classes according to the probable nature of their energy sources, as follows:

(1) Non-thermal: This class includes Seyfert nuclei, most N-type radio galaxies, and also very luminous Zwicky compact objects. The existence of high-velocity clouds, and a wide variety of densities and temperatures appear to be common characteristics of these objects. QSOs are considered to be related, with similar but more extreme properties.

(2) Hot, massive stars: This class includes some small galaxies, parts of galaxies or appendages to galaxies, which are usually much less luminous than class (1) objects.

(3) Stellar: Some compact galaxies having only absorption line spectra fall into this class.

A variety of morphological forms exists amongst class (1) and (2) objects.

Emission line spectra of QSOs are reviewed, and related to the physical conditions in the emitting regions.

1. Introduction

The widespread interest that exists today in compact extragalactic objects has developed in the past seven years or so, and stems largely from:

(a) the discovery of the quasi-stellar objects;

(b) the survey work of Zwicky (1964, 1966; 1964–68) which disclosed that an abundant population of compact objects exists in space, following the survey work of Haro (1956) and Markarian (1967, 1969) which selected objects with abnormally strong UV continua and which yielded some of the same kind of compact galaxies;

(c) the continuing work on identification and spectroscopy of radio galaxies, which include compact objects.

The earliest work on compact objects was of course that by Seyfert (1943), in his original isolation of what are now known as the Seyfert galaxies. As we shall see, the term 'compact' is a loose descriptive term, and the compact objects in Zwicky's definition comprise compact parts of galaxies as well as compact galaxies, e.g. the 'Ambartsumian knot' in NGC 3561 (Zwicky and Humason, 1961; Zwicky, 1966; Stockton, 1968). His definition, however, excludes the QSOs, because he requires the objects to be just distinguishable from stars on the 48-in. Palomar Schmidt plates.

In any case, Zwicky's lists contain a heterogeneous collection of objects. We shall take it that the definition of 'compact' has partly an operational significance, signifying a galaxy or part of a galaxy with a surface brightness exceeding some assigned value, and partly a deductive physical significance, following from properties such as short-term variability and implied physical conditions. Thus the family of compact objects comprises QSOs, Seyfert nuclei, N-type galaxies, and compact galaxies or parts of galaxies as categorized by Zwicky.

Both Morgan (1970) and Ambartsumian (1970) have emphasized the importance of a morphological approach to the study of these objects, so that the important and physically meaningful correlations between the various types of observations can be found and so that ultimately a satisfactory theory embracing all the data can be formulated. Morgan (this volume, p. 97) has described a classification scheme for these objects, with criteria for classification and selected standards for the form classes, and the present paper is intended to be read with that scheme kept always in mind as a basic morphological classification which minimizes observational selection effects due to distance.

I want to make a simple division of my subject matter along physical lines, and first I divide the compacts in Zwicky's sense into three classes:

(1) The first class consists of objects with an energy source of very small dimensions, having a large and often variable energy output generally believed to be of non-thermal origin. The ultimate energy source is probably gravitational energy, but its nature is not fully understood, and other hypotheses, e.g. that by Ambartsumian (1958, 1965), have been advanced. With a non-thermal source, high-energy particles provide the means of transferring energy from the source to what is seen – in our case, either short-wavelength radiation which ionizes the gas and/or the line emission itself. This class includes the Seyfert nuclei and most N-type radio galaxies, and it has a direct link with the QSOs which have similar but more extreme properties. The Seyfert galaxy II Zw $0430+05 \equiv 3C$ 120 is an example. The class is distinguished partly by its line spectrum, which has strong, broad emission lines, arising from ions going up to a high ionization potential. We shall, however, see later that some N-type radio galaxies have fairly narrow emission lines, yet the radio emission requires a non-thermal source and places them in this class.

(2) The second class consists of small galaxies, parts of galaxies, or appendages to galaxies, and is distinguished from class (1) by the line spectra, which display sharp emission lines resembling bright high-excitation H_{II} regions like the Orion Nebula. The physical distinction from class (1) is that the energy source appears to be hot, massive stars – gigantic O-associations. There is no evidence for non-thermal emission.

(3) Some compact galaxies have only absorption-line spectra, with no emission lines to indicate the presence of hot gas. Such objects must consist only of stars, and the star density must be high. It is not known how common this class is, and little quantitative work has been done on them.

The division of the Zwicky compacts and those Markarian and Haro galaxies which fall in this category into three classes has been made possible by the slit spectroscopic or objective prism studies by Haro, Markarian, and Zwicky themselves, by Arp *et al.* (1968), DuPuy (1968), Weedman and Khachikian (1968, 1969), Sargent (1970a, b, c), and Arakelian *et al.* (1970).

Section 2 will cover the classical Seyfert galaxies, N-type galaxies, and compacts of class (1); compacts of classes (2) and (3) will be covered in less detail in Sections 3 and 4, respectively. Finally, Section 5 will discuss rather briefly the emission-line spectra only of the QSOs; the paper by Lynds (this volume, p. 127) will cover the absorption-line spectra of QSOs.

2. Seyfert and N-Type Galaxies; Compacts of Class (1)

2.1. SEYFERT GALAXIES

A fuller account of these than is possible here can be found in Chapters 5, 8, and 9 of the review article by G. Burbidge (1970). Following the original paper by Seyfert (1943), the first general discussion of the observations which attempted to put them in a physical framework was that by Burbidge *et al.* (1963), on violent events in the nuclei of galaxies. Here the phenomena were related to other evidence for explosive activity in radio galaxies, etc. Many more recent papers and a useful bibliography were published in the Proceedings of a symposium on Seyfert galaxies (Pacholczyk and Weymann, 1968).

Table I lists the bright galaxies known to fall into Seyfert's category, and fainter more distant compact objects found by the various surveys, and discussed in Section 2.3. Radio galaxies classified as N-type are not included here as they are listed in Table III.

Seyfert galaxies are defined by the following properties:

(a) They have a bright star-like nucleus, unresolved optically. Danielson *et al.* (1968) found that the main luminosity in NGC 4151 is coming from a region with an upper limit of $0''.18$ for its half-intensity angular diameter;

(b) The spectra show broad Balmer emission lines, often having a narrower core and very broad wings; the forbidden lines can be broad or narrow;

(c) The emission lines include transitions in ions of higher ionization potential than those found characteristically in H II regions in galaxies, e.g. [Ne v].

Even this subdivision of the class (1) of compact objects is itself a somewhat heterogeneous grouping; de Vaucouleurs and de Vaucouleurs (1968) distinguished between Seyferts with spiral-arm outer parts and those without, and even Seyfert's original list contains three distinctly different types, exemplified by the best-studied objects, NGC 1068, NGC 1275, and NGC 4151. Table I demonstrates this diversity in another way, in the several different kinds of surveys which have yielded Seyfert galaxies: as well as those already mentioned there are the atlases and catalogues by Vorontsov-Velyaminov (1959) and Arp (1966) and the radio-source catalogues.

Since NGC 1068, 1275, and 4151, from Seyfert's classical list, have been much studied and demonstrate distinct differences from one another, it is worthwhile describing their properties in more detail. For a more complete bibliography covering these and other objects, see G. Burbidge (1970).

2.1.1. NGC 1068

This galaxy has a bright main body and well-developed spiral arms, and its rotation has yielded a mass within 2 kpc of the center of $3 \times 10^{10} M_\odot$ (Burbidge *et al.*, 1959). NGC 3227 and NGC 7469 resemble it in having fairly bright spiral arms, and the masses of these are similar – $3.5 \times 10^{10} M_\odot$ (Rubin and Ford, 1968) and $1 \times 10^{10} M_\odot$ (Burbidge *et al.*, 1963), respectively. The nucleus of NGC 1068 is not as concentrated as in some Seyferts; spectra show absorption lines presumably due to the stellar

TABLE I

Seyfert galaxies and extragalactic objects with compact nuclei and broad emission lines

No.	z	No.	z
NGC		**Markarian**	
1068	0.0038	9	0.038
1275	0.0180	10	0.029
1566[1]	0.0039	34	0.0507
3227	0.0034	42	0.024
3516	0.0092	50	0.023
3783[2]	0.0093	69	0.076
4051	0.0023	79	
4151	0.0033	105	
5548	0.0166	106	
6814[3]	0.0053	110	
7469	0.0166	124	
Zwicky		141	
III 0008 + 10	0.089	142	
0039.5 + 4003[4]	0.1026	205[8]	0.070
I 0051 + 12	0.061	231	
II 0119 − 01	0.054	273	
I 1535 + 55[5]	0.0386	279	
II 2130 + 09	0.061	290	
Vorontsov-Velyaminov (1959)		**Radio**	
144[6]	0.021	3C 120[9]	
150[7]	0.027	OQ 208[10]	

References to Table I

NGC galaxies mostly in Seyfert (1943) and redshift catalogues. Published references for Zwicky and Markarian galaxies given at end of Section 1; unpublished identifications kindly provided by E. Khachikian and M.-H. Demoulin, to be published elsewhere. Additional references:

1. G. de Vaucouleurs: 1961, *Mem. Roy. Astron. Soc.* **68**, 69.
2. T. Page: 1967, *Astron. J.* **72**, 821.
3. M.-H. Demoulin, private communication.
4. F. Zwicky *et al.*: 1970, *Publ. Astron. Soc. Pacific* **82**, 93.
5. J. deVeny and R. Lynds: 1969, *Publ. Astron. Soc. Pacific* **81**, 535.
6. G. and M. Burbidge: 1964, *Astrophys. J.* **140**, 1307.
7. G. and M. Burbidge: 1961, *Astron. J.* **66**, 541.
8. Weedman, D. W.: 1970.
9. M. Burbidge: 1967, *Astrophys. J. Letters* **149**, L51; W. Sargent: 1967, *Publ. Astron. Soc. Pacific* **79**, 369.
10. This paper, Section 2.4, p. 120.

content, as well as the Balmer and forbidden emission lines which are both broad, though not as broad as the H lines in NGC 4151.

Osterbrock and Parker (1965) and Dibai and Pronik (1965) studied the physical conditions in the gas giving rise to the line emission. These are:

$T_e \simeq 9000 \, \text{K}$
$N_e > 4 \times 10^3$ from [O II]
$> 4 \times 10^4$ from [S II]
$\approx 2 \times 10^5$ from [Ar IV].

As in all such objects that have been studied in sufficient detail to reveal their physical properties, it has been found that very strong density fluctuations must be present in the emitting gas. Dense, cooler condensations are embedded in a more tenuous hotter gas. Even lines from neutral gas are seen: [O I] $\lambda 6300$ (which is a very characteristic feature in spectra of Seyfert nuclei and related galaxies) and [N I] $\lambda 5199$ (this is similar to the [O II] $\lambda 3727$ transition, $^4S° - {}^2D°$).

The Bowen fluorescence lines of permitted O III in the ultraviolet (excited by He II $\lambda 304$ and seen in normal planetary nebulae) are definitely absent, and Osterbrock and Parker deduced from this fact that there must be an intimate mixture of neutral and ionized matter, so that He II radiation is absorbed in photoionization before it can reach the main body of gas. This can occur only if there are many sources of ionizing radiation, e.g. shock fronts or thin sheets between cooler regions. The ionization is high but not extreme; [Ne V] and [Fe VII] are seen but not [Fe X].

A very interesting feature is the character of the emission lines as seen under moderately high dispersion. Walker (1966, 1968) found that immediately outside the very small-diameter continuum source, the broad emission lines break up into a number of distinct separate structures – named by him 'haystack profiles' – which indicate that there are separate cloud complexes moving at several hundred km s^{-1} relative to one another.

2.1.2. NGC 1275

This Seyfert galaxy, the brightest member of the Perseus cluster, is the well-known strong radio source with a very interesting radio structure, including a component of extremely small angular diameter at the center, and outer structures that involve nearby radio-source galaxies in the Perseus cluster. NGC 1275 is also famous for having two distinct emission-line redshifts, discovered by Minkowski (1957). The main body of the galaxy has a recession velocity of 5300 km s^{-1}, and an outer filamentary structure on one side of the object has a velocity of some 8300 km s^{-1}. The velocity field was mapped out by Burbidge and Burbidge (1965) and was interpreted as being due to an explosive event $\sim 10^6$ yr ago in the nucleus. This ejected a large body of gas asymmetrically. Narrow-band filter photography by Lynds (1970) showed that, in addition to the gas ejected at about 3000 km s^{-1}, the gas at velocities close to the systemic velocity is distributed around the galaxy in a filamentary structure, looking very like a gigantic Crab Nebula.

Dibai and Pronik (1966) studied the physical conditions in the line-emitting gas. The mass of the gas ejected at 3000 km s^{-1} is some $10^8 \, M_\odot$. Although the main galaxy has often been called an elliptical, Minkowski (1968) showed that there are

Balmer absorption lines in its spectrum, so that early-type stars must be present and the stellar population must thus be younger than that characteristic of ellipticals.

2.1.3. NGC 4151

This object is representative of the commonest type of Seyfert galaxy. Although classified as a spiral, its spiral arms are rather irregular and ill-defined, and are much fainter relative to the nucleus than is the case in NGC 1068. The Balmer emission lines are very broad, with a half-intensity half-width of some 3000 km s^{-1}, and they have a narrower core. The forbidden lines have profiles like the hydrogen cores. These cores show the 'haystack' profiles demonstrating the existence of fast-moving cloud complexes of ionized gas (Walker, 1968).

The most detailed spectrophotometric study is by Oke and Sargent (1968). The gas producing the emission lines is hotter than that in NGC 1068, with $T_e \simeq 18\,000$ K and $N_e \simeq 5 \times 10^3$ cm^{-3}. In this object also, extreme density and temperature fluctuations must exist, because lines of [O I] and of [Fe VII], [Fe X] are present, and Oke and Sargent suggested the identification of [Fe XIV] $\lambda 5303$ also. This might arise in a hot low-density gas at $T_e \simeq 10^6$ K, $N_e \simeq 10^2$ cm^{-3}, which would give pressure equilibrium with the cooler, denser regions. Table II lists the lines identified by Oke and Sargent.

The extreme range of ionization poses a problem. Williams and Weymann (1968) and Osterbrock (1969) discussed the line identifications and strengths, and deductions therefrom. Shock waves produced in collisions of high-velocity clouds will produce high-temperature regions; Shklovsky (1966) suggested that there might be another component of ionizing radiation, in addition to the far UV tail of the known UV component, which might give a rise in the continuum flux in the X-ray region and which could produce the ions of highest ionization potential.

Williams and Weymann pointed out that the Bowen fluorescence lines of [O III], which are not seen in NGC 1068, do appear in NGC 4151, but are much weaker than in planetary nebulae. Regarding the abundances of the elements, the line intensities calculated by these authors were derived on the assumption of 'normal', solar-neighborhood, relative abundances. Osterbrock (1970) noted, however, that helium appears to be underabundant in this nucleus.

2.1.4. Emission-Line Widths in Spectra of Seyfert Nuclei

The question of the line-broadening mechanism that produces the great widths of the emission lines in spectra of Seyfert nuclei is still not settled. These could be due to electron scattering or to Doppler broadening by either random or organized motions. Electron scattering was suggested by Burbidge *et al.* (1966) for the broadening of emission lines in spectra of QSOs, and for NGC 4151 it was suggested by Oke and Sargent (1968). Williams and Weymann (1968), however, found difficulties with this hypothesis.

It is true that the 'haystack' profiles in NGC 1068 and 4151 demonstrate that part, at least, of the broadening is due to large random velocities of the gas. The very extended wings in the permitted lines in some Seyfert nuclei, however, are quite

TABLE II

Emission lines in nucleus of NGC 4151: Identifications, intensities, and comparison with the planetary nebula NGC 7027, according to Oke and Sargent (1968)

λ(Å)	Identification	Equivalent width (Å)	Flux at source (units of 10^{40} erg s^{-1})	Strength relative to $H\beta = 100$	
				NGC 4151	NGC 7027
10830.2	He I	163.5	5.98	81	87
10049.4	Pζ	30.0:	1.19:	16:	5
7329.9 7330.7	[O II]	11.0	0.62	8	32
6731.3	[S II]	33.3	2.08	28	6
6717.0	[S II]	27.7	1.75	24	
6583.6	[N II]	29.0	1.83	25	90
6562.8	Hα wings	360.0	22.75	307	290
	Hα core	34.0	2.15	29	
6548.1	[N II]	6.2	0.39	5	30
6374.5	[Fe X]	2.8:	0.18	2:	–
6363.9	[O I]	4.7	0.31	4	6
6300.2	[O I]	20.6	1.36	18	20
6085.3	[Fe VII]	7.6	0.52	7	–
5875.6	He I	4.4	0.31	4	11
5754.8	[N II]	3.0:	0.22:	3:	8
5720.9	[Fe VII]	4.4	0.32	4	–
5303.6	[Fe XIV]	1.0:	0.08	1:	–
5006.8	[O III]	188.0	15.80	214	1460
4959.9	[O III]	62.0	5.23	70	480
4861.3	Hβ wings	72.0	6.07	82	100
	Hβ core	16.0	1.35	18	
4799.5	[Fe III]	1.0:	0.10:	1:	–
4740.3	[A IV]	1.7	0.15	2	10
4711.4	[A IV]	1.7	0.15	2	8
4685.7	He II	21.7	1.88	25	46
4658.1	[Fe III]	5.9	0.51	7	–
4471.5	He I	1.0:	0.10:	1:	4
4363.2	[O III]	5.4	0.48	7	26
4340.5	Hγ wings	21.4	1.92	26	47
	Hγ core	5.6	0.50	7	
4243.0	?	0.5:	0.04:	1:	–
4228.0	?	0.5:	0.04:	1:	–
4101.7	Hδ wings	5.4	0.51	7	26
	Hδ core	3.7	0.35	5	
4076.2	[S II]	2.3	0.22	3	16
4068.6	[S II]	3.5	0.34	5	
3970.1 3968.5	Hε [Ne III]	10.3	1.03	14	52
3889.1 3888.6	Hζ He I	4.3	0.47	6	20
3869.7	[Ne III]	19.3	2.12	29	120
3728.9 3726.2	[O II] [O II]	28.7	3.75	51	35
3425.8	[Ne V]	14.5	2.46	33	130

smooth. In the case of NGC 3516, in which Seyfert found very great broadening, with a half-width of some 4000 km s^{-1}, Souffrin (1969) suggested that the very broad H wings might be produced by Doppler broadening in an inner region of higher density ($N_e \geqslant 10^8$ cm^{-3}), in which the (less broad) forbidden lines cannot appear because of the high density. [OIII] $\lambda 4363$, of intermediate width, would be produced in an intermediate region with lower turbulent velocities where $N_e \simeq 10^6$ cm^{-3}, and the other, narrower, forbidden lines would arise in an outer region where $N_e \approx 10^4$ cm^{-3}.

Whatever the main broadening agent is found to be, extreme lack of homogeneity in the emitting region must be an important factor, and all those who have worked on the physical conditions in Seyfert nuclei have arrived at this conclusion. Weymann (1970) suggested that N_e as high as 3×10^9 cm^{-3} might exist in an inner core of radius 7.5×10^{15} cm, in which only the Balmer wings would be produced, and here electron scattering might indeed be important.

2.1.5. *Absorption Lines in Gas in NGC 4151*

One of the very interesting discoveries, and one that links the Seyfert nuclei with QSOs, has been the multiple-redshift absorption lines of H and HeI found by Anderson and Kraft (1969) in coudé spectra of the nucleus of NGC 4151, and their variability discovered by Cromwell and Weymann (1970). Anderson and Kraft found that HeI $\lambda 3889$ appeared with three narrow components, giving velocities (relative to the nucleus itself) of -280, -550, and -840 km s^{-1}; weaker components were also visible in the Balmer lines, the largest velocity differences being -970 km s^{-1} at Hβ. These lines can best be interpreted as arising in gas ejected from the nucleus of NGC 4151, which is known to be variable in optical light. Further, Cromwell and Weymann found that these components are variable and can appear and disappear with a time scale of order one year. Osterbrock (1970) estimated that the gas outflow suggested by these observations amounts to about $\frac{1}{2} M_\odot$ per year.

2.1.6. *Emission-Line Variability*

Variability on a longer time scale was found by Andrillat and Souffrin (1967) in the emission-line spectrum of NGC 3516. Their spectra, obtained in 1967, showed that a definite change had occurred since the observations by Seyfert in 1943. The Balmer lines had almost disappeared, and the forbidden lines had strengthened. They suggested that matter ejected by the active inner nucleus had expanded and become less dense. It will be very interesting to carry out long-term monitoring of the spectra of a variety of Seyfert nuclei, and see whether others vary in this way, because an idea of the size of the variable regions can be provided by detailed study of such variations.

2.2. N-GALAXIES

The original isolation and classification of N-type galaxies was the form classification in Morgan's classical paper of 1958. He defined these galaxies as "systems having small brilliant nuclei superposed on a considerably fainter background". In that paper he classified NGC 4051 and 4151, two of the original Seyfert galaxies, as N systems. Later, in the paper on radio galaxies by Matthews *et al.* (1964), N-type galaxies were defined as "galaxies having brilliant star-like nuclei containing most of the luminosity of the system. A faint nebulous envelope of small visible extent is observed". It was realized in that paper that many radio galaxies fall into this category, and it was also pointed out that these objects may be related to the compact galaxies discovered by Zwicky.

In practice, the term 'N-galaxy' has been most often used for radio galaxies conforming to the description by Matthews *et al.* Now that a specific classification scheme has been described by Morgan (this volume, p. 97) with 3 basic classes and 2 subclasses, it is to be hoped that there will be more precision in the allotment of new objects into appropriate classes, and that this will replace the arbitrary naming of some objects as 'Seyfert galaxies' and some as 'N-galaxies', which has led to some confusion.

As spectroscopic information has accumulated for objects that have been given a form classification as N-type systems, it has become apparent that a wide variety of types of line spectra are possible. These range from objects with strong and very broad emission lines, just like those described in Section 2.1, objects with narrower but still intensely strong emission lines, and objects with much weaker emission lines, and even with absorption lines only. Lynds (1968a) has discussed this variety of spectra. It is probable that a more careful form classification of objects with absorption lines only would place them in Morgan's class C, and then my physical division would allot them to type (3) of the compacts, to be described in Section 4.

Two good examples of N-type radio galaxies are 3C 234 and 3C 445, and the emission lines and their estimated relative strengths were given by Schmidt (1965a) in his paper on redshifts of 31 radio galaxies. Table III, taken from G. Burbidge (1970), lists the spectrum lines seen in 18 N galaxies identified with radio sources.

An interesting question concerns the relative contribution to the emitted light coming from stars, from hot gas, and from non-thermal radiation. Since my discussion concerns only the line spectra, the important question is whether or not absorption lines coming from the stellar component are visible in the spectrum. Since the non-thermal component of the continuum emission may be variable, as in the N-galaxy 3C 371, absorption lines may be visible at times and disappear when the starlight is much exceeded by the thermal radiation from hot gas and synchrotron radiation.

Two N-type galaxies have been found to display a phenomenon which may relate to the double redshift of NGC 1275. We describe these objects next.

3C 227 and 3C 390.3

These two objects are listed as N-galaxies by Wyndham (1965); 3C 227 was described by Matthews *et al.* (1964) and Zwicky had informed Wyndham that 3C 390.3 was very compact, only just discernible as extended with the 200-in. telescope, having a redshift velocity of about 17000 km s^{-1}. Lynds (1968b) described their spectra,

TABLE III

Spectroscopic properties of N galaxies identified with radio sources

Object	z	Lines
MSH 05−43 (Pictor A)	0.0342	[OII]λ3727, [NeIII]λ3869, Hδ, Hγ, Hβ, [OIII]$\lambda\lambda$4363, 4959, 5007.
3C 371	0.0508	[OII]λ3727, [OIII]$\lambda\lambda$4959, 5007, CaII H and K abs.
3C 445	0.0568	[NeV]λ3426, [OII]λ3727, [NeIII]$\lambda\lambda$3869, 3968, [SII]$\lambda\lambda$4068, 4076, Hδ, Hγ, Hβ, [OIII]$\lambda\lambda$4363, 4959, 5007. No abs.
3C 390.3[a]	0.0569	[NeV]λ3426, [OII]λ3727, [NeIII]$\lambda\lambda$3869, 3968, Hδ, Hγ, Hβ, [OIII]$\lambda\lambda$4363, 4959, 5007, Hα. No abs.
PKS 0521−36	0.061	[NeV]$\lambda\lambda$3346, 3426, [OII]λ3727, [NeIII]$\lambda\lambda$3869, 3968, Hδ, Hγ, [OIII]$\lambda\lambda$4959, 5007.
3C 227[b]	0.0855	[NeV]λ3426, [OII]λ3727, [NeIII]$\lambda\lambda$3869, 3968, Hγ, Hβ, [OIII]$\lambda\lambda$4363, 4959, 5007, HeI λ5876, Hα. No abs.
PKS 1417−19	0.1192	[OII]λ3727, [NeIII]λ3869, [OIII]$\lambda\lambda$4363, 5007, Hγ, Hβ, Hα. No abs.
PKS 2300−18	0.129	[NeIII]λ3869, Hγ, Hβ, [OIII]$\lambda\lambda$4363, 4959, 5007.
PKS 1340+05	0.1333	[OII]λ3727, CaII H and K abs.
PKS 2349−01	0.174	[OII]λ3727, Hδ, Hγ, Hβ, [OIII]$\lambda\lambda$4363, 4959, 5007.
3C 234	0.1846	[NeV]$\lambda\lambda$3346, 3426, [OII]λ3727, [NeIII]$\lambda\lambda$3869, 3968, Hζ, Hδ, Hγ, [OIII]$\lambda\lambda$4363, 4959, 5007, HeII $\lambda\lambda$3203, 4686, Hβ. No abs.
3C 287.1	0.2156	[NeV]λ3426, [OII]λ3727, [NeIII]$\lambda\lambda$3869, 3968, Hβ, [OIII]$\lambda\lambda$4959, 5007.
3C 17	0.2201	[OII]λ3727, [NeIII]λ3869, Hβ, [OIII]$\lambda\lambda$4959, 5007.
3C 459	0.2205	[NeV]λ3426, [OII]λ3727, [NeIII]λ3869, [OIII]$\lambda\lambda$4959, 5007. Weak em.; no abs.?
3C 171	0.2387	[OII]λ3727, [NeIII]$\lambda\lambda$3869, 3968, [OIII]$\lambda\lambda$4363, 4959, 5007, Hβ.
3C 79	0.2561	[NeV]$\lambda\lambda$3346, 3426, [OII]λ3727, [NeIII]$\lambda\lambda$3869, 3968, Hγ, [OIII]$\lambda\lambda$4363, 4959, 5007, HeII 4686, Hβ.
3C 109	0.3056	[NeV]λ3426, [OII]λ3727, [NeIII]$\lambda\lambda$3869, 3968, Hγ, Hβ, [OIII]$\lambda\lambda$4363, 4959, 5007.
3C 177	no publ. z	No emission lines, CaII H, K, and G band abs.

[a] Balmer emission lines very broad, with two maxima.
[b] Balmer emission lines double.

and pointed out that 3C 227 has hydrogen emission lines that are clearly double, and the lines in 3C 390.3 are fully as wide as the total redshift. The forbidden lines in 3C 227, among which [NeV] λ3426 is very prominent, are single and agree in redshift with the longer-wavelength component of the hydrogen lines.

In 3C 390.3, we have found that the extremely broad emission lines of H are also double, although the two maxima are not as distinct as in 3C 227. Again, the forbidden lines are single and give the same redshift as the longer-wavelength component of hydrogen. Measurements on our Lick spectrograms give:

Mean recession velocity, longward components of Hα, Hβ + forbidden lines (this component of Hγ is blended with [OIII] λ4363)... = 16 800 km s^{-1}; Mean recession velocity, shortward components of Hα, Hβ, Hγ... = 12 250 km s^{-1}. Thus the velocity difference is 4550 km s^{-1}, exceeding the difference of 3000 km s^{-1} seen in NGC 1275. Possibly this phenomenon may represent an early stage in the development of structures that later may resemble what is seen in NGC 1275. The absence of forbidden lines in the shortward component suggests that the electron density is at present higher in the gas giving rise to this emission.

2.3. COMPACTS OF CLASS (1)

We turn now to my physical grouping, class (1) of the Zwicky compact objects. After the objects are picked on the basis of their compact nature, this grouping is made solely according to the spectrographic information, and objects falling into this category resemble classical Seyfert galaxies in having very broad strong hydrogen lines in their spectra. However, they have larger redshifts than the classical Seyfert galaxies. Known examples are listed in Table I.

Three good examples of this class are III Zw 0008 + 10, for which Arp (1968) published a direct photograph and the spectrum ($z=0.089$), I Zw 0051 + 12, with $z=0.061$, studied by Sargent (1968a), and II Zw 2130 + 09, with $z=0.061$, also studied by Sargent (1968b). The spectrum of III Zw 0008 + 10 has very broad hydrogen lines, and narrow forbidden lines of [O II], [O III], and [Ne III]. In view of the clearly-defined doubling of the hydrogen lines in 3C 227 and 3C 390.3, it is interesting that Arp noted that doubling in Hβ and Hδ was discernible in the spectrum of III Zw 0008 + 10, and Hγ could be seen on microphotometer tracings to be double.

I Zw 0051 + 12 and II Zw 2130 + 09 are interesting because they both have strong broad permitted lines of Fe II in their spectra. The former has no forbidden lines at all; the electron density must be greater than 10^6 cm^{-3} and Sargent's spectrophotometric measurements then yielded $R \leqslant 5 \times 10^{17}$ cm for the dimension of the line-emitting region. The object II Zw 2130 + 09 does have forbidden lines, but they are weaker than usual so the electron density is probably fairly high. The Fe II emission lines in these two objects provide an interesting link between QSOs and compact galaxies of this class; these same Fe II lines were seen in the spectrum of the QSO 3C 273, which is known to have a high electron density because the forbidden lines of [O III] are not seen in its spectrum.

Finally, we may note that conversion of the redshifts of these compact galaxies to distances, using a Hubble constant of 75 km s^{-1} Mpc^{-1}, yields photographic absolute magnitudes of -22.6, -23.2, and -22.8, respectively. This is a considerably higher luminosity than that of normal bright elliptical galaxies.

2.4. REDSHIFTS AND THEIR DISTRIBUTION IN THE SKY FOR COMPACT OBJECTS

We may ask whether there is anything unusual in the distribution of redshifts of those compact objects characterized by non-thermal emission, or whether there is anything in their distribution in the sky in relation to other galaxies, to indicate that their redshifts might not be due to expansion of the Universe.

The existence of a peak at $z=0.061$ in a histogram plotting numbers of objects at steps of 0.01 in redshifts was pointed out by Burbidge (1968) and Burbidge and Burbidge (1969). This might mean fluctuations in the recent past in the occurrence of such objects, or a non-cosmological origin for the redshifts, but no explanation or interpretation is available. We await the measurement of more redshifts to see whether the peak is maintained, whether it grows, or whether it disappears.

The occurrence of compact objects near galaxies can also be looked for. The radio source OQ 208 was thought to be identified with a star-like object 10" from a brighter galaxy (Ryle and Pooley, 1969), but spectroscopic observations at various observatories, including my own observations at Lick, revealed that the star-like object is in fact a normal star, and the spectrum of the galaxy is a good example of the Seyfert or compact class (1) type, with broad strong emission lines and a redshift $z = 0.077$.

Weedman (1970) found that Markarian 205 is a compact galaxy with a spectrum like that of typical Seyfert galaxies and a redshift $z = 0.070$, and it lies only 40" from the center of the normal spiral galaxy NGC 4319, well within the spiral arms of that galaxy. Lick spectrograms of NGC 4319 show that it has a recession velocity of some $+1500$ km s^{-1}, i.e. $z = 0.005$.

We are reminded of the fact that two QSOs also lie very near spiral galaxies: 3C 275.1 is near NGC 4651 (Sandage *et al.*, 1965), and the radio-quiet object QS 1108+285 ($z = 2.192$) lies about 1 arc min from NGC 3561 (Stockton 1969), the galaxy long known to have a compact appendage in the form of the 'Ambartsumiam knot'.

3. Compacts of Class (2)

3.1. GENERAL DISCUSSION

The spectroscopic observations of Zwicky compacts by Sargent (1970b) show that objects with sharp emission lines are most numerous. Also the Markarian and Haro galaxies, with strong UV continua, are mostly of this type (see references given in Section 1). The range of ionization and excitation exhibited by the emission lines is like that of bright H II regions in our own and other galaxies, and the presumption is strong that the source of energy is a large number of massive young OB stars.

Objects of this type have a great range in size, luminosity, and morphological appearance. Direct photographs of several, and a typical spectrum, are reported by Sargent (1970b). Some objects have jets; some are in the form of rings with no central nucleus; most are irregular.

An interesting example of an extremely small compact double galaxy with sharp emission lines in its spectrum was described by Arp (1965). The recession velocity is only $+1326$ km s^{-1}, yet the apparent photographic magnitudes of the two components are 17.8 and 17.9, and their separation and respective diameters are only about 1 arc sec. Thus the absolute magnitudes are only -12.7 and -12.8 (for $H = 100$ km s^{-1} Mpc^{-1}) and the diameter of each component is only 70 pc. The nature and evolutionary history of such an object is very puzzling, and since it was discovered by chance on a plate taken with the Palomar 200-in. telescope, the space density of galaxies of this sort might be quite high. A similar faint galaxy was discovered by Kinman (1965) in an investigation of the fainter Haro-Luyten blue stellar objects; it has sharp emission lines, a small redshift, and an absolute B magnitude of about -14 and was just detectable as non-stellar on Kinman's photograph taken with the Lick 120-in. telescope.

3.2. Possible Young Galactic Nuclei

Sargent (1970c) drew attention to a particular kind of dwarf compact galaxy with sharp emission lines and 21-cm radiation, which he and Searle are studying and which they think might be young recently-formed objects. The three examples of this type for which they have observations are II Zw 0553+03, I Zw 0930+55, and I Zw 1531+46. These have small redshifts and values of M_p fainter than -17. The 21-cm observations are by Chamaraux et al. (1970); see also Heidmann (this volume, p. 264). The structures suggest that these objects may consist of giant O-associations embedded in neutral hydrogen, although at the moment the 21-cm detection is certain in only one of the three objects. Sargent and Searle conclude that they are either young or have only recently formed massive stars (the observations do not exclude the possibility that a substructure of faint red stars may be present). Because of the way in which these galaxies were discovered, it is clear that they may be very frequent in space.

4. Compacts of Class (3)

In this class of compact galaxy the spectra display only absorption lines, typically Ca II, the G-band, and the Mg b feature, though a few show Balmer lines as well as Ca II. The colors are like those of elliptical galaxies, and it is clear that the objects are dense groupings of stars of the population type normally taken to be some 10^{10} yr old. Their evolutionary history is therefore of great interest: what has caused them to have a different structure from that of normal elliptical galaxies?

The best-known example of this kind of galaxy is NGC 4486-B, the companion of M87. Rood (1965) measured its brightness distribution and, using the large velocity dispersion measured by Minkowski (1961), found that it has a mass of $5 \times 10^{10} M_\odot$ and a high mass-to-light ratio of 80. He suggested that its small compact size might be due to tidal limitation by the extensive and very massive M87.

There are other compact galaxies of this kind, however, for which the suggestion of tidal limitation cannot be made. Zwicky (1968) discovered two clusters of compact galaxies, Zw Cl 0152+33 and Zw Cl 1710+64, and the question of their relation to clusters of normal galaxies is not at present understood. Sargent (1971) has studied the virial relation between kinetic and potential energy for Zw Cl 0152+33.

5. Quasi-Stellar Objects

I shall cover only the emission-line spectra of QSOs; Lynds will discuss the very interesting absorption-line spectra, and I have time only for a brief and condensed account. A fuller description of both emission and absorption line spectra was given earlier this year (E. M. Burbidge, 1970).

5.1. IDENTIFICATION OF EMISSION LINES

The emission lines found in spectra of QSOs are those expected to be produced in a hot gas at fairly low density, with a chemical composition similar to that of the Sun and the stars and gaseous nebulae in the solar neighborhood. This was shown by the first line identifications made by Schmidt (1965b), in which emission lines predicted to be strong in the far ultraviolet spectral region in planetary nebulae (Osterbrock, 1963) were found to be present, with very large redshifts, in the spectra of QSOs.

Table IV, taken from E. M. Burbidge (1970), lists most of the emission lines that have been found with rest wavelengths in the range 1200–5000 Å (weaker features seen rarely are not included). A large range of ionization is found, from 7.6 eV

TABLE IV
Principal emission lines seen in spectra of QSOs, 1200–5000 Å

Ident.	λ(A)	Ident.	λ(A)
Lγ-α	1216[a]	Mg II	2798[a,b]
N V	1240[b]	[Ar IV]	2854, 2869
Si IV	1397[b]	[Mg V]	2931
O IV]	1406	He II	3203
C IV	1549[a,b]	[Ne V]	3346, 3426
He II	1640	[O II]	3727[b]
O III]	1664	[Ne III]	3869, 3968
C III]	1909[a]	[O III]	4363, 4959, 5007[a]
C II]	2326	Balmer series	4102, 4340, 4861[a]

[a] Strongest transitions.
[b] Blends of doublets.

which is required for single ionization of Mg, to 97 eV required to produce Ne^{+4} or 109 eV for Mg^{+4} (or even 186 eV for Mg^{+6} since [Mg VII] has been identified in spectra of a few QSOs. A useful table which groups the lines according to electron configuration in the emitting ions was published by Lynds (1968b). This lists all possible transitions between about 1000 and 7000 Å, in the various ionization stages of all elements having a solar-neighborhood ('normal') relative abundance greater than 10^{-6} hydrogen, and those transitions which have actually been observed are noted. Burbidge and Burbidge (1967) gave a compilation of emission lines observed in 60 QSOs, with eye estimates of intensity, and the results for recent measurements of line intensities in 28 QSOs were given by Oke et al. (1970).

5.2. PHYSICAL CONDITIONS IN THE LINE-EMITTING REGION

Following the first study of the physics of the line-emitting region in 3C 273 and 3C 48 (Greenstein and Schmidt, 1964), and the study of average conditions in a composite of several QSOs (Osterbrock and Parker, 1966), it has become increasingly apparent that, as in the Seyfert galaxies, the line-emitting region must be non-homogeneous and probably consists of a combination of stratified layers and denser clouds or filaments in a hotter, lower-density background. In these earlier studies a level of ionization was assumed. Bahcall and Kozlovsky (1969), however, calculated the ionization equilibrium in 3C 273 and 3C 48, assuming that the energy for ionization comes from a continuum with a power-law distribution of the flux, $F(v)$, as a function of frequency, v.

Osterbrock (1969) published an updated list of calculated line strengths for the elements H, He, C, N, O, Ne, and Mg in 'normal' abundances, and also a further list of weaker lines from Mg, Si, S, Ar, some of which have been observed. A result which was apparent in the analysis of several QSOs by Osterbrock and Parker, and also in the study of 3C 273 by Bahcall and Kozlovsky where the level of ionization was calculated and not assumed, was that He appears to be underabundant in these objects, whereas the other elements, relative to hydrogen, appear normal. Further work on this is desirable.

The widths of the emission lines are generally not as great as in the spectra of Seyfert nuclei; typically they are about 50 Å, although the resonance lines Ly-α and C IV λ1549 may be much wider. Often the forbidden lines are narrower than the permitted lines, as in some Seyfert nuclei, and this led Burbidge *et al.* (1966) to propose a model in which electron scattering is the broadening agent, in a stratified medium where the various lines are produced in different regions. It is not known whether this mechanism, or Doppler broadening by random or organized motion, is responsible for the line broadening.

The discovery by Lynds and Wills (1970) of 4C 5.34, a QSO with the very large redshift of $z = 2.877$, has made possible the study of the Ly-α/Ly-β intensity ratio, a quantity pointed out by Bahcall (1966) to be of interest, and the Lyman continuum. Ly-β can also be studied in 5C 2.56, which has $z = 2.388$; this latter QSO, which had become very faint, recently brightened so that I was able again to obtain spectrograms at Lick and to confirm a tentative earlier measurement of its redshift.

Finally, I would like to draw attention to the fact that, despite their similarities, there are distinct spectroscopic differences between the QSOs, as was seen in the first quantitative study of 3C 273 and 3C 48. For example, the emission line N V λ1240 may not appear at all in some QSOs of large redshift, while in others it may be almost as strong as Ly-α emission, as is the case in PHL 3424 (Sandage and Luyten, 1967) and in 5C 2.56. Quantitative study of these and other differences can now be undertaken.

Acknowledgement

Extragalactic research at UCSD is supported by the National Science Foundation and by NASA under grant NGL-05-005-004.

References

Ambartsumian, V. A.: 1958, in *Solvay Conference on Structure and Evolution of the Universe*, (ed. by R. Stoops, Brussels), p. 241.

Ambartsumian, V. A.: 1965, in *Solvay Conference on Structure and Evolution of Galaxies*, Interscience, John Wiley and Sons Ltd., London, p. 1.
Ambartsumian, V. A.: 1970, *Nuclei of Galaxies*, Study Week, Pontifical Academy of Sciences, in press.
Anderson, K. S. and Kraft, R. P.: 1969, *Astrophys. J.* **158**, 859.
Andrillat, Y. and Souffrin, S.: 1967, *Astrophys. Letters* **1**, 111.
Arakelian, M. A., Dibai, E. A., and Yesipov, V. F.: 1970, *Astrofizika* **6**, 39.
Arp, H. C.: 1965, *Astrophys. J.* **142**, 402.
Arp, H. C.: 1966, *Atlas of Peculiar Galaxies*, Cal. Tech. Book Store, Pasadena.
Arp, H. C.: 1968, *Astrophys. J.*, **152**, 1101.
Arp, H. C., Khachikian, E. E., Lynds, C. R., and Weedman, D.: 1968, *Astrophys. J. Letters* **152**, L103.
Bahcall, J.: 1966, *Astrophys. J.* **145**, 684.
Bahcall, J. and Kozlovsky, B. Z.: 1969, *Astrophys. J.* **155**, 1077; *Astrophys. J.* **158**, 529.
Burbidge, E. M.: 1970, *Nuclei of Galaxies*, Study Week, Pontifical Academy of Sciences, in press.
Burbidge, G. R.: 1968, *Astrophys. J. Letters* **154**, L41.
Burbidge, G. R.: 1970, *Ann. Rev. Astron. Astrophys.* **8**, 369.
Burbidge, G. R. and Burbidge, E. M.: 1965, *Astrophys. J.* **142**, 1351.
Burbidge, G. R. and Burbidge, E. M.: 1967, *Quasi Stellar Objects*, W. H. Freeman, San Francisco.
Burbidge, G. R. and Burbidge, E. M.: 1969, *Nature* **222**, 735.
Burbidge, G. R., Burbidge, E. M., Hoyle, F., and Lynds, C. R.: 1966, *Nature* **210**, 774.
Burbidge, G. R., Burbidge, E. M., and Prendergast, K. H.: 1959, *Astrophys. J.* **130**, 26.
Burbidge, G. R., Burbidge, E. M., and Prendergast, K. H.: 1963, *Astrophys. J.* **137**, 1022.
Burbidge, G. R., Burbidge, E. M., and Sandage, A. R.: 1963, *Rev. Mod. Phys.* **35**, 947.
Chamaraux, P., Heidmann, J., and Lauqué, R.: 1970, *Astron. Astrophys.*, in press.
Cromwell, R. and Weymann, R. J.: 1970, *Astrophys. J. Letters* **159**, L147.
Danielson, R. E., Savage, B. D., and Schwarzschild, M.: 1968, *Astrophys. J. Letters* **154**, L117.
Dibai, E. A. and Pronik, V. I.: 1965, *Astrofizika* **1**, 78.
Dibai, E. A. and Pronik, V. I.: 1966, *Izv. Krymsk. Astrofiz. Obs.* **35**, 87.
DuPuy, D. L.: 1968, *Publ. Astron. Soc. Pacific* **80**, 29.
Greenstein, J. L. and Schmidt, M.: 1964, *Astrophys. J.* **140**, 1.
Haro, G.: 1956, *Bol. Obs. Tonantzintla Tacubaya* **2**, No. 14, 8.
Kinman, T. D.: 1965, *Astrophys. J.* **142**, 1241.
Lynds, C. R.: 1968a, *Astron. J.* **73**, 888.
Lynds, C. R.: 1968b, in S. Bashkin (ed.), *Proc. Conf. on Beam-Foil Spectroscopy*; Gordon and Breach, New York, p. 539.
Lynds, C. R.: 1970, *Astrophys. J. Letters*, **159**, L151.
Lynds, C. R. and Wills, D.: 1970, *Nature* **226**, 532.
Markarian, B. E.: 1967, *Astrofizika* **3**, 55.
Markarian, B. E.: 1969, *Astrofizika* **5**, 443, 581.
Matthews, T. A., Morgan, W. W., and Schmidt, M.: 1964, *Astrophys. J.* **140**, 35.
Minkowski, R.: 1957, in H. C. van de Hulst (ed.), 'Radio Astronomy', *IAU Symp.* **4**, 107.
Minkowski, R.: 1961, in G. C. McVittie (ed.), 'Problems of Extragalactic Research', *IAU Symp.* **15**, 112.
Minkowski, R.: 1968, *Astron. J.* **73**, 842.
Morgan, W. W.: 1958, *Publ. Astron. Soc. Pacific* **70**, 364.
Morgan, W. W.: 1970, *Nuclei of Galaxies*, Study Week, Pontifical Academy of Sciences, in press.
Oke, J. B., Neugebauer, G., and Becklin, E.: 1970, *Astrophys. J.* **159**, 341.
Oke, J. B. and Sargent, W. L. W.: 1968, *Astrophys. J.* **151**, 807.
Osterbrock, D. E.: 1963, *Planetary Space Sci.* **11**, 621.
Osterbrock, D. E.: 1969, *Mem. Soc. Roy. Sci., Liège. Ser. V.* **16**, 391.
Osterbrock, D. E.: 1970, *Nuclei of Galaxies*, Study Week, Pontifical Academy of Sciences, in press.
Osterbrock, D. E. and Parker, R. A. R.: 1965, *Astrophys. J.* **141**, 892.
Osterbrock, D. E. and Parker, R. A. R.: 1966, *Astrophys. J.* **143**, 268.
Pacholczyk, A. G. and Weymann, R.: 1968, *Astron. J.* **73**, 836.

Rood, H. J.: 1965, *Astron. J.* **70**, 689.
Rubin, V. C. and Ford, W. K.: 1968, *Astrophys. J.* **154**, 431.
Ryle, M. and Pooley, G. G.: 1969, *Astrophys. Letters* **4**, 137.
Sandage, A. R. and Luyten, W. J.: 1967, *Astrophys. J.* **148**, 767.
Sandage, A. R., Véron, P., and Wyndham, J.: 1965, *Astrophys. J.* **142**, 1307.
Sargent, W. L. W.: 1968a, *Astrophys. J. Letters* **152**, L31.
Sargent, W. L. W.: 1968b, *Astron. J.* **73**, 893.
Sargent, W. L. W.: 1970a, *Astrophys. J.* **159**, 765.
Sargent, W. L. W.: 1970b, *Astrophys. J.* **160**, 405.
Sargent, W. L. W.: 1970c, *Nuclei of Galaxies*, Study Week, Pontifical Academy of Sciences, in press.
Sargent, W. L. W.: 1971, *Astrophys. J.*, in press.
Schmidt, M.: 1965a, *Astrophys. J.* **141**, 1.
Schmidt, M.: 1965b, *Astrophys. J.* **141**, 1295.
Seyfert, C. K.: 1943, *Astrophys. J.* **97**, 28.
Shklovsky, I. S.: 1966, *Soviet Astron.* **9**, 683.
Souffrin, S.: 1969, *Astron. Astrophys.* **1**, 305, 414.
Stockton, A. N.: 1968, *Astron. J.* **73**, 887.
Stockton, A. N.: 1968, *Astrophys. J. Letters* **155**, L141.
de Vaucouleurs, G. and de Vaucouleurs, A.: 1968, *Astron. J.* **73**, 858.
Vorontsov-Velyaminov, B. A.: 1959, *Atlas and Catalogue of Interacting Galaxies*, Moscow.
Walker, M. F.: 1966, in M. Arakeljan (ed.), 'Non-Stable Phenomena in Galaxies', *IAU Symp.* **29**, 21.
Walker, M. F.: 1968, *Astrophys. J.* **151**, 71.
Weedman, D. W.: 1970, *Astrophys. J. Letters* **161**, L113.
Weedman, D. W. and Khachikian, E. K.: 1968, *Astrofizika* **4**, 587.
Weedman, D. W. and Khachikian, E. K.: 1969, *Astrofizika* **5**, 113.
Weymann, R. J.: 1970, *Astrophys. J.* **160**, 31.
Williams, R. E. and Weymann, R. J.: 1968, *Astron. J.* **73**, 456.
Wyndham, J.: 1965, *Astrophys. J.* **144**, 459.
Zwicky, F.: 1964, *Astrophys. J.* **140**, 1467.
Zwicky, F.: 1966, *Astrophys. J.* **143**, 192.
Zwicky, F.: 1964–1968, privately circulated lists of 2000 compact galaxies available on request to the California Institute of Technology.
Zwicky, F.: 1968, *Compt. Rend. Acad. Sci. Paris* **266**, 103.
Zwicky, F. and Humason, M. L.: 1961, *Astrophys. J.* **133**, 794.

Discussion

Pastoriza: A drastic change in the strength of Hβ emission in NGC 1566 was detected at Córdoba in 1969. When the Seyfert characteristics were first observed in the nucleus of NGC 1566 by de Vaucouleurs at Mt. Stromlo in 1956, Hβ was broad ($\sim$ 3000 km s^{-1}) with a narrow core and the line was much stronger than the [O III] lines. The situation was still very much the same in 1952 when Shobbrook confirmed this result, also at Mt. Stromlo. However, in several spectrograms taken at Córdoba in 1969, Hβ has almost vanished while the [O III] emission lines are still strong, indicating that the variation in Hβ cannot be explained by an increase in the continuum. The new observations are reported in Astrophysical Letters (in press).

Tovmasjan: Recently Arakelian, Dibai, Yesipov and Markarian have observed 120 galaxies from Markarian's second and third lists and discovered 18 galaxies with spectra characteristic of Seyfert galaxies – broad emission hydrogen lines and narrow forbidden lines. Results for some of them (including Markarian 205) have already been published.

Weymann: The [Fe XIV] identification in NGC 4151 should be regarded as uncertain. On plates taken by Mme. Souffrin, the feature was identified as [Ca V] 5309, and the same identification was made independently by Alloin and Weymann at Steward Observatory.

NGC 1068 should not be regarded as a case of an object having 'both broad hydrogen *and* broad forbidden lines' since the extremely broad hydrogen wings characteristic of most of the Seyfert galaxies are lacking in this object. I would like to ask Mrs Burbidge if she knows of any cases where there are very broad hydrogen wings as well as equally broad wings on lines which are definitely [O III] rather than [Fe II].

Mrs Burbidge: I know of several cases where the widths of the hydrogen lines and the forbidden lines are comparable, and similar to NGC 1068 (and indeed similar to the QSOs, in which the majority show rather broad lines of both permitted and forbidden lines, but *not* the extremely extended wings shown by the H lines in e.g. NGC 3516 and 4151). In these cases of moderately broad profiles for both permitted and forbidden lines, the widths are of the sort shown by Walker's 'haystack' profiles in NGC 1068. I think one can say definitely that no cases are known where the *forbidden* lines show the extremely broad wings seen in the Balmer lines in NGC 4151.

THE ABSORPTION-LINE QSOs

C. R. LYNDS

Kitt Peak National Observatory, Tucson, Ariz., U.S.A.

Abstract. Many features of the distribution of emission and absorption line redshifts of QSO spectra and of the distribution of the differences between emission and absorption line redshifts may be due to observational selection. However, it seems significant that all six of the absorption line objects of largest redshift have multiple absorption systems. Observations seem to favor more strongly the hypothesis that the multiple absorption line systems are due to clouds associated with the QSO rather than in the intervening space. The multiple absorption line systems of the QSOs PKS 0237-23, Ton 1530 and 4C 05.34 are discussed. The persistence of the Lyman-α absorption feature and identification of multiple-absorption line systems strongly suggests that unidentified absorption lines shortward of the Lyman-α emission also arise from the Lyman-α transition.

I have collected together all of the information available to me on the spectroscopic characteristics of QSOs relevant to the occurrence of absorption lines. Figure 1 shows the emission-line redshift distribution for 236 out of a sample of 237 objects. The QSO omitted, although confirmed to be extragalactic, is only known to have absorption lines. The spectra of 60 objects in Figure 1 show absorption lines with sufficient certainty that it is possible to make proposals for their identification and redshift. These objects are indicated by the total cross-hatched area in Figure 1. A dozen or more additional objects are suspected of having absorption lines. The spectra of many QSOs appear to indicate the presence of more than one absorbing cloud. Such objects are indicated in Figure 1 by additional cross-hatching.

Superficially, the redshift distribution for all QSOs shows a steep increase out to a redshift of approximately 0.4 and a steady decline until what appears to be a statistically significant steep decline at a redshift of about 2.1. If we were dealing with a sample that was complete over the entire range of redshift, the numbers would increase steadily – sharply at first but slowing somewhat toward larger redshift. The initial steep rise in the observed distribution may indicate that the sample is to some extent representative out to a redshift of about 0.3, but the subsequent decline illustrates the enormous incompleteness in the sample. The incompleteness at a redshift of 2.1 is so severe that physical interpretations of the steep decline should probably await a more complete understanding of the factors that may influence observational selection.

The apparent redshift distribution for the absorption-line objects is somewhat different from that for all objects taken together. There is no evidence for a steady decline with redshift, and there is what appears to be a peak at a redshift of 2.0. The detection of absorption lines in objects depends strongly on the Mg II resonance doublet ($\lambda 2798$) out to a redshift of about 1.1. The fact that this feature cannot be observed for redshifts less than about 0.2 explains why there are no absorption-line objects with redshifts less than this value. For redshifts larger than 1.1 the C IV resonance doublet ($\lambda 1549$) plays an increasingly important role in the detection of absorption, and there are several reasons to expect that this feature is more persistent – that is, can be more easily detected when small quantities of material are involved – than the Mg II lines. This effect may explain why the redshift distribution does not show the decline that is present for all of the QSOs taken together. It is also important

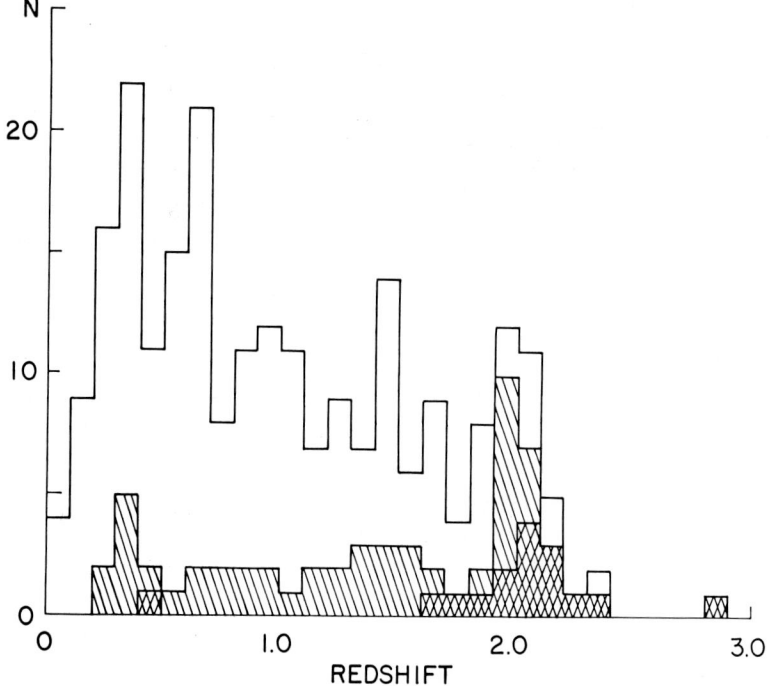

Fig. 1. The distribution of emission-line redshift for 236 QSOs. The entire cross-hatched area represents all absorption-line QSOs, and the area having additional cross-hatching denotes those objects that have multiple absorption systems.

to note that there is a considerable body of evidence indicating that Lyman-α of hydrogen is usually the most persistent feature arising in the clouds responsible for the absorption spectra. This, together with the fact that the region of Lyman-α cannot be really well observed for redshifts smaller than about 1.9, may explain, at least in part, the prevalence of absorption lines for objects having redshifts near 2.0. However, it should be remembered that many of the large-redshift objects not noted as having absorption lines are very faint and difficult to observe sufficiently well to rule out the presence of absorption lines comparable in strength to those found in the other sources.

The redshift distribution of objects having *multiple* absorption redshifts may well be influenced by the factors just mentioned, but the fact that all six of the absorption-line objects of largest redshift have *multiple* absorption systems appears to me to be significant. Furthermore, the three objects with redshifts greater than 2.2 appear to have many more than one or two absorption redshift systems.

It is instructive to look at the distribution of redshift residuals, $z_{abs} - z_{em}$ for the absorption systems in all of the absorption-line QSOs. This distribution is shown in the lower part of Figure 2. (Actually, certain absorption systems characteristic of the largest-redshifts objects have been excluded because they would have dominated the distribution. These absorption systems will be discussed later.) There is seen to be a

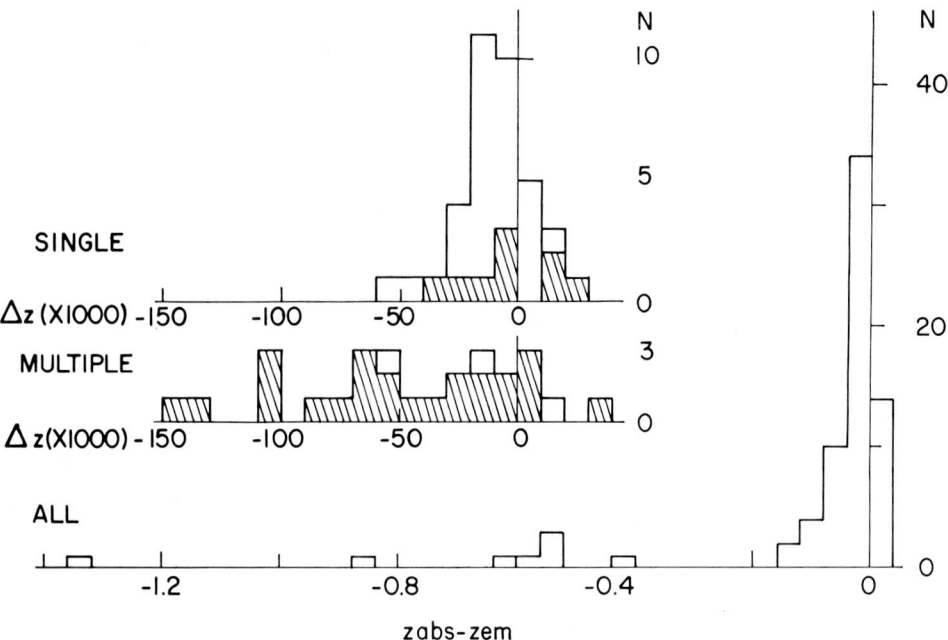

Fig. 2. The distribution of absorption redshift residuals, $z_{abs} - z_{em}$, for all absorption-line QSOs and for single-system and multiple-system QSOs separately. The cross-hatched area indicates those objects having emission-line redshifts greater than 1.8.

strong tendency for the absorption-line redshifts to be near the emission-line value or, more exactly, 0.02 less than the emission-line value. The distribution is quite compact and the largest positive residual is 0.031. There are, however, eight systems showing very large negative residuals, and these eight systems are present in the spectra of only four objects, one of which accounts for five of the systems. These four objects all have more than one absorption system and have at least one system with a redshift near the emission-line value.

From the appearance of the distribution of residuals in Figure 2, one might speculate that there are two classes of absorption-line QSO: those with single absorption systems having redshifts always near the emission-line value and those with multiple systems distributed over a much larger range of redshift. The two upper parts of Figure 2 show, on an expanded scale, the distribution of residuals for each of the two classes of QSO. It is seen that there appears to be a major difference between the two distributions, with the single-system QSOs showing a much stronger tendency for small residuals than do the multiple-system objects. However, the bulk of the single-system objects have redshifts smaller than about 1.8 and depend on transitions (Mg II $\lambda 2798$ and C IV $\lambda 1549$) presumably less effective than Lyman-α in revealing the presence of multiple systems. Therefore, it is more germane to the question to look at the distribution of residuals for only those objects having redshifts greater than 1.8. These objects are indicated by cross-hatching in the illustration. It is seen that there remains some indication that single absorption systems tend to have small residuals, but the significance of the effect is placed somewhat in doubt by the fact that only two, or at most four, of the nine single-system objects can be considered to have been intensively studied.

The distribution of redshifts for the absorption systems – in both absolute value and relative to the emission-line redshifts of the objects – is of particular importance to the question of the location and origin of the media responsible for the absorption lines. For example, the strong tendency for QSOs to show small absorption-redshift residuals, for at least the more prominent absorption systems, strongly implies that the absorbing media are located in the immediate vicinity of the objects. On the other hand, those QSOs having multiple absorption systems tend to fit more naturally into the notion that the absorbing material is distributed in discrete clouds throughout the intervening space between such objects and the observer and that the redshifts are cosmological in origin. There is no *a priori* reason why both situations cannot obtain, with the observed absorption lines having both an intrinsic and an extrinsic origin, as they do in the case of galactic stars. Finally, there is the possibility that all of the observed absorbing material is uniquely associated with the particular QSOs in which they are observed and that the clouds showing very large redshift residuals actually have their origin and their large differential motions as a consequence of an energetic expulsion of material from the QSO. This latter interpretation has some difficulty, not necessarily insurmountable, in accounting for the extreme narrowness of the observed absorption lines in comparison with the inferred expulsion velocity.

In connection with the possibility that *some* of the absorbing clouds, especially those showing large redshift residuals, are cosmological in nature and not associated with the particular objects in which they are observed, it is important to realize that there should be a correspondence between the space density of such clouds and the probability of finding clouds at particular redshifts in the spectrum of a QSO. Thus, if the clouds were uniformly distributed throughout the Universe, there should be observed no statistically significant deviation from uniformity in the observed redshift distribution of the clouds. The indication from the present sample of multiple redshift QSOs is that the distribution of redshift is more-or-less uniform with a probably significant tendency for there to be more absorbing clouds located at large redshift.

This interpretation is, however, confronted with some evidence of a striking contradiction. If the absorbing clouds are presumed to be cosmological in distribution, any of the QSOs with emission-line redshifts larger than a particular value should serve equally well – observational factors being equal – as probes in revealing the presence of the clouds. Therefore, it is noteworthy that, of the 44 objects having redshifts greater than 1.7 and thus capable of revealing the presence of clouds of smaller redshift, only three objects do reveal such clouds. These three objects have multiple absorption systems and show seven clouds, all having redshifts considerably less than the emission-line redshifts of the objects. This would seem to represent a rather large statistical fluctuation within the framework of the hypothesis that such clouds are cosmological in distribution. In fact, if the case were any stronger, one might be compelled to conclude that observed clouds are in every case peculiarly associated with the particular objects in which they are observed. As a note of caution, however, it must be confessed that the three QSOs in question have been more intensively studied than almost all of the other QSOs.

The distilled characteristics of absorption-line QSOs that have been presented indicate, in spite of the many necessary observational reservations, that the attribute of absorption lines shows a variation of degree as a function of redshift; absorption lines become an increasingly dominant characteristic as one passes to objects of very large redshift. This impression is considerably strengthened by a familiarity with the nature of the original observational material. As I have had the privilege of working with a considerable amount of such material, it occurs to me that it may be of interest to present in the time remaining some of this along with my views as to its interpretation. In particular, the entire discussion will pertain to a hypothesis used as a basis for a likely interpretation of the multitude of 'unidentified' absorption lines found in objects of large redshift.

The immense complexity of the absorption spectra of objects of large redshift requires an appreciation of the formidable observational problem and the importance of spectral resolution. Some of the effects of improving spectral resolution are illustrated in Figure 3, where a portion of the very complex spectrum of Ton 1530 is reproduced from six spectrograms ranging in resolving power from 10 Å to 1 Å. (As a technical note, all spectrograms except the one of highest dispersion have been

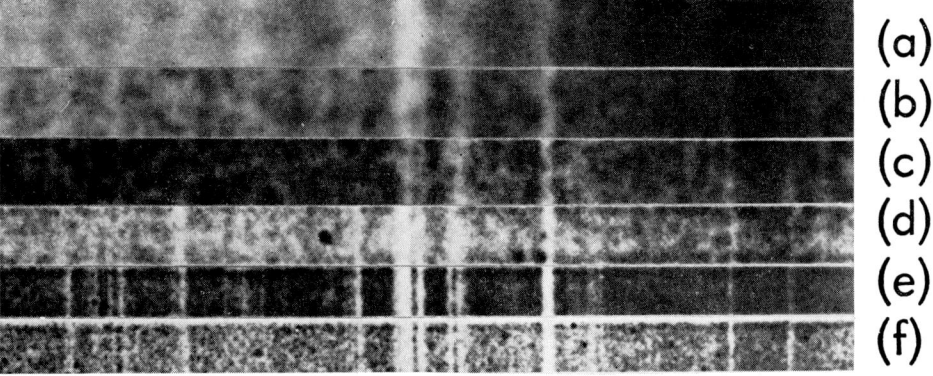

Fig. 3. The effects of spectral resolution as shown by six spectrograms of Ton 1530 ranging in resolution from 10 Å for spectrum (*a*) to 1 Å for spectrum (*f*).

photographically 'folded' so that the information content per resolution element is roughly the same for all spectrograms reproduced. This exacting process was accomplished by Mr Hanna of the Kitt Peak National Observatory photographic laboratory.) It is seen that high resolution is not only important for revealing weak and narrow lines and resolving closely-positioned lines, but that it is essential for clarifying the true character of complex blends involving both weak *and* strong lines. The narrowest lines in spectrum (f) of Figure 3 are saturated and unresolved with a width of the order of 1 Å – or 0.3 Å in the rest frame of this particular source. Because of the faintness of most QSOs and the consequent fact that a successful effort at the highest possible dispersion usually requires nearly perfect observing conditions, we seldom have the luxury of spectrograms of the quality of spectrum (f) and must in many cases be satisfied with spectrograms resembling spectrum (a).

The difficulties of obtaining adequate spectral resolution, when compounded with the presence of multiple redshifts and the possibility of spectrum variations, tend to place added importance on physical reasonableness as a consideration in making line identifications. In my opinion it is also prudent to preserve a certain amount of 'continuity' in the approach to line identification in new and especially in difficult objects; the interpretations that have been found to be successful and well-reasoned in previously studied objects should form much of the basis for the analysis.

One of the first absorption-line QSOs to be discovered and perhaps the one that can inspire the most confidence in the subsequent analyses of other objects is 3C 191. The spectrum of this object contains broad emission lines that can with considerable confidence be identified with Lyman-α of hydrogen, the resonance transitions of Si IV (λ1397) and C IV (λ1549), the 'Balmer-α' transition in He II (λ1640), and the lowest inter-system transition in C III (λ1909). A portion of the confidence in these identifications stems from the early demonstration by Schmidt (1965) of continuity of spectroscopic characteristics in going from the firmly-established interpretations of small-redshift objects to the large redshift (2.012) of 3C 9. Furthermore, the line identifications in 3C 191 are reasonable within the framework of normal element abundances and self-consistent excitation conditions.

If any additional support for the validity of the emission-line identifications was required, it was forthcoming in the spectrum of 3C 191. This evidence takes the form of numerous narrow absorption lines that may be conclusively identified with ground-state transitions in hydrogen, Si II, Si III, Si IV, N V, C II, and C IV at a redshift very nearly equal to the emission-line value. A previously published spectrogram of 3C 191 is reproduced in Figure 4. Notice particularly that the proximity between emission and absorption lines is perfectly natural within the framework of an implied physical and geometrical relationship between emitting and absorbing material. The evidence for such a relationship is strengthened in the present case by the fact that some of the observed absorption lines arise from excited fine-structure levels within the ground states, suggesting that the emitting body may provide the necessary excitation. Among the many absorption-line systems in subsequently studied objects, I know of none

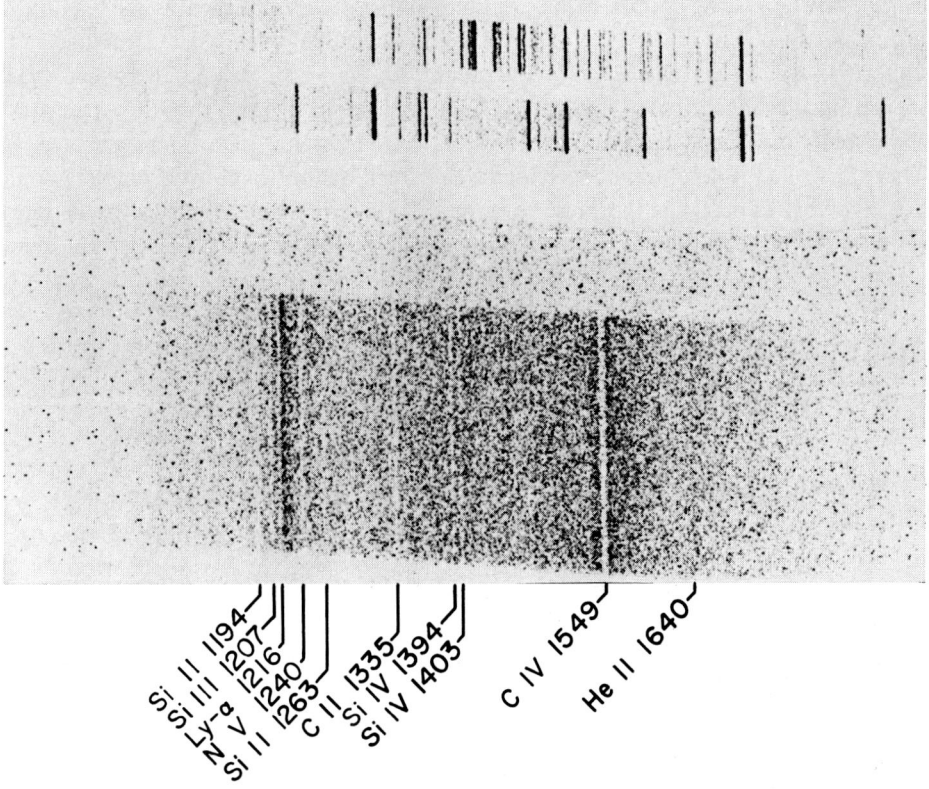

Fig. 4. A spectrogram showing the well-developed absorption spectrum of 3C 191.

that gives clear evidence for population of excited fine-structure levels whenever the absorption-line redshift differs greatly from the emission-line value.

The proximity between absorption and emission lines or, more generally, an apparently special kinematic relationship between absorbing and emitting material may form a tentative basis for identification of an absorption feature, even when there is only one. Thus, in one of the first absorption-line QSOs discovered, Sandage (1965) was quite justified in suggesting that the absorption line occurring in the middle of the C IV λ1549 emission in BSO 1 was due to the same transition in absorption. This point is especially relevant in view of the fact that a large fraction of absorption-line QSOs display only one known absorption line. In essentially all cases such a line is found to be in or near an emission line due to Lyman-α, C IV λ1549, or Mg II λ2798. The provisional assumption that the identification of the absorption line is the same as the associated emission line is very likely to be correct, for in several cases I have checked by obtaining improved observations, only one (PHL 938) was found to be erroneous.

Because the spectrum of 3C 191 is exceptional and can be misleading, it should be mentioned that in other QSOs one seldom finds the lines of the ions of silicon and other low-abundance species to be as prominent – if they are seen at all – as they are in the spectrum of 3C 191. Such lines may be absent even when Lyman-α and C IV λ1549 are strong. This effect is consistent with the considerable body of evidence indicating that most absorption lines (at least the stronger ones) in QSOs are saturated and consequently insensitive indicators of the amount of absorbing material. There is also considerable evidence that when small quantities of absorbing material are involved, Lyman-α is the most persistent of the commonly observed transitions (at least for large-redshift objects). However, this should not be taken to mean that

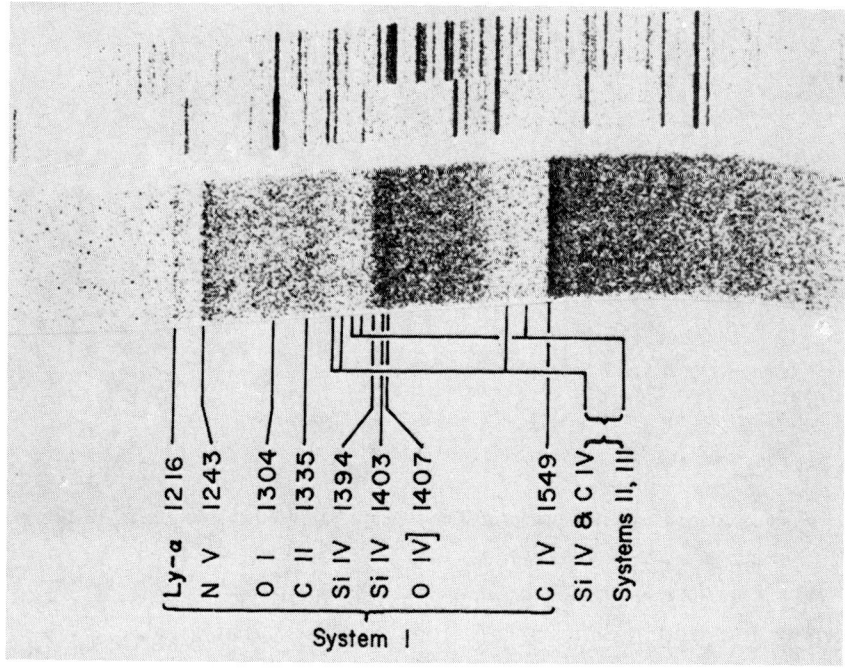

Fig. 5. A spectrogram showing the broad absorption bands in PHL 5200. Systems II and III refer to structure that appears superficially as emission.

there is no evidence for small variations of element abundances or level of excitation in passing from one object to another.

The probable relevance of a special relationship between the redshift distributions for absorbing and emitting material is further illustrated by the spectrum of PHL 5200 illustrated in Figure 5. Emission lines, unquestionably due to the resonance transitions of hydrogen, N V, Si IV, and C IV, are bordered on their violet sides by extremely broad absorption bands. Not only does the identification of the absorption and emission features with the same transitions follow naturally from the obvious kinematic relationship, but the interpretation is also supported by the coherence of the structure within the Si IV band with that within the C IV band. Implicit in this interpretation of the spectrum of PHL 5200 is the distinct possibility of a generic connection between objects of this type and the multiple-absorption-redshift QSOs. Thus, a simple reduction in the amount of absorption at all redshifts would probably transform PHL 5200 into an object with relatively distinct absorption-line redshifts.

The spectrum of PHL 5200 tends to lend credibility to the possibility of multiple absorption-line redshifts, but it did not completely foretell the extreme situation

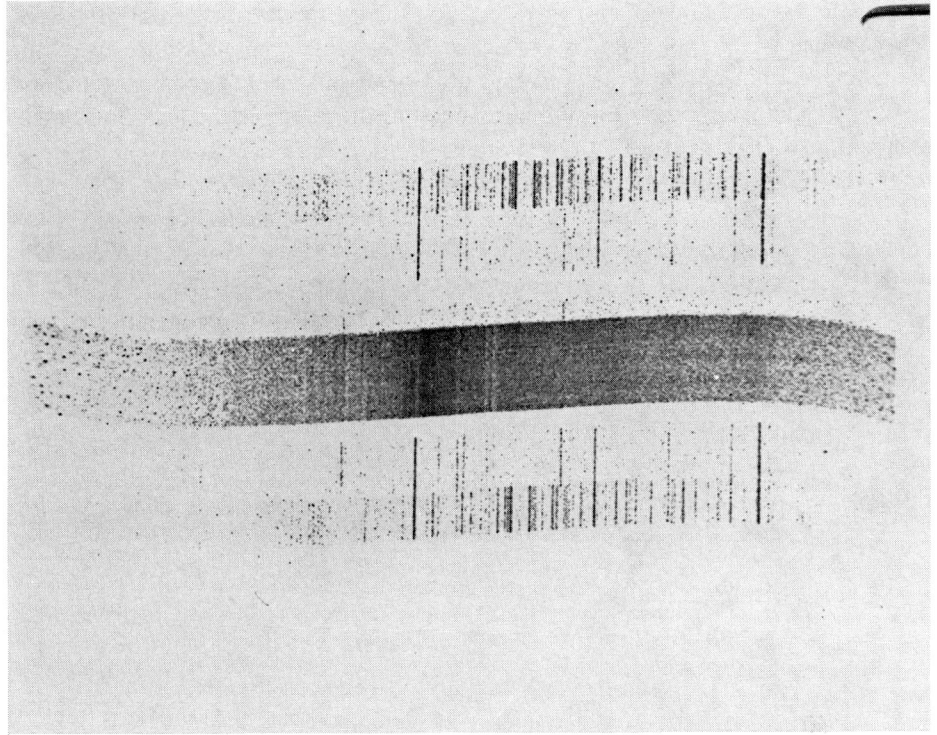

Fig. 6. A relatively low dispersion spectrogram of PKS 0237-23. Many of the absorption features are shown to have fine structure and many additional weak lines are revealed by higher dispersion spectrograms.

found in PKS 0237-23. A low-dispersion spectrogram of this object reproduced in Figure 6 shows the strong and very broad Lyman-α emission line, a less prominent C IV λ1549 emission line, and numerous absorption features. The application of higher dispersion reveals the presence of many additional lines. The initial attempts to account for the lines in one or two redshift systems were not particularly successful in explaining the more prominent features. However, the analysis has now reached the state where several investigators would probably agree to the presence of approximately six redshift systems. These include a system with a redshift near the emission-line value of 2.22 and several systems with considerably smaller redshifts, the smallest of which is 1.364 and is well-established by the presence of several prominent and important transitions.

These absorption systems were essentially all that could be established by what might simply be described as a 'pattern-recognition' method, in spite of the fact that many lines were left without identifications. However, almost all of the prominent unidentified lines occur on the short-wavelength side of the Lyman-α emission line. This fact must be recognized as representing a semblance of order and regularity in the data and, as such, can form the basis of a hypothesis by means of which it is possible to advance a reasonable interpretation for the unidentified absorption lines. The hypothesis is simply that a large fraction of the most prominent lines arise from the Lyman-α transition in clouds that are too poor in material to produce detectable features arising from other transitions. Implicit in the basis for the hypothesis is the assumption, well supported by observation, that Lyman-α can usually be expected to be the most persistent absorption feature in large-redshift systems. This interpretation is really only a minor extension of ideas that have already been successfully applied in the interpretation of QSO spectra, and it will be strongly supported by the characteristics of the two remaining QSOs to be discussed, Ton 1530 and the large-redshift object 4C 05.34. (The absorption systems that have been inferred only from the prevalence of absorption lines near and shortward of Lyman-α emission in the large-redshift objects are the ones that were referred to as having been excluded from the data in Figure 2.)

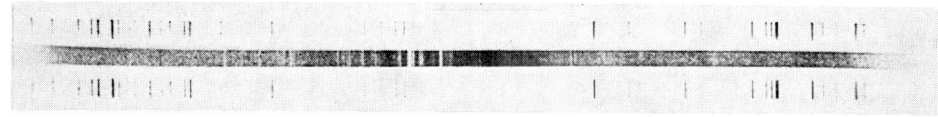

Fig. 7. A high dispersion spectrogram of Ton 1530. A spectrogram of still higher dispersion clearly shows the most prominent absorption line to be triple.

Ton 1530 has already been the subject of two published investigations, but the present discussion is based on new observational material. One of several recent spectrograms of the object is reproduced in Figure 7. The Lyman-α emission line in the center of the spectrum is seen to be flanked on the violet side by numerous narrow absorption lines. Only a few absorption lines have been found longward of Lyman-α (out to about 5000 Å). By the technique of 'pattern-recognition' it has been possible to establish the presence of six or seven absorption-line redshifts (two of the systems have nearly the same redshift), one of which is much smaller than the emission-line value – 1.486, as compared with 2.046. These redshifts serve to identify nearly all detectable absorption lines longward of Lyman-α but leave most of the lines shortward of Lyman-α without identifications. The distribution of the unidentified lines represents the same type of regularity as was found in PKS 0237-23 and indicates that a large fraction of the lines may reasonably be interpreted as arising from Lyman-α.

The QSO 4C 05.34 shows an extreme version of the same type of absorption spectrum as PKS 0237-23 and Ton 1530 (as well as several other sources for which the

Fig. 8. A spectrogram of the large-redshift object 4C 05.34. The emission-line redshift is 2.877, and the emission line in the center of the spectrum is Lyman-α.

analysis is not yet complete) and, perhaps significantly, has the largest redshift, 2.877, known for a QSO. A recent spectrogram of this object is reproduced in Figure 8. This and several other spectrograms show few prominent absorption lines longward of the Lyman-α emission, located in the center of the spectrogram, but a very large number of lines shortward of Lyman-α. Approximately 100 reasonably definite absorption lines have been measured in this object. By the procedure of 'pattern-recognition' at least five absorption-line redshifts have been established; the largest is 2.875, near the emission-line value, and the smallest is 2.473. These systems account for all of the prominent lines longward of Lyman-α but only a small fraction of those shortward of Lyman-α. The very special arrangement of the unidentified absorption lines with respect to emission in a transition having the astrophysical importance of Lyman-α leaves little doubt in my mind that the correct interpretation of the lines is that they also arise from Lyman-α.

Acknowledgements

I wish to acknowledge helpful discussions with T. D. Kinman and W. H. McCrea during the preparation of this paper, which was largely accomplished while visiting the Astronomy Centre, University of Sussex.

References

Sandage, A.: 1965, *Astrophys. J.* **141**, 1560.
Schmidt, M.: 1965, *Astrophys. J.* **141**, 1275.

Discussion

Mrs Burbidge: You mentioned the saturation of the absorption lines – this is indeed quite strong. A student, Y-W. Chan at La Jolla, has applied the Strömgren line-pair method to Lick spectrograms of the C IV doublet in the strongest absorption-line system in Ton 1530 and in two of the intermediate-redshift systems, about $z = 1.6$, in PKS 0237-23, as well as to the Mg II doublet in the $z = 0.6$ redshift system in PHL 938. The lines are definitely narrower than the Lick instrumental profile; the velocity dispersions are about 75 km s^{-1} in the two line pairs in PKS 0237-23, about the same (73 km s^{-1}) in Ton 1530, and only 49 km s^{-1} in PHL 938. This latter is in good agreement with the curve-of-growth determination of the same quantity by Demoulin and Doras.

Dent: Do you have any evidence for changes in these multiple Lyman-alpha absorption lines in these QSOs?

Lynds: I see apparent changes but am not yet certain as to their reality.

Terzian: What is the redshift range of the absorption lines in individual QSOs? Is it not possible that some of this absorption is due to intergalactic clouds which are not associated with the QSOs?

Lynds: As I mentioned, the distribution of absorption-line redshift for objects having multiple absorption systems fits naturally into the interpretation that the absorbing clouds are intergalactic. However, I may have neglected to point out that there is evidence of a striking contradiction. There are 44 QSOs in the sample having redshifts greater than 1.7 that should be equally effective as probes in revealing the presence of absorption by intergalactic clouds with redshifts less than 1.7. In fact, however, only three objects do so, and these three objects have multiple absorption systems with at least one absorption system near the emission-line redshift which is 1.955 in the object with the smallest emission redshift. Furthermore, these three objects show seven absorption redshifts less than 1.7, five of which are revealed by only one of the three objects. This would seem to be a rather large statistical fluctuation within the framework of a hypothesis that the absorbing clouds are cosmological in distribution. It must be confessed that the three QSOs in question are perhaps the most intensively studied of all QSOs. On the other hand, this reservation does not appear to be particularly strong and the weight of evidence may well compel us to conclude that absorbing clouds are in every case peculiarly associated with the particular objects in which they are observed.

Mrs Rubin: What can you say about the absorption line velocity systems which are more positive than the emission line velocity systems in a single object?

Lynds: The largest positive redshift residual in my sample is 0.031, corresponding to about 3000 km s^{-1} for redshifts near 2.0. This is comfortably within acceptable velocity dispersions for the corresponding cosmological epoch.

Ozernoy: If the absorbing gas is in the QSOs themselves, what is your estimate of the total mass of gas clouds?

Lynds: I have made no estimate.

Roeder: I have just finished using the statistical test of Bahcall and Peebles for the randomness of the absorption-line systems, applied in several different General Relativity models of the universe. If one uses all absorption-line systems, one finds that they are strongly non-random. If one uses only the three or four objects with narrow absorption lines, one finds again that the absorption redshift systems seem to be non-random, although of course this conclusion is not extremely strong because of the small number of objects.

SPECTRAL ENERGY DISTRIBUTIONS OF NUCLEI OF PECULIAR GALAXIES

J. B. OKE

Hale Observatories, Pasadena, California, U.S.A.

1. Introduction

Studies of galaxies during the last seventy years have turned up a large array of peculiar objects. Vigorous work on these was rare until a few years ago when quasars were discovered. It then became clear that an understanding of quasars might come from an understanding of the less-extreme peculiar galaxies. It was further clear that even the physical processes occurring in nuclei of peculiar galaxies were far from easy to understand.

There are several groups of objects with which I will be concerned in this paper:

(1) Seyfert Galaxies. NGC 1068 was recognized as peculiar by Fath (1908) in 1908. The first studies of NGC 4151 were made in 1918 by Campbell and Moore (1918) at the Lick Observatory. Seyfert (1943) made the first quantitative study of a substantial number of galaxies with small bright nuclei. These 'Seyfert' galaxies are nearly all very close objects, and, with the possible exception of NGC 1275, are spiral galaxies. They all have tiny, bright nuclei.

(2) Vorontsov-Velyaminov and Krasnogorskaya (1962) and Vorontsov-Velyaminov and Arhipova (1963, 1964, 1968) have published a catalog of peculiar galaxies. Many of these turn up in other lists such as those of Markarian (see below).

(3) Haro (1956) also has published a list of peculiar galaxies, of which a few have been studied.

(4) Zwicky Compact Galaxies. Since 1964 Zwicky has distributed seven lists of galaxies which have high surface brightness per square sec of arc. He also includes parts of galaxies with this same property. They are all nonstellar on the 48-in. Schmidt survey and, therefore, should be extragalactic objects. Their spectra have been studied mainly by Zwicky (1964, 1966, 1967) and Sargent (1970b). Many of these galaxies are very distant.

(5) Markarian Galaxies. Markarian has published three lists of peculiar galaxies (Markarian 1967, 1969a, 1969b). They have been discovered from objective prism spectra and are characterized by strong ultraviolet radiation as compared with normal galaxies. These objects are relatively close. Spectra have been studied by Weedman and Khachikian (1968, 1969), Arakelian, Dibai, and Yesipov (1970), and Sargent (1970a).

(6) Radio Galaxies. Many galaxies identified with radio sources are not peculiar in appearance. Other radio galaxies, usually referred to as N galaxies, (Matthews *et al.*,

1964) are peculiar and have tiny, very bright nuclei. Spectra have been studied mainly by Schmidt (1965), Sandage (1966), and M. Burbidge (1967).

(7) Quasars. In this paper quasars are discussed in terms of their similarities to the nuclei of peculiar galaxies.

2. Observational Techniques

In the visual spectral range at least three kinds of observations can be made:

(1) *Direct Photographs.* These can be used to classify the objects to determine, for instance, the relationship of the bright nucleus to the background galaxy. This type of observation has already been discussed in this Symposium by Dr Morgan.

(2) *Slit Spectra.* These indicate for the nuclear region what lines are present, what the line shapes are, what velocity fields exist, and the red shift. Analysis of line ratios may help to give the electron density N_e and electron temperature T_e.

(3) *Photoelectric Spectrometer Observations.* These give absolute line intensities over the whole spectral range, and line intensities in the far red and near infrared which is not usually accessible with slit spectrographs. The technique is also the best method for finding very broad, low-contrast emission and absorption features. Finally, the absolute spectral energy distribution of the continuous spectrum can be obtained. From this one can hope to separate the radiation from stars in the galactic nucleus from the other contributors to the radiation.

(4) *Infrared Photometry.* These measurements can be put on an absolute basis and extend by a large factor the wavelength range covered by normal spectrophotometry mentioned in (3) above. Dr Neugebauer will discuss these kinds of observations in a later paper.

In this paper I will discuss, almost exclusively, data obtained with photoelectric spectrometers.

3. Spectral Properties

Peculiar galaxies of the sort described above have a wide range of properties insofar as their spectra are concerned. Many have very strong emission lines. Others may show only Hα and λ3727 in emission, while occasionally no emission lines are seen at all. In some objects all the emission lines are narrow and similar in breadth. Others have the 'Seyfert characteristic', namely very broad permitted lines and narrower forbidden lines. Most objects discussed in this paper have this latter characteristic. In a few compact galaxies the forbidden lines are weak or absent; usually in these objects emission lines of permitted FeII are observed, indicating densities of 10^6 or 10^7 electrons cm^{-3} or more.

The continuous spectra of the reddest objects are very similar to those of normal galaxies; broad absorption features caused by the MgI lines, the G-band and the H and K lines of CaII are usually present. The bluer objects have smooth continuous spectra with little or no evidence of absorption features. In these objects the ultraviolet radiation may be abnormally high.

4. Sources of Continuous Radiation

(1) The stars in the background galaxy will contribute to the total observed radiation. It will be assumed that the absolute energy distribution of any stellar contribution is similar to that observed in the nucleus of M 31 (Oke and Sandage, 1968). This may not be a valid assumption since Wampler (1971) has shown that the stellar radiation near the nucleus of NGC 1068 is significantly bluer than that from the nucleus of M31.

(2) Bound-free and free-free radiation from hydrogen. This must be present since Balmer emission lines are often very strong. If this contribution is large, a Balmer jump in emission may be observed as in NGC 4151 (Oke and Sargent, 1968). Two-photon emission could occur although most estimates of the electron density in these objects rule it out.

(3) Non-thermal radiation, such as synchrotron and inverse Compton radiation may be present. It is assumed that any nonthermal contributor has a spectrum of the form $F_\nu \propto \nu^\alpha$ where ν is the frequency and α the spectral index.

(4) Large amounts of hot interstellar dust could radiate substantially in the infrared. In such a case, any variability of this component must have time scales of many years.

All of the above radiation sources may be affected by interstellar reddening. Corrections for insterstellar reddening in our own Galaxy are usually easily made, since most peculiar galaxies which have been studied are at high galactic latitudes. Interstellar reddening within the central region of the observed galaxy is much harder to estimate. Wampler (1968b) has studied the ratio of infrared to blue sulphur forbidden lines in Seyfert galaxies and finds strong evidence for reddening. A comparison of Paschen and Balmer line intensities and, in principle at least, the Balmer decrement, can also be used to determine the amount of reddening.

5. Distance Scale Problems and Selection Effects

Ideally, one would like to always observe only the bright nucleus of a peculiar galaxy out to some limiting distance, say 25 or 50 parsec. This can never be done in practice because (a) one always observes along a line-of-sight which penetrates the whole galaxy, and (b) galaxies are at different distances and only in a few cases is a resolution of 50 parsec possible. The classical 'Seyfert' galaxies are all close and it is possible to isolate the tiny, bright nucleus quite well. But even in NGC 4151, Wampler (1971) has shown that the stellar contribution within a few sec of arc of the center is appreciable and variable with wavelength. More information can be obtained about the stellar background by measuring radiation through apertures of different sizes, but even in this circumstance some model for the light distribution is needed to predict the contribution at the center.

In the case of a distant Zwicky compact galaxy, the observing aperture may include not only light from the stellar nucleus, but also light from most of the background galaxy as well. Such a galaxy is discovered only because the nucleus is exceedingly bright compared with the whole galaxy. A galaxy such as NGC 4151, which has only a moderately bright nucleus, would probably never be detected as a peculiar galaxy if it were at a large distance. As a result, galaxies with only mild activity in the nucleus are all nearby objects. Those objects which have violent activity are seen only at very large distances, presumably because they are very rare.

Because the amount of background stellar radiation depends so much on distance, it is necessary to separate this component of the radiation from the other contributors. Only if this can be done will it be possible to study adequately the conditions in the nuclei of peculiar galaxies.

6. Observational Material

Data will be presented which are representative of spectrophotometric observational results for 12 Markarian galaxies, 14 Zwicky compact objects, and 16 radio galaxies.

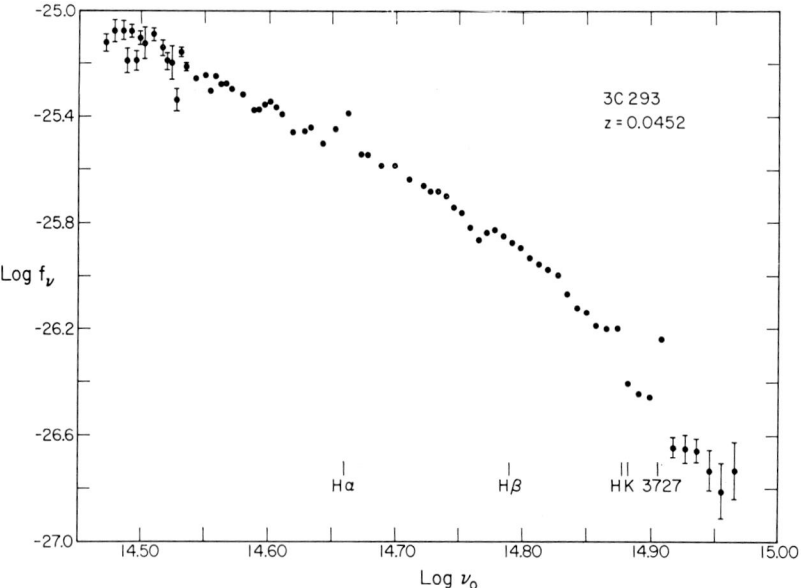

Fig. 1. The absolute spectral energy distribution of the radio galaxy 3C 293. The logarithm of the apparent flux f_ν erg s^{-1} cm^{-2} Hz^{-1} is plotted against the rest frequency ν_0. Positions of several emission lines and the H and K lines of Ca II are marked. Standard deviations are shown if they are greater than 0.02 in log f_ν. The red shift is indicated.

The results are shown and described in Figures 1 to 7. The object 3C 293 is a radio galaxy of normal appearance. The spectral energy distribution is almost identical with that of ordinary elliptical galaxies except for the presence of Hα and λ3727 of [O II] in emission. The spectral index, α, which will be defined over the spectral range

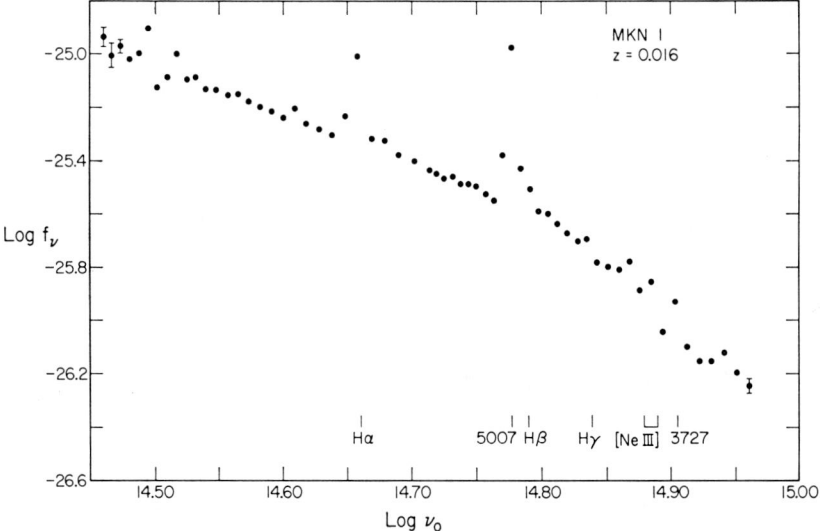

Fig. 2. Markarian No. 1.

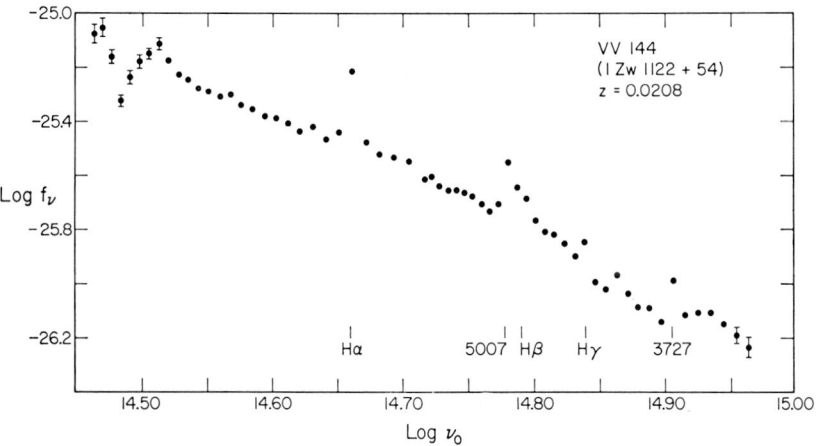

Fig. 3. Vorontsov-Velyaminov object VV 144.

from 4000 Å to 10000 Å is −2.7 and typical of the central regions of normal galaxies.

Markarian No. 1 (which is also NGC 449) and the Vorontsov-Velyaminov object VV 144 (also the Zwicky compact I Zw 1122+54) have energy distributions which are mixtures of a background galaxy and some non-thermal source. The spectral indices are both near −2.0 and there is strong evidence that the H and K lines of Ca II are present in absorption.

The radio galaxy 3C 382 also has a spectral index of −2.0 from 4000 Å to 10000 Å,

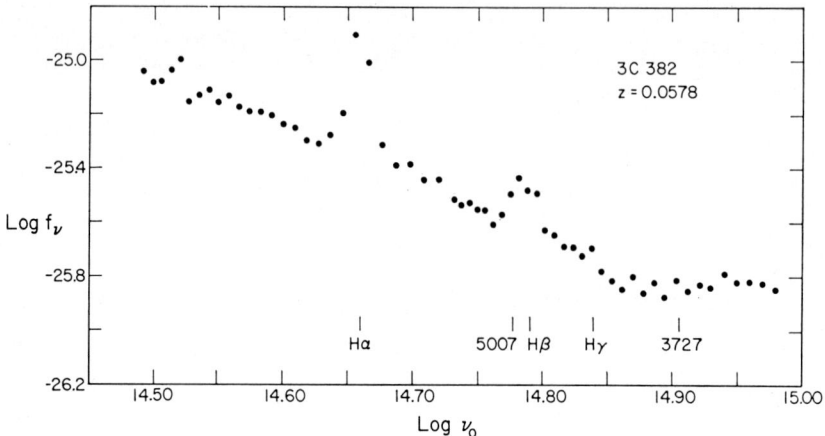

Fig. 4. Radio Galaxy 3C 382.

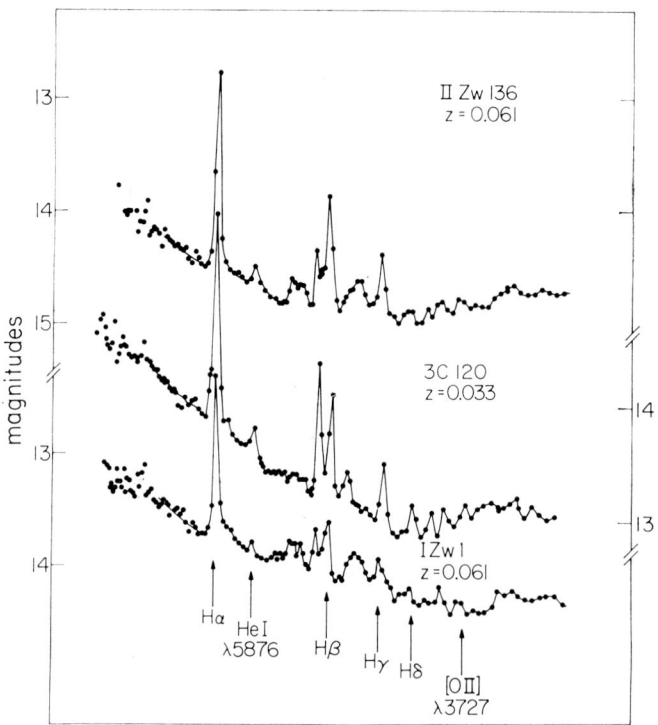

Fig. 5. Three objects II Zw 136 (II Zw 2130 + 09), 3C 120, and I Zw 1 (I Zw 0051 + 12). The vertical scale is $-2.5 \log f_\nu +$ const. and the horizontal scale is $1/\lambda$ where λ is the observed wavelength in microns. The emission on either side of Hβ and λ5007 of [O III] is due to Fe II.

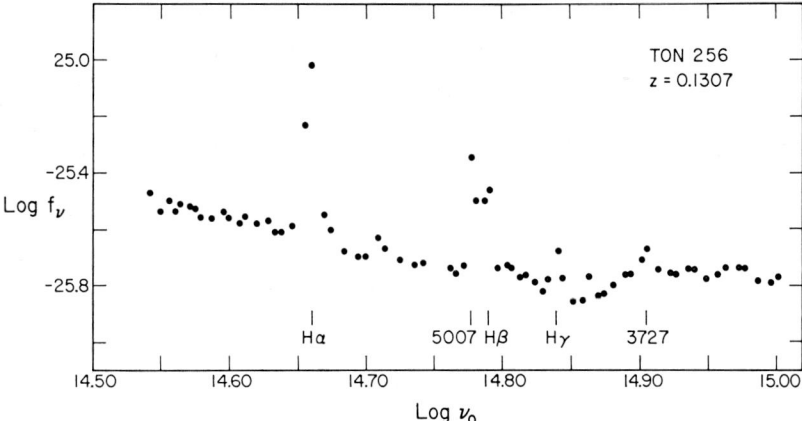

Fig. 6. Tonanzintla object Ton 256. Axes are the same as in Figure 1. The permitted lines are very broad.

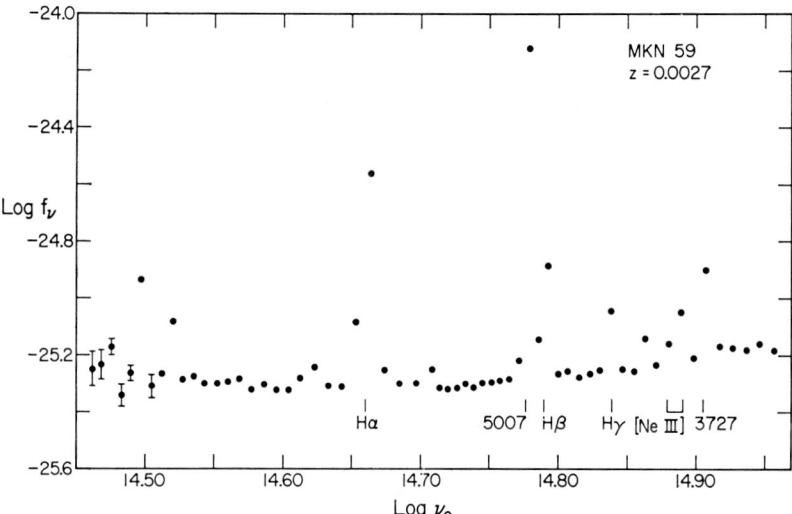

Fig. 7. Markarian No. 59. The emission lines are very strong and sharp. The object is probably a very large H II region.

but in this object the spectrum becomes flat below 4000 Å. This could be caused by a Balmer jump in emission or by many unresolved emission lines in this spectral range. Many quasi-stellar sources have energy distributions similar to that in this object. Schmidt (1965) has obtained a spectrum of this object and finds no evidence for absorption lines.

The three objects II Zw 2130+09, 3C 120, and I Zw 0051+12 have been observed by Sargent and Oke and have spectral indices of -1.13, -1.93, and -1.24 respectively. Both Zwicky objects have strong Fe II lines (Sargent, 1968a, 1968b) and weak or absent forbidden lines. It is not certain whether the N-galaxy 3C 120 has Fe II lines. As in 3C 382, all the spectra are flat in the ultraviolet and there is no evidence for absorption features. Ton 256 has a spectral energy distribution very much like those of the two Zwicky objects mentioned above.

Markarian No. 59 is one of only three objects so far observed in which the spectral index is positive, being +0.2. The forbidden and permitted lines are very strong and sharp. It is suspected that this object is a very large H II region.

Absolute spectral energy distributions similar to those described above for peculiar galaxies have also been obtained for quasi-stellar sources by Oke (1966, 1967a) Wampler (1967a, 1967b, 1968a), and by Oke *et al.* (1970). A comparison of energy distributions of quasi-stellar sources with those of galaxies in which absorption lines are not seen, shows some striking similarities.

(a) The range of spectral indices is similar.

(b) Many galaxies, such as those mentioned above and quasars such as 3C 345, PKS 0405-12, and 3C 334, have spectral energy distributions which become flat below a rest wavelength of 4000 Å.

(c) Some quasi-stellar sources, such as 3C 323.1, have energy distributions, line strengths, and line breadths which are almost identical with those of peculiar galaxies such as the Braccesi object B 264 and Ton 256.

7. General Properties of Spectral Energy Distributions of Peculiar Galaxies and Quasars

It has been shown by Sandage, for example, (Sandage, 1967b) that there is a regular progression of colors as one goes from quasi-stellar sources, through N-galaxies to normal elliptical galaxies. This progression is, however, somewhat masked in the broad-band color measurements by the strong emission lines often present in the spectrum. It is, therefore, preferable to use the spectral index α of the continuous energy distribution as a measure of color. As indicated above, α is here defined between 4000 Å and 10 000 Å and the region below 4000 Å is neglected.

Another quantity which is relevant for these objects is the absolute flux F_{ν_0} erg s^{-1} Hz^{-1} of the continuum at a fixed rest frequency ν_0; ν_0 is chosen to correspond to 4000 Å. Distances are calculated using a Hubble constant of 100 km s^{-1} Mpc^{-1} and a cosmological constant $q_0 = +1$. The results for the sample of objects being discussed in this paper are shown in Figure 8 where the spectral index α is plotted against log F_{ν_0}. A few specific objects are marked. Quasars, assumed to be at cosmological distances, are shown by crosses. If they are not at cosmological distances, they shift horizontally far to the left. G marks the location of the brightest cluster galaxy when measured out to an isophote corresponding to 25 mag. arc sec^{-2}.

The family of curves shows the result of combining the spectral energy distribution of a bright galaxy with various contributions of a power-law spectrum. The bright galaxy is assumed to be somewhat fainter than the brightest cluster galaxy, since it

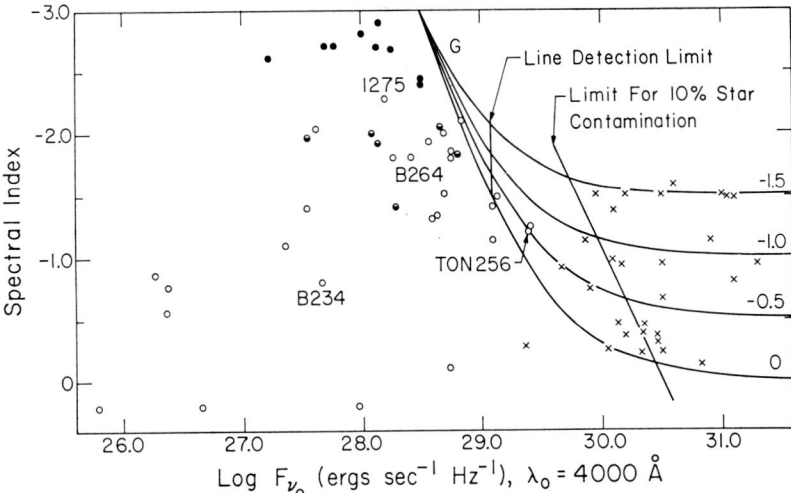

Fig. 8. The spectral index α defined over the range from 4000 Å to 10000 Å plotted against the absolute flux at a rest wavelength of 4000 Å. Quasars are plotted as crosses. G is the location of the brightest cluster galaxy.

is usually measured with an aperture which does not include all the light. Each curve goes from a pure galaxy spectrum to a pure power-law spectrum. The oblique straight line shows the limit beyond which radiation from even a very bright galaxy will not contribute more than 10% to the total radiation at any wavelength. It is clear that most, but not quite all, quasi-stellar sources have energy distributions which are virtually unaffected by any background galaxy.

Another limit which can be placed in this figure is the point beyond which the H and K lines of CaII would not be detected. Because observational data are still not very precise, and there are often emission lines near H and K, it is estimated that the radiation at H and K can be diluted by approximately a factor 3 and still leave the lines detectable. This limit is shown by the vertical line in Figure 8; to its right, the H and K lines would probably not be detected using data of typical accuracy at the present time. This same vertical line also represents the limit of detectability of the MgI lines at $\lambda 5180$. This feature is weaker than H and K, but is in a more favorable spectral region for detection.

The family of curves shown in Figure 8 applies only for a contributing galaxy of fixed luminosity. If the galaxy is made fainter, the curves all shift horizontally to the left. When allowance is made for this, the following general conclusions can be reached:

(1) For $\alpha < -2.1$, the H and K lines should always be visible in the observed spectral energy distribution.

(2) For $-2.1 < \alpha < -1.4$, the H and K lines may or may not be seen, depending on the galaxy luminosity and the spectral index of the non-thermal source.

(3) For $\alpha > -1.4$, the H and K lines should not be seen.

It should be emphasized that the above remarks assume a modest accuracy for the observations. As observational accuracy and wavelength resolution improve, the limits will shift. It also should be pointed out that it is possible to make detailed composite energy distributions. When these are compared with observations, somewhat more definite conclusions can be reached concerning the relative importance of nonthermal and galaxy radiation.

In Figure 8, the solid dots denote galaxies where H and K lines are seen. The half-filled circles denote those galaxies where H and K may be present. H and K are not seen in those objects shown by open circles. With the exception of NGC 1275, for which better observations are needed, the various domains defined above are filled with the expected kinds of objects.

The three objects with positive spectral indices are all like Mkn 59 which was shown in Figure 7. They are all probably H II regions. The three objects, B 234, B 264, and Ton 256 which have often, in the past, been called quasi-stellar objects, have been shown by Arp (1970b) to be galaxies.

8. Variability

Another tool for separating the stellar and non-thermal radiation is available when the object is variable, since it is reasonable to assume that any light variability is associated only with the non-thermal radiation. If it could be assumed that the non-thermal spectral index is independent of brightness, then the separation of the non-thermal and galaxy contribution would be trivial. In practice, this assumption is almost certainly incorrect and the whole problem becomes more complicated, and other data such as absorption line strengths are desirable.

A good example of a variable object which has been studied in some detail is 3C 371 (Oke, 1967b; Sandage, 1967a). When the object is faint, features such as the H and K lines of Ca II and the $\lambda 5180$ band of Mg I are just detectable. When the object is bright, these are not seen. Furthermore, the continuous energy distribution has a different slope depending on the brightness, as is expected of a composite spectrum when the two components are quite different. A reasonable separation of the galaxy and non-thermal contributions can be made. Arp (1970a) has shown that this object is a large galaxy with the outer regions being very faint.

Much more work should be done to find variable galaxies and study their spectral changes.

9. Conclusions

All of the objects which have been discussed in this paper, apart from quasars, are extended and must be galaxies. Since the galaxy is visible, it is clear that the observed spectral energy distribution must have some contribution from stars in addition to the non-thermal component. It is important to separate these components where possible in order (a) to study the non-thermal radiation (b) to investigate the nature of the stellar component in the nucleus, and (c) to determine how much energy is coming from each source of radiation and consequently to learn what kinds of galaxies produce what kinds of non-thermal radiation.

I have attempted to show that progress can and is being made in these directions. It is clear that more objects should be observed spectrophotometrically. In particular, emphasis should be placed on variable galaxies. It will probably be useful to study objects by measuring colors and magnitudes through various apertures. Finally, spectrophotometric data should be improved by increasing the accuracy of the observations and also by increasing to some extent the wavelength resolution. If these improvements are made, it will be possible to detect absorption features such as the H and K lines, the G-band and the MgI band in many more peculiar galaxies.

Acknowledgement

The author wishes to thank Mr. B. Turnrose who helped to reduce and analyse much of the data presented in this paper. This work was supported in part through grant No. NGR 05-002-134 from the National Aeronautics and Space Administration.

References

Arakelian, M. A., Dibai, E. A., and Yesipov, V. F.: 1970, *Astrofizika* **6**, 39.
Arp, H. C.: 1970a, *Astrophys. Letters* **5**, 75.
Arp, H. C.: 1970b, *Astrophys. J.* **162**, 811.
Burbidge, E. M.: 1967, *Astrophys. J. Letters* **149**, L51.
Campbell, W. W. and Moore, J. H.: 1918, *Pub. Lick Obs.* **13**, 122.
Fath, E. A.: 1908, *Lick. Obs. Bull.* **5**, 71.
Haro, G.: 1956, *Bol. Obs. Tonantzintla Tacubaya* **2**, 8.
Markarian, B. E.: 1967, *Astrofizika* **3**, 55.
Markarian, B. E.: 1969a, *Astrofizika* **5**, 443.
Markarian, B. E.: 1969b, *Astrofizika* **5**, 581.
Matthews, T. A., Morgan, W. W., and Schmidt, M.: 1964, *Astrophys. J.* **140**, 35.
Oke, J. B.: 1966, *Astrophys. J.* **145**, 668.
Oke, J. B.: 1967a, *Astrophys. J.* **147**, 901.
Oke, J. B.: 1967b, *Astrophys. J. Letters* **150**, L5.
Oke, J. B., Neugebauer, G., and Becklin, E. E.: 1970, *Astrophys. J.* **159**, 341.
Oke, J. B. and Sandage, A.: 1968, *Astrophys. J.* **154**, 21.
Oke, J. B. and Sargent, W. L. W.: 1968, *Astrophys. J.* **151**, 807.
Sandage, A.: 1966, *Astrophys. J.* **145**, 1.
Sandage, A.: 1967a, *Astrophys. L. Letters* **150**, L9.
Sandage, A.: 1967b, *Astrophys. J. Letters* **150**, L177.
Sargent, W. L. W.: 1968a, *Astron. J.* **73**, 893.
Sargent, W. L. W.: 1968b, *Astrophys. J. Letters* **151**, L31.
Sargent, W. L. W.: 1970a, *Astrophys. J.* **159**, 765.
Sargent, W. L. W.: 1970b, *Astrophys. J.* **160**, 405.
Schmidt, M.: 1965, *Astrophys. J.* **141**, 1.
Seyfert, C. K.: 1943, *Astrophys. J.* **97**, 28.
Vorontsov-Velyaminov, B. A. and Arhipova, V.: 1963, *Trudy Gos. Astron. Inst. Sternberga* **33**.
Vorontsov-Velyaminov, B. A. and Arhipova, V.: 1964, *Trudy Gos. Astron. Inst. Sternberga* **34**.
Vorontsov-Velyaminov, B. A. and Arhipova, V.: 1968, *Trudy Gos. Astron. Inst. Sternberga* **38**.
Vorontsov-Velyaminov, B. A. and Krasnogorskaya, A.: 1962, *Trudy Gos. Astron. Inst. Sternberga* **32**.
Wampler, E. J.: 1967a, *Astrophys. J.* **147**, 1.
Wampler, E. J.: 1967b, *Astrophys. J. Letters* **148**, L101.

Wampler, E. J.: 1968a, *Astrophys. J.* **153**, 19.
Wampler, E. J.: 1968b, *Astrophys. J. Letters* **154**, L53.
Wampler, E. J.: 1971, *Astrophys. J.* **164**, 1.
Weedman, D. W. and Khachikian, E. Ye.: 1968, *Astrofizika* **4**, 587.
Weedman, D. W. and Khachikian, E. Ye.: 1969, *Astrofizika* **5**, 113.
Zwicky, F.: 1964, *Astrophys. J.* **140**, 1467.
Zwicky, F.: 1966, *Astrophys. J.* **143**, 192.
Zwicky, F.: 1967, *Adv. Astron. Astrophys.* **5**, 267.

Discussion

Wray: I would like to draw attention to the fact that the strength of N_1 and N_2 exceeds that of Hβ in some of the nuclear regions and inner H II regions in ordinary galaxies, as shown by Dr. Rubin as well as generally in H II regions in the outer parts of ordinary spiral galaxies, as observed by Searle. Perhaps it is possible that the relative strength of N_1, N_2 compared with the inner core of the Hβ line (neglecting the broad wings) occurring in a number of Seyfert galaxies and compact galaxies is due in some degree to physical processes similar to those occurring in nuclear regions and inner H II regions of some ordinary galaxies, a likely mechanism being excitation by a component of OB stars. It is interesting to note that these sharp hydrogen line cores are absent from spectra of a number of quasi-stellar objects and that a number of these objects, as you have shown, most probably do not contain significant, if any, underlying stellar populations. This suggests that the broad wings of the hydrogen lines in Seyfert galaxies also are not likely to be due to stellar excitation. In short it appears that two excitation mechanisms may be operating; OB stars for objects ranging from ordinary galactic H II regions to at least a part of the Seyfert galaxy and compact galaxy emission radiation, and a second source responsible for broad emission lines in objects ranging from Seyfert galaxies and some compact galaxies through quasi-stellar objects.

THE PHYSICAL PROPERTIES OF THE ABSORPTION ENVELOPES OF TWO QSOs

JEFFERY D. SCARGLE

Lick Observatory, Board of Studies in Astronomy and Astrophysics, University of California, Santa Cruz, and Ames Research Center, NASA, Moffett Field, Calif., U.S.A.

LAWRENCE J. CAROFF

Ames Research Center, NASA, Moffett Field, Calif., U.S.A.

and

PETER D. NOERDLINGER

New Mexico Institute of Mining and Technology, Socorro and Ames Research Center, NASA, Moffett Field, Calif., U.S.A.

Abstract. The wide absorption troughs of certain resonance lines in the spectra of the QSOs PHL 5200 and RS 23 are produced by pure scattering and must therefore be compensated by an equal amount of emission. Analysis of the transfer of resonance-line radiation through a differentially expanding atmosphere shows that the absorption-emission profiles can be quantitatively understood on this basis.

We have been studying the transfer of radiation by resonant scattering in an expanding nebula with application to QSOs having absorption wings or troughs to the blue of emission. The study was prompted by the absorption troughs running from the resonance lines of C IV, Si IV, N V, and Ly-α to a velocity $c/30$ blueward in the object PHL 5200 (Lynds, 1967). (The necessary figures to accompany this talk appear in Lynds (1967), Burbidge (1968, 1969, 1970) and our own paper (1970).) In Burbidge (1968, 1969), tentative evidence is presented for changes in the Si IV trough in about 9 months. These changes must involve material comprising $>1\%$ of the envelope. If they are real, the envelope must contain such a component <1 pc in diameter.

By averaging tracings of three spectra kindly loaned by E. M. Burbidge, we can estimate that the optical depth in the Si IV and C IV troughs in PHL 5200 must be at least $\tau = 3$, whatever happens to the light. If we use

$$\tau = N\sigma/\delta v,$$

where N = column density, σ = integrated scattering cross-section and $\delta v \approx v_0/30$ = = width of trough, we may estimate the column densities and get minimum mass values, by using cosmic abundance ratios. The result is

$$M > 5 M_\odot (R_{pc})^2 (\Omega/4\pi),$$

where Ω is the solid angle subtended by the envelope at the central source.

Since the envelope is known to have existed for three years, this gives $M > 5 M_\odot$ unless Ω is small.

We can show collisional de-excitation is small, and photo-ionization out of the upper state is negligible at distances greater than 3×10^{14} cm from the center. Therefore, the light 'absorbed' in the troughs must all come out. If the envelope is a jet directed toward us, the light can go to the sides and not be seen. However, there is a symmetric CIII] 1909 emission line, whose width is about that of the troughs, and so it seems reasonable that this line comes from the envelope. If so, a jet is excluded. In addition, our more spherical models produce good fits to the lines.

If the envelope is spherical but too diffuse to be seen against the sky, it must be ⩾25 arc sec in diameter. This implies $R \approx 100$ kpc and leads to very large masses.

Therefore, we believe the envelope is unresolved. Radiation from the central object that is scattered to us from the back side of the nebula is shifted from the blue of the line to the red side of the line (all frequencies here being defined in the QSO rest frame). Light scattered by material out towards the side is seen at the line center, and so on. Thus the 'missing' light has been redistributed in frequency.

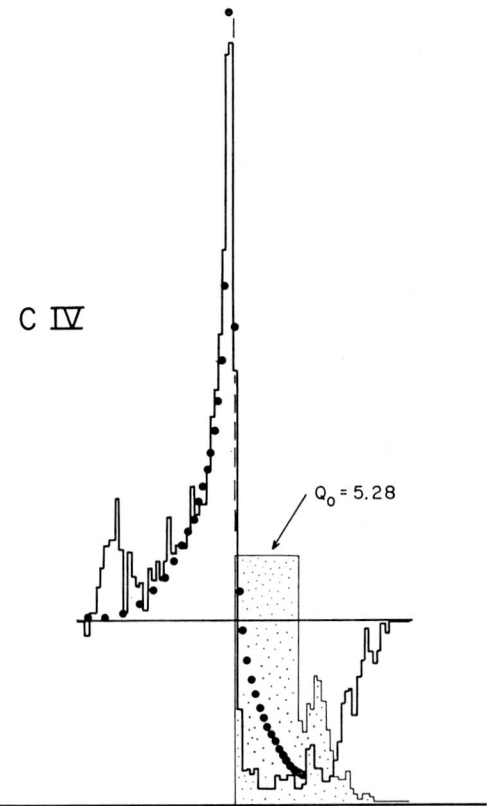

Fig. 1. Comparison of the theoretical (filled circles) and observed (heavy lines) profiles for the CIV line in PHL 5200. The observed curve has been referred to the estimated continuum level. The profile has been force-fit as well as possible to the blue of the zero-velocity point (vertical dashed line), and the theoretical curve is not plotted where the agreement is exact. The redward profile is determined completely independently of the data. Dotted histogram, run of Q_0 in the absorption trough; vertical scale is indicated by the maximum value (5.28 is an artificial cutoff and the true value is indeterminate).

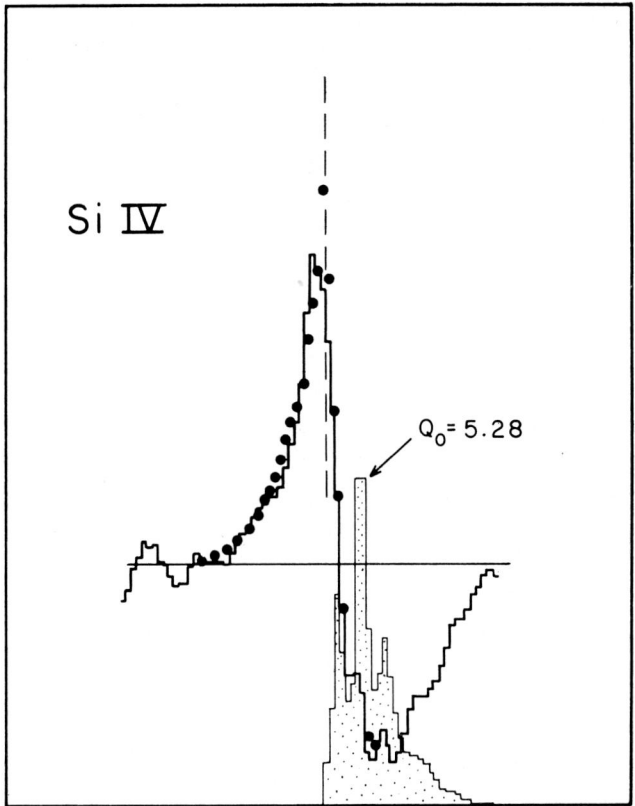

Fig. 2. Same as Figure 1 for the Si IV profile in PHL 5200, except that the observed curve has been slightly smoothed.

Our Monte Carlo program traces photons through an expanding nebula under these assumptions:

(a) uniform, isotropic expansion;
(b) thermal width small compared with width of trough;
(c) there is no physical boundary, although the density can be set equal to zero in velocity space. Thus, the photons see material always moving away from them, and they eventually escape in velocity space, being unable any longer to find material of the correct velocity to scatter them.

There is only one parameter left at any frequency, an effective optical depth

$$Q_0 = \frac{n\sigma}{Kv_0}$$

n = number density, $K = c^{-1} \, dv/dr$, where v = radial velocity of expansion and r is distance from the central source.

Here n is a function of v, and $v = v_0 (1 - v/c)$ so Q_0 is a function of v or ν. The

Monte Carlo program tells us how light from frequency v in an absorption trough is redistributed over the range $(2v_0 - v, v)$ extending an equal distance on either side of rest frequency v_0.

When Q_0 is small (of the order 1–3) the redistribution is rather uniform, and at higher values, up to our largest value $Q_0 = 12$ it develops a moderate preference toward the red.

Using our redistribution profiles, we start at the blue end of an absorption trough, and trace redward. At each frequency, we allow residual (unscattered) light $I_0 \exp(-Q_0)$ and we add in all light scattered from larger frequency. The local Q_0 value must be adjusted to bring the total

$$I_{\text{scattered}} + I_0 \exp(-Q_0)$$

equal to the observed intensity. When too much light has been scattered from previously considered frequencies (further in the blue), no fit can be obtained.

Our published graphs (Scargle et al., 1970) show good fits obtained by this method, except that we cannot get the C IV trough as deep as observed. Even in this case, the redward profile is good.

The trough can be made deeper by taking a non-uniform expansion, $v/r > dv/dr$. This throws more light into the line and the red. Exact equality is still preserved between integrated absorption and re-emission. A recent scan by Oke (1970) suggests that our continuum may have been taken too low in the red. If so, absorption exceeds emission and the envelope is probably asymmetric, although other possibilities exist.

The QSO known as RS 23 (Burbidge, 1970) shows similar troughs, but cannot be fitted by our present models, because emission exceeds absorption. It would appear interesting to consider models with some emission by collisional excitation in the envelope.

References

Burbidge, E. M.: 1968, *Astrophys. J. Letters* **152**, L777.
Burbidge, E. M.: 1969, *Astrophys. J. Letters* **155**, L43.
Burbidge, E. M.: 1970, *Astrophys. J. Letters* **160**, L533.
Lynds, C. R.: 1967, *Astrophys. J.* **147**, 396.
Oke, J. B.: 1970, private communication.
Scargle, J., Caroff, L., and Noerdlinger, P.: 1970, *Astrophys. J. Letters* **161**, L115.

Discussion

Rees: The light travel time across your assumed envelope may be thousand of years and your calculations also involve the further assumption that the continuum strength has remained constant over this period. Otherwise, you have another free parameter available.

Noerdlinger: We assumed a steady continuum. If it had increased during the light travel time across the nebula, this could explain the possible imbalance between emission and absorption, since the redward redistributed light comes from an earlier period in the life of the envelope. Incidentally, we had very few free parameters. The zero of velocity had to be picked just right, or the fits came out very badly. Also, Q_0 had to go zero at zero velocity to give good fits; there was no in-falling material.

THE BALMER LINES IN THE SEYFERT GALAXIES
NGC 5548 AND NGC 4151

R. WEYMANN and R. CROMWELL

Steward Observatory, Tucson, Ariz., U.S.A.

Abstract. The profiles of the Balmer lines in NGC 5548 as reported by Dibai *et al.* (1968) were somewhat asymmetric, whereas those reported by Anderson (1970) are smooth and symmetric. We present profiles which are strongly asymmetric, resembling those of Dibai *et al.* Evidently electron scattering is not the sole principal broadening agent and we must deal with velocities ∼2500 km s⁻¹ in a very small volume.

The transient nature of the P-Cygni type profiles in the Balmer lines of NGC 4151 has previously been noted (Cromwell and Weymann, 1970). These lines have since disappeared, at the resolution available to us, in a time of 3 months. A model in which frequent outbursts of shells or filaments produce transient features in the Balmer lines, while the accumulated material from past outbursts produces the relatively stable He I λ3889 feature, seems the most plausible.

This contribution deals with two items: the marked asymmetry observed in the Hβ profile of NGC 5548, and the appearance and subsequent disappearance of displaced absorption cores in the Balmer lines of NGC 4151.

There have been two recent attempts to interpret the wide wings characteristic of the Balmer line profiles in Seyfert galaxies in terms of electron scattering occurring in very dense, very small regions of the nuclei of these objects (Weymann, 1970; Mathis, 1970).

According to profiles recently obtained by Anderson (1970), the Hβ profile in NGC 5548 is quite symmetrical and smooth, and can be matched quite well by the theoretical line profiles. This success in matching the profile thus supported the reality of the electron-scattering model.

During the interval December 1969–June 1970 however, we obtained several spectrograms of NGC 5548, *all* of which show a rather marked asymmetry. A representative tracing from one of these spectrograms is shown in Figure 1. Also included is the profile derived by Anderson, as well as a tracing of a spectrogram taken in May 1966 by Dibai *et al.* (1968). The same asymmetry is evident in this tracing, (though not as clearly as on our spectrograms) and Dibai *et al.* explicitly drew attention to it. We cannot say whether the differences in the profiles obtained by ourselves and Dibai *et al.* and by Anderson are real. No convincing systematic changes between December 1969 and June 1970 were noted. Dibai *et al.* also called attention to some apparent finer structure in the profiles; there is a suspicion of this in our material, but it is by no means confirmed.

The main conclusion to be drawn from this asymmetry is simply that electron scattering cannot be the sole principle broadening agent; mass motions of at least 2500 km s⁻¹ are involved. Dibai *et al.* proposed that radial expansion was involved, and, assuming an electron density in this region of 10^7 cm⁻³ deduced a characteristic time of 30 yr. The density could very well be as high as 10^9 cm^{-3} however, in which case the time would be only 1 yr. If the gas is bound by some massive object it is hard to see how the motion could persist, unless it is fairly well ordered, which suggests rotation. The asymmetry could then be due either to obscuring material or to clumpi-

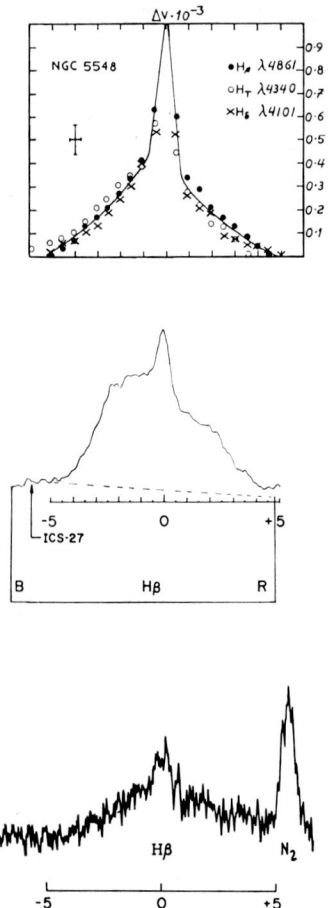

Fig. 1. Profiles of Hβ in the Seyfert galaxy NGC 5548. The top profile is based on work of Anderson (1970), the middle profile is an intensity tracing of a spectrogram taken by the present authors in January 1970, and the bottom profile is from a spectrogram of Dibai *et al.* taken in May 1966.

ness. In either case, one anticipates changes in the spectrum on a time scale of perhaps a few months to 30 yr.

Anderson and Kraft (1969) have discussed the displaced absorption cores of He I 3889 and the Balmer lines in NGC 4151 and derived from these features very high mass ejection rates. Subsequent observation by us (Cromwell and Weymann, 1970) in December 1969 and January 1970 showed that the Hβ absorption was quite strong, and must be transient. This fact, together with the presumption that these features were formed closer to the nucleus than was assumed by Anderson and Kraft, led us to an estimate of the mean mass ejection rate very much lower than Anderson and Kraft's.

By late April, this absorption core had weakened to the point where, with our resolution, it had disappeared. Figure 2 shows microphotometer tracings of plates

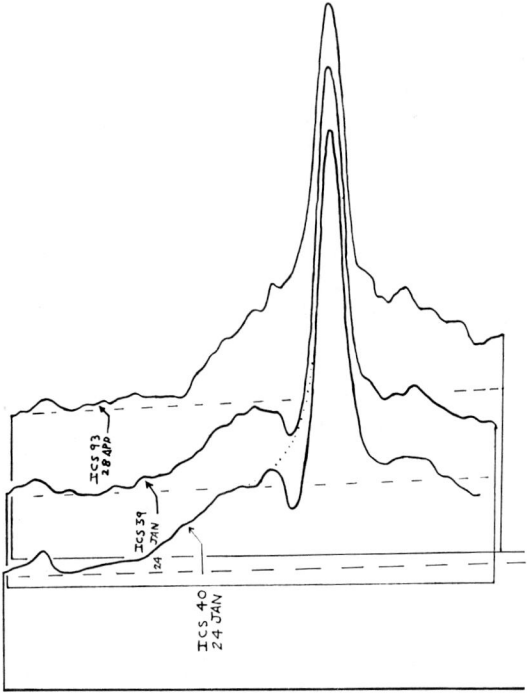

Fig. 2. Profiles of Hβ in the Seyfert galaxy NGC 4151. The bottom two profiles were taken on the same night (January 24) and show the displaced absorption core. The top profile was taken April 28, by which time, at our resolution, the absorption core had disappeared.

taken in January, and one in April. This displacement of the core from the line center is 1000 km s^{-1} so that during these three months the material would have moved only 10^{15} cm.

The interpretation is complicated by the fact that the nucleus was abnormally faint during December and January. (Babadzhanjanz *et al.*, 1970). This is also evident from the spectrograms at this epoch, which show a large number of faint emission lines normally undetectable. We consider now the following possibilities for the rapid change, in order to see whether this very small dimension has any physical significance.

(a) The feature is permanently present, but is normally masked by the non-thermal continuum;

(b) a distant small cloud passed in front of the nucleus;

(c) the optical depth in Hβ of an intervening cloud changed due to changes in excitation and ionization associated with the fluctuation in brightness of the nucleus;

(d) a shell of gas was expelled from the nucleus.

We cannot exclude (a) with certainty, but feel it is not likely for the following reasons: The absorbing material would have to be seen projected against only the broad emission wings, rather than the continuum, requiring the continuum and broad emission to arise from quite different regions of space. Moreover, the limited material at our disposal suggests that the ratio of the continuum intensity to the broad emission wing intensity remained roughly constant during December–April, so that 'filling in' would not have occurred anyway. Possibility (b) is highly unlikely: the cloud would have to be of the order 10^{15} cm and very dense, but ionized, to produce the required optical depth. In order not to have the passage of such a cloud across the nucleus be exceedingly unlikely, there must be enormous numbers of such clouds, in which case the emission from them would exceed by many orders of magnitude that actually observed.

Possibilities (c) and (d) have been considered together by solving the ionization equilibrium and statistical equilibrium equations governing the population of the $2s$ and $2p$ states in hydrogen and the $2s^3 S$ state of neutral helium. We *assume* photoionization from a point source with a power law extrapolation of the visible radiation into the far UV. The details will be published elsewhere, but the main conclusions are as follows:

(1) Of the four processes considered for controlling the rate at which electrons leave the $2s$ and $2p$ states (namely: two photon decay from $2s$; $2s$–$2p$ collisions followed immediately by Ly-α escape; $2s$–sp equilibrium with occasional Lγ-α escape; photoionization from the $n=2$ level) we find that only the third leads to acceptable time scales, and suggests a thin shell of radius $R=10^{16}$ cm, density 10^{10} cm^{-3} during the outburst. For a thin, expanding shell it can be shown in this case that the optical depth in the Balmer lines varies roughly as R^{-8}.

(2) An *increase* in the optical depth of Hβ in a shell due to a *decrease* in the intensity of the ionizing flux over a three month period is plausible only in the Lyman-α escape regime described above. In particular, for lower values of N_e and slightly larger values of R where the rate of escape of electrons from the $2s$ state is governed by the rate of $2s$–sp collisions, increasing the ionizing flux does not affect the optical depth of Hβ.

(3) A slab with $N_e \sim 10^{10}$ cm^{-3}, $R \sim 10^{16}$ cm and thickness such that $\tau(H\beta) \sim 1$ would have $\tau(3889) \ll 1$.

(4) Material which accumulates in a fairly narrow strip characterized by $N_e \sim 10^7$, $R \sim 3 \times 10^{17}$ cm down to $N_e \sim 10^4$, $R \sim 10^{19}$ cm can have $\tau(3889) \sim 1$ with $\tau(H\beta) \sim \frac{1}{10}$.

A model for this spectroscopic behavior which seems not grossly to contradict the observed facts thus involves the frequent expulsion of shells or filaments, each of which produces a short lived displaced Hβ core, but whose cumulative effect is to produce a relatively more stable He I 3889 displaced core. The mass loss rate associated with this picture is probably intermediate between that suggested by Anderson and Kraft and by Cromwell and Weymann.

References

Anderson, K.: 1970, *Astrophys. J.* **162**, 743.
Anderson, K. and Kraft, R. P.: 1969, *Astrophys. J.* **158**, 859.
Babadzhanjanz, M. K., Hagen-Thorn, V. A., and Ljutyi, V. M.: 1970, *Astron. Cirk. Izdav. Bjuro Astron. Soobshch. Acad. Sci. U.S.S.R.*, No. 544.
Cromwell, R. and Weymann, R.: 1970, *Astrophys. J. Letters* **159**, L147.
Dibai, E. A., Yesipov, V. F., and Pronik, V. I.: 1968, *Soviet Astron.* **11**, 553.
Mathis, J.: 1970, *Astrophys. J.* **162**, 761.
Weymann, R.: 1970, *Astrophys. J.* **160**, 31.

PHYSICAL CONDITIONS IN SEYFERT TYPE GALAXIES MARKARIAN 9, 10 AND 42

E. YE. KHACHIKIAN

Byurakan Astrophysical Observatory, Armenia, U.S.S.R.

I would like to present some new data concerning the Seyfert type galaxies Markarian 9, 10 and 42 (Khachikian, 1968; Arp *et al.*, 1968; Khachikian and Weedman, 1969).

A. MARK. 9

The large scale photograph of this galaxy shows that it consists of three parts: (a) A very bright and compact nucleus which is a little elongated and has a size 4".5 by 5"; (b) A very bright ring the plane of which is perpendicular to the direction of elongation of the nucleus. The diameter of the ring is about 11".6; (c) A faint diffuse envelope is elongated in the same direction as the nucleus and surrounds both (*a*) and (*b*) parts. The envelope diameter is approximately 20".

B. MARK. 10

This is an Sb spiral with a very bright and star-like nucleus. The nucleus is a little elongated with size 5".1 by 4".4. Mark. 10 is a giant galaxy with diameter ~ 55 kpc.

C. MARK. 42

The galaxy has a condensed nucleus with nebulous envelope. We have no large scale photograph of this galaxy.

Figure 1 shows slit spectra tracings of Mark. 9, 10 and 42 (Cassegrain image-tube spectrograph of 200-in. Hale telescope, dispersion 85 Å/mm). The relative line-intensities and equivalent widths of identified lines of these galaxies are shown in Table I.

Photometric data for these galaxies are listed in Table II (Weedman and Khachikian, 1968 a, b).

Conclusion

As seen from Figure 1 the profiles of the Balmer lines in Mark. 9 and 10 are very smooth without sharp cores superposed as in NGC 4151 (Oke and Sargent, 1968). This fact is incompatible with a two-zone model of the nuclei of Seyfert galaxies (Dibai and Pronik, 1968). At the same time if the broadening mechanism of Balmer lines is due to Doppler effect it is difficult to imagine how both hydrogen and forbidden lines can arise in the same region because of large differences between their line-widths.

It is also seen from Figure 1 that the H-lines of Mark. 9 and 10 have components shifted toward short wave-lengths by 0.005 λ and 0.011 λ respectively, which means that in the nuclei of these galaxies there possibly took place an asymmetrical ejection

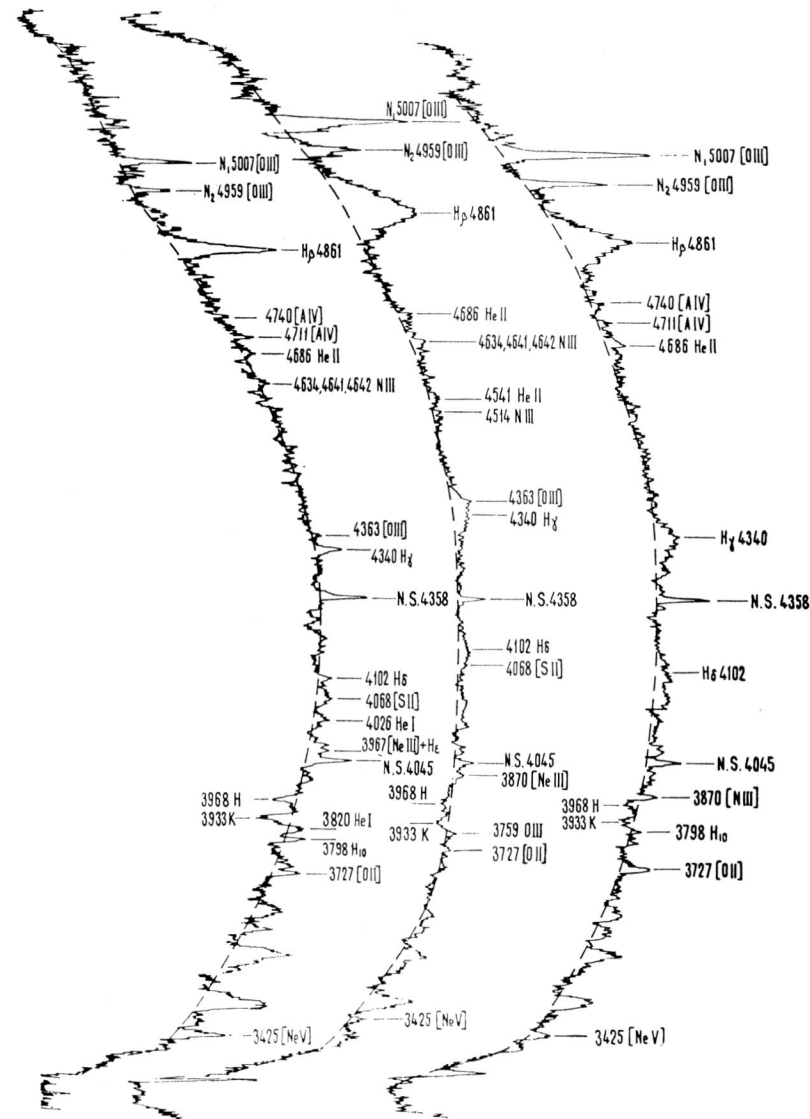

Fig. 1. Slit spectrum tracings of Mark. 10, 9 and 42 (Cassegrain image-tube spectrograph of 200-in. Hale telescope, dispersion 85 Å mm^{-1}) (from the right to the left).

of gas with radial veolocities of about 1500 km s^{-1} and 3000 km s^{-1} repectively.

As for Mark. 42 it seems that the Hβ line of this galaxy has broad wings with a very strong central peak.

There are no absorption line features in the spectra of these galaxies except the H and K lines of Ca II which are not shifted and have Galactic origin.

A more detailed discussion of these galaxies will be published elsewhere.

TABLE I

Ion	λ_0	Mark. 9		Mark. 10		Mark. 42	
		$I/I_{H\beta}$	W_λ	$I/I_{H\beta}$	W_λ	$I/I_{H\beta}$	W_λ
[OIII]	5007	6.1	18.3	10.3	25.2	3.5	3.7
[OIII]	4959	2.1	6.0	3.0	7.0	1.4	1.6
H_β	4861	10.0	37.6	10.0	33.7	10.0	16.5
[AIV]	4740			0.5	1.4		
[AIV]	4711			0.4	1.2	1.0	1.1
HeII	4686	0.6	2.2	1.3	3.8	1.5	1.8
CIII	4668			1.0	3.5		
NIII	4634, 4641, 4642	0.7	2.3				
HeII	4541	0.5	1.8				
NIII	4514	0.6	2.0				
[OIII]	4363	0.8	2.8				
$H\gamma$	4340	3.8	13.1	3.8	12.2	2.9	4.0
$H\delta$	4102	2.1	9.0	1.9	7.2	1.2	1.2
[SII]	4076					1.7	2.0
[SII]	4068	0.4	1.6				
HeI	4026					1.4	1.7
$H\varepsilon$	3970					1.2	1.6
[NeIII]	3968						
[NeIII]	3870	0.7	2.2	0.4	1.6		
HeI	3820					1.8	2.2
H_{10}	3798			0.3	1.1		
H_{11}	3771			0.2	0.7		
OIII	3759	0.4	1.3				
[OII]	3727	0.2	0.7	1.4	4.0	1.3	3.2
[NeV]	3425	0.9	1.1	1.5	1.8		

TABLE II

Mark. No.	B	B−V	U−B	z	M_B	R_{Mpc}
9	14.77	+0.40	−0.68	0.039	−21.2	156
10	14.71	+0.47	−0.70	0.029	−20.7	116
42	16.24	+0.79	−0.19	0.024	−18.8	96

References

Arp, H. C., Khachikian, E. Ye, Lynds, C. R., Weedman, D. W.: 1968, *Astrophys. J. Letters* **152**, L103.
Dibai, E. and Pronik, V.: 1968, in M. Arakeljan (ed.), 'Non-Stable Phenomena in Galaxies', *IAU Symp.* **29**, 83.
Khachikian, E. Ye.: 1968, *Astron. J.* **73**, 891.
Khachikian, E. Ye. and Weedman, D. W.: 1969, *Astron. Cirk. Izdav. Bjuro. Astron. Soobshch. Acad. Sci. U.S.S.R.* No. 506.
Oke, J. B. and Sargent, W. L. W.: 1968, *Astron. J.* **73**, 894.
Weedman, D. W. and Khachikian, E.: 1968a, *Astrofizika* **4**, 587.
Weedman, D. W. and Khachikian, E.: 1968b, *Astrofizika* **5**, 113.

INFRARED RADIATION FROM COMPACT OBJECTS

G. NEUGEBAUER

California Institute of Technology, Pasadena, Calif., U.S.A.

Abstract. Current infrared observations of Seyfert galaxies, QSOs, and compact galaxies in the lists of Markarian and Zwicky have been reviewed. The bright Seyfert galaxies generally show similar infrared excesses at the longer wavelengths accessible from the ground. Only NGC 1068 has been observed, by F. J. Low, in the 100μ region; if the spectral distributions of all Seyfert galaxies are similar to that of NGC 1068, the intrinsic luminosities vary from 1 to 100×10^{44} erg s^{-1}. Measurements by Kleinmann and Low, Gillett and Stein, Pacholczyk, and Penston and Neugebauer at 10 and 2μ apparently confirm variability of NGC 1068 and NGC 4151 on a time scale too short to allow the infrared radiation to come predominantly from dust shells.

The published infrared observations of QSOs have been limited to wavelengths shorter than 3.5μ except for 3C 273. The energy distributions either show a power law fall-off, a flat spectrum, or a combination of these two. Those four QSOs which show large variability all have steep power law spectra. If one accepts that the distance of QSOs are cosmological, the extrapolated luminosity at 2μ of several QSOs exceeds that of 3C 273. The integrated luminosity of 3C 273 depends critically on the spectrum in the unobserved $10-1000\mu$ region, but is probably in the range 10^{47} to 10^{48} erg s^{-1}.

The infrared observations of the compact Markarian and Zwicky objects show a correlation between the infrared excess and the broad optical emission lines. Presumably the infrared can be used to differentiate between the stellar and non-thermal components of the radiation from these objects.

THE RELATION BETWEEN THE OPTICAL AND CENTRIMETRIC POLARIZED EMISSION IN BL LAC AND OTHER QSOs

T. D. KINMAN

*Kitt Peak National Observatory, Tucson, Ariz., U.S.A.**

Abstract. The polarized optical and centrimetric fluxes from BL Lac have their maxima at position angles which are roughly orthogonal. A similar tendency is found in other QSOs. If these fluxes are of synchrotron radiation then the magnetic fields associated with the optical and centrimetric radiation are likely to be highly inclined with respect to each other. The implications of this are discussed.

Discussion

Lasker: Have power spectra been computed for your results, and if so, what is the logarithmic slope?

Kinman: No.

---

* Operated by the Association of Universities for Research in Astronomy, Inc., under contract with the National Science Foundation.

ACTIVITY IN THE NUCLEI OF SEYFERT GALAXIES IN THE VISUAL AND INFRARED

A. G. PACHOLCZYK

Steward Observatory, Tucson, Ariz., U.S.A.

Abstract. This paper reports (1) a possible 5.1 yr period in the optical variations of the nucleus of NGC 4151; and (2) variations with a time-scale of a few days in the 2.2 micron flux density of NGC 1068, implying that the size of the infrared emitting region is not larger than a few times 10^{15} cm.

I would like to communicate two results of photometric investigations, concerning the activity in the nuclei of the Seyfert galaxies.

First is the possibility of presence of a 5.1 yr period in the optical variations of the nucleus of NGC 4151. This possibility is not incompatible with the analysis of data

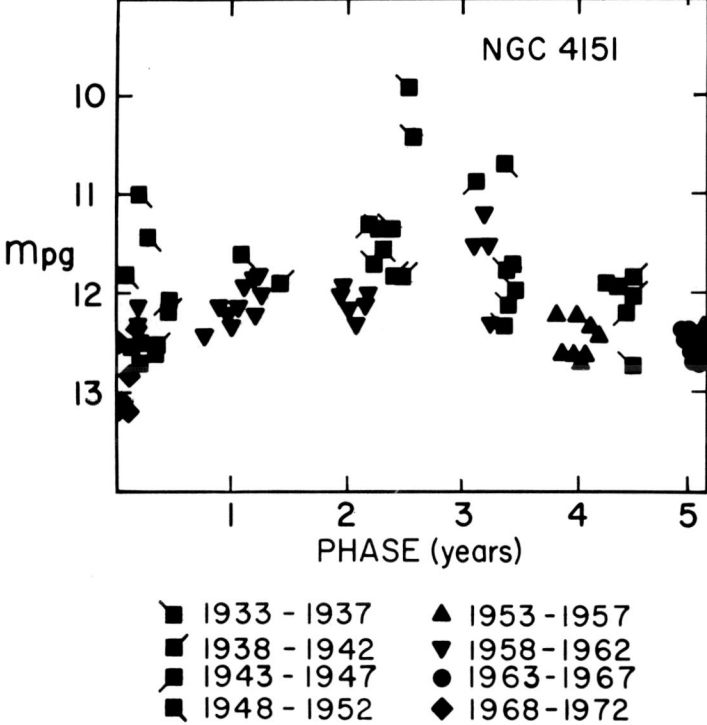

Fig. 1. The photographic magnitude of NGC 4151 as a function of phase within a 5.1 yr period.

from Harvard patrol plates taken during the years 1932–1952 and from Steward plates taken between the years 1956 and 1968. Detailed description of the photometric data, procedure, errors and results is given in the paper by Pacholczyk (1971). Figure 1 represents the photographic magnitude of NGC 4151 as a function of phase within a 5.1 yr period. Note a fairly well defined lower envelope of points. The amplitude of this long term change is about 1.5 mag. On this long term variation are superimposed short term changes with a similar amplitude but a time-scale of months (observed photoelectrically since 1967). The available data, summarized in Figure 1, are, of course, insufficient to make a definite statement about periodicity, and any more sophisticated statistical analysis of those data will not be meaningful at the present stage except, perhaps, for pointing out that the average magnitude during half-cycles between the minima is 11.64 ± 0.12 while the magnitude during the half-cycles around the minima is 12.34 ± 0.05. The Harvard plates used for measurements reported in Figure 1 were exposed $1-\frac{1}{2}$ h or more and contained a star-like image of NGC 4151 well above the plate limit. Efforts are being currently made to extend the measurements to shorter exposure plates, and to plates of other than RH series, which are abundant in the Harvard collection, and which will permit an analysis of the variability as far back as 1898.

The 5.1 yr period might be a manifestation of the presence of a rotating single massive object in the central region of the nucleus of NGC 4151. The possibility of the existence of 'spinars' in quasi-stellar objects and galactic nuclei was pointed out by Morrison (1969), Woltjer (1970), Cavaliere *et al.* (1969) while 'magnetoids' were invoked by Ozernoy (1966) and Piddington (1970). The object in NGC 4151, with a mass of the order of $2 \times 10^9 \, M_\odot$, rotating with the angular velocity Ω of 4×10^{-8} s, would have a radius larger than the Schwarzschild radius of 5×10^{13} cm but smaller than $\Omega/c = 8 \times 10^{17}$ cm. The emission of variable optical radiation is clearly not in the form of pulses, as in the quasistellar source 3C 345, but exhibits secondary changes with a time-scale of months due to some mechanism not necessarily connected with the rotation.

The second result is the presence of infrared variations of flux at 2.2 microns (K photometric system) of the Seyfert galaxy NGC 1068 with a time-scale of a few days, implying that the size of the infrared emitting region is not larger than a few times 10^{15} cm. Figure 2 summarizes the observations; Figure 3 describes the magnitude of photometric errors involved by displaying standard deviations of an observation of a comparison star and of NGC 1068. The deviations for NGC 1068 are significantly higher than if they were solely due to photometric errors. A detailed description of the photometric data and procedures, as well as a discussion of the accuracy of

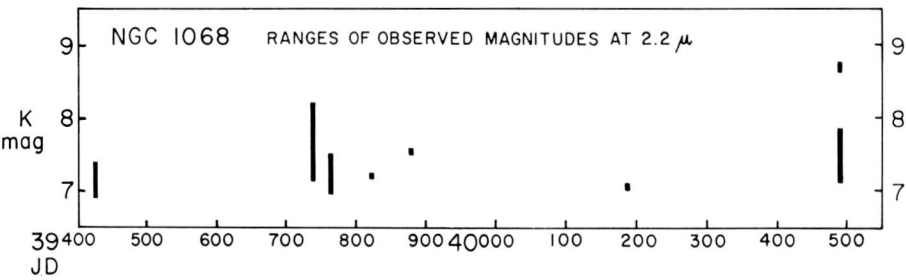

Fig. 2. Ranges of magnitudes of NGC 1068 at 2.2μ. Time is expressed in Julian days.

measurements and of differential extinction and other effects possibly contributing to the large scatter of data for NGC 1068, is contained in the paper by Pacholczyk (1970). A discussion of the role of the diaphragm centering errors, based on recent data on the distribution of brightness at 2.2μ in NGC 1068 obtained by G. Neugebauer, is given in the Appendix. Observations of NGC 4151 do not rule out possible variability of the infrared flux at 2.2μ (Figure 3); it is however clear that observations with more sensitive instrumentation are needed definitely to establish the variability of NGC 4151.

The limit on the size of the emitting region in NGC 1068, imposed by infrared variability, is incompatible with the interpretation that the observed infrared radiation is emitted by dust grains which absorb energy from a central compact optical or ultraviolet source. The 2.2μ emitting region must be at a distance of the order of 10^{18} cm from the central source and have a size of the same order of magnitude for any type of dust grains with an absorption efficiency not far below unity at ultraviolet and optical frequencies. A homogeneous synchrotron model, or even more, a model consisting of a large number of synchrotron sources ('irtrons'), cannot reconcile an emitting region as small as a few times 10^{15} cm with the lack of a self-absorption feature observed up to 100μ without incurring inverse Compton losses substantially exceeding the electron energy losses due to synchrotron radiation. The infrared characteristics of NGC 1068 could be explained in terms of a non-uniform synchrotron model with the field decreasing outwards, somewhat similar to that discussed by Rees and Sciama (1966) for 3C 273. The infrared radiation of a given wavelength would originate predominantly in a region at a certain distance from the center of a spherically-symmetric model; at smaller distances the source would be optically thick

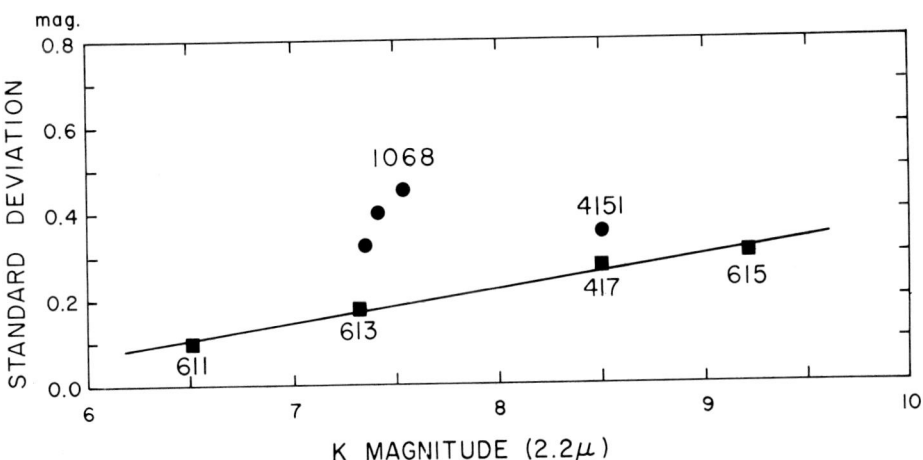

Fig. 3. Standard deviation from mean value of a single observation consisting of 20 differences between measurements of 'star' and of 'sky', each measurement employing an integration time of 15 sec. Squares represent observations of comparison stars of various magnitudes at 2.2μ. Dots refer to measurements of the Seyfert galaxies NGC 1068 (the lower dot refers to all 33 observations of NGC 1068, the upper one represents observations on JD 39737, and the middle observations on JD 40487–90) and NGC 4151.

to radiation of that and longer wavelength; radiation of longer wavelength would be emitted mainly at larger distances. The details of the model are being computed by W. G. Fogarty in his dissertation.

Acknowledgement

The work reported here was supported by the National Science Foundation under grant GP-9616.

References

Cavaliere, A., Pacini, F., and Setti, G.: 1969, *Astrophys. Letters* **4**, 103.
Morrison, P.: 1969, *Astrophys. J. Letters* **157**, L73.
Ozernoy, L. M.: 1966, *Astron. Zh.* **43**, 300.
Pacholczyk, A. G.: 1970, *Astrophys. J. Letters* **161**, L207.
Pacholczyk, A. G.: 1971, *Astrophys. J.* **163**, 449.
Piddington, J. H.: 1970, *Monthly Notices Roy. Astron. Soc.* **148**, 131.
Rees, M. J. and Sciama, D. W.: 1966, *Nature* **211**, 805.
Woltjer, L.: 1970, *Columbia Astrophys. Lab. Contr.* No. 24.

Appendix

Diaphragm Centering Errors in K Photometry of NGC 1068

Dr G. Neugebauer kindly communicated the results of his observations of NGC 1068 at 2.2μ with diaphragms of sizes ranging from 2″ to 20″ obtained in December 1969.

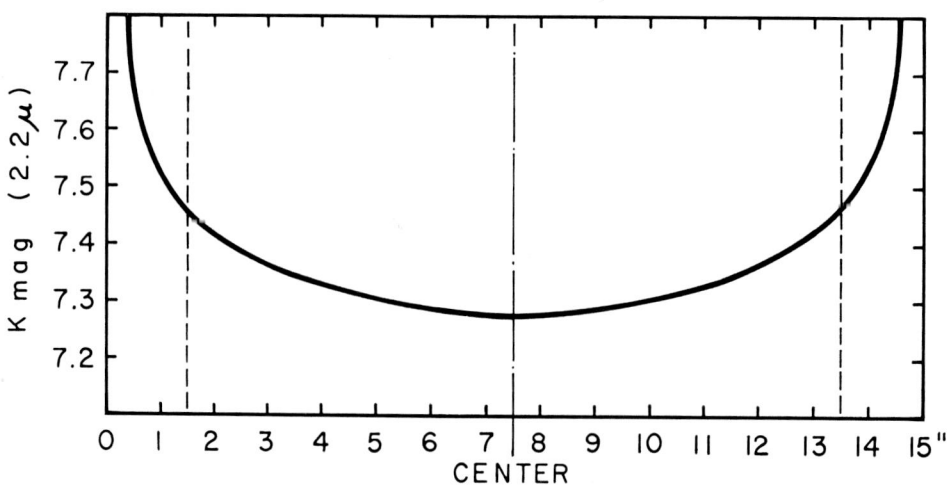

Fig. 4. The K magnitude of NGC 1068 as a function of the position of the center of the galaxy along the diameter of a circular diaphragm of 15″. The difference in the K magnitudes for any position between the dotted lines and the central position is smaller than the total photometric error of measurement of the K magnitude of a point-like object (star) of comparable magnitude.

His data permitted us to calculate the photometric errors due to improper centering of this object within the photometer diaphragm. In the paper 'Infrared Variability of the Seyfert Galaxy NGC 1068' (Pacholczyk, 1970) variations in the 2.2μ flux from this object are reported; the measurements were done with a 15" diaphragm. Figure 4 represents the K magnitude (2.2μ) of NGC 1068 as a function of the position of the center of the galaxy along the diameter of a circular diaphragm of size of 15", computed by Mr. L. Thompson on the basis of Dr. Neugebauer's results (the K magnitude of NGC 1068 decreased more or less linearly with the logarithm of the diaphragm size from 8.28 mag. at 2" to 7.12 mag. at 20"). It can be seen from Figure 4 that for any position of the center of NGC 1068 within a 12" circle concentric with the diaphragm, the difference in the K magnitude is smaller than the total photometric error of measurement of the K magnitude of a point-like object (star) of com-

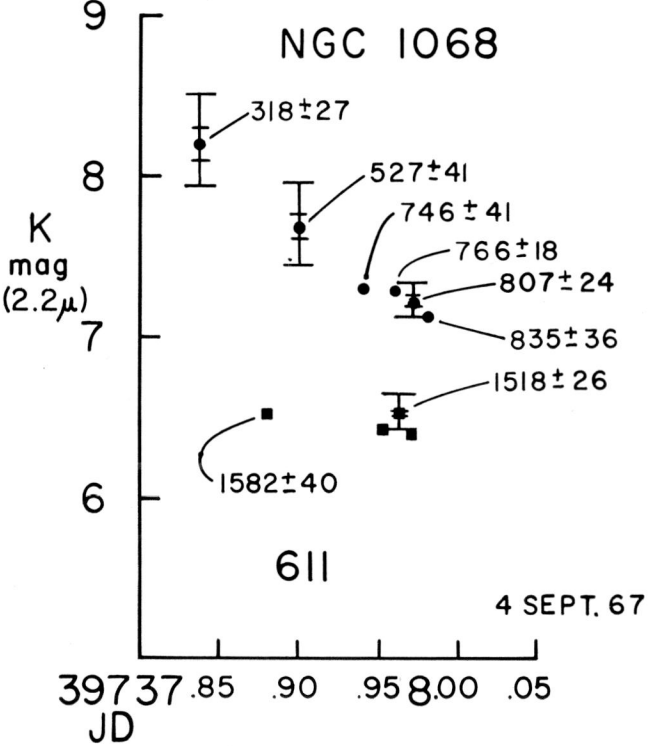

Fig. 5. Observations of NGC 1068 at 2.2μ on the night of September 4, 1967. The numbers are proportional to average voltages of the photometer signals and their probable errors. The error bars correspond to one (inner bars) and three (outer bars) times the total error of the photometric measurement (in magnitudes).

parable magnitude. Centering errors do not therefore affect significantly the measurements of NGC 1068 and by no means can be responsible for the observed variations of K magnitudes of this object.

Figure 5, representing an example of observations of NGC 1068 on the night of

September 4, 1967, confirms the conclusion that the centering errors are not significant. The numbers next to every point in Figure 5 are proportional to average voltages of the photometer signals (each point is an average of 20 differences of readings on 'star' and on 'sky') and their probable errors. The errors are not increasing with decreasing voltage, indicating that the lower average voltages of the upper points are not due to a larger scatter of individual readings (as it would be the case if centering errors were responsible for the lower average), but systematically to lower individual readings. The errors of voltage measurements for NGC 1068 are not larger than those for a point source (comparison star 611, squares in Figure 5). The error bars in Figure 5 represent one (inner bars) and three (outer bars) times the total error of the photometric measurements (in magnitudes, and therefore increasing with increasing magnitude of the object), and include contributions due to extinction and zero point determination.

Note added in proof. In a recent paper Neugebauer *et al.* (*Astrophys. J. Letters* **166**, L45, 1971) point out that, according to their measurements with diaphragm sizes ranging from 2" to 20", the 2.2 micron flux of NGC 1068 can be interpreted as a superposition of a point-like source and a diffuse component. They note that 56% of the 2.2 micron flux measured by them with a 15" diaphragm falls outside a 2" diaphragm and conclude that this is compatible with the variations of flux, observed with a 15" diaphragm, with an amplitude not larger than roughly a factor of two. This limit on amplitude is based on a small number of measurements and its validity rests entirely on the assumption that centering errors at small apertures are negligible. In fact, any errors in centering of such small apertures as 2" will tend to overestimate the percentage of flux in question and are much more likely to occur with a small diaphragm than with a 15" diaphragm, even under conditions of good seeing.

OPTICAL VARIABILITY OF TWENTY-TWO QUASI-STELLAR OBJECTS

R. J. ANGIONE* and H. J. SMITH

University of Texas, Austin, Texas, U.S.A.

Abstract. Light fluctuations have been found in all 22 QSOs studied by measurement of plates from the Harvard collection, which cover the last eighty years. The conclusions of this study are: (i) There appear to be at least three general classes of variation: (a) erratic, small-amplitude variations, (b) erratic, large-amplitude variations, and (c) slow quasi-periodic variations, e.g. as in 3C 273; (ii) No significant differences were detected between the rates of rise and decline of luminosity; (iii) Definite secular trends over at least 50 years were found in 5 QSOs; (iv) No simple clearcut periods greater than one year have been found; (v) There may be a trend of decreasing amplitude of fluctuations in apparent magnitude with increasing luminosity.

Twenty-five objects, of which 22 proved to be QSOs, were selected for historical photometric investigation, solely on the basis of their relatively bright apparent magnitudes (brighter than $m = 16.6$). The Harvard historical plate collection covering the last eighty years was searched for plates containing images of any of these objects; where possible, B magnitudes of the QSOs were measured with an iris photometer against photoelectrically-determined comparison star sequences in the immediate vicinity of each (Angione, 1970), or estimated visually against the sequence if iris photometry was not feasible on the particular plate. Various criteria establish the standard deviation of the photographic B magnitudes so determined as about 0.10 mag. An average of 74 plates per object was available.

These historical data can be used to ask at least six important questions concerning the behavior and nature of QSOs:

(1) What proportion of QSOs vary 'significantly' (by more than 0.1 mag.) over time scales ranging up to decades?

(2) Do the types of variations show any consistent differences or groupings?

(3) Do QSOs really 'flare'?

(4) Do QSOs show secular trends – individually or collectively?

(5) Is periodic behavior indicated?

(6) Do QSOs collectively demonstrate an amplitude-luminosity relation?

The remainder of this note presents brief summaries of our new information bearing on each of these questions. For detailed data and discussions of these results, see Angione (1970).

1. Percentage of QSOs Found to Vary

All 22 of the true QSOs in this sample were found to vary significantly over the 50–70 yr of available data. Standard deviations of individual QSO magnitudes about their means ranged from 0.13 to 0.62 mag. (3C 48 and 3C 273, both well-known variables,

---

* Now at San Diego State College.

had respectively $\sigma = 0.15$ and 0.21). After the completion of the data-taking we found that the original list of objects studied in the Harvard plate stacks contained two 'QSOs' (Ton 883 and 3C 68.2W) which were reported to be mis-identified stars, presumably constant in brightness. It is gratifying that these objects did prove to be

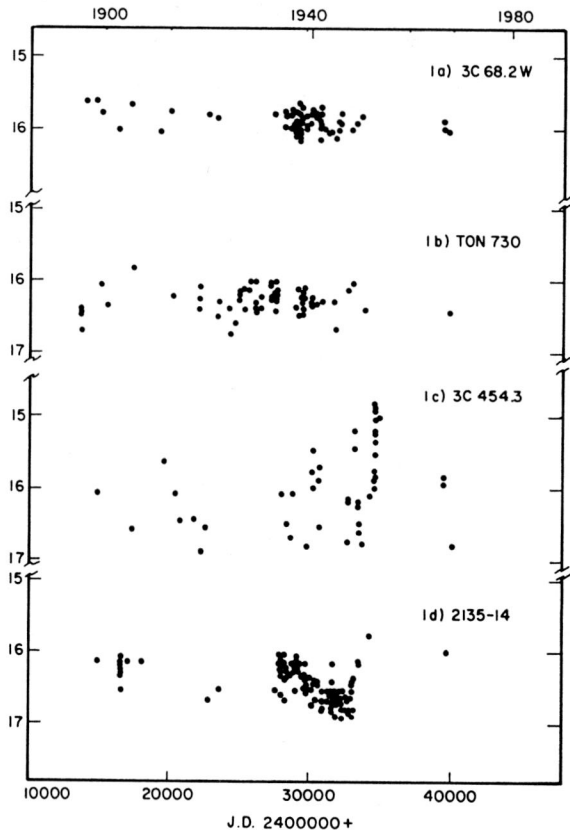

Fig. 1. Typical photometric data; 1a: A non-variable star, indicating the size and nature of the random observational errors in the material. 1b: A low-amplitude QSO. 1c: A large-amplitude 'violent' QSO. 1d: A 'coherently' variable QSO.

constant in the photometry of this program, with individual-measurement standard deviations of about 0.10 mag.; the 'light curve' of 3C 68.2W is shown in Figure 1a.

2. Types of Variations

There appear to be at least three general classes of variation, at least to the resolution available in these data: (a) relatively rapid, erratic, small-amplitude variations (Figure 1b); (b) rapid, erratic, large-amplitude variations (Figure 1c); (c) relatively slow, correlated, quasi-periodic variations like those of 3C 273 (Figure 1d), with more erratic behavior apparently superimposed. In our data there are no well-defined connecting links between classes a and b; the distribution may be truly bi-modal (see also Figure 4).

3. Reality of 'Flares'?

To test whether the rates of increase of brightness are significantly different from those of decline, the frequencies of observed changes by a given magnitude in a given interval of time were collected from the entire body of data. Figure 2 shows the

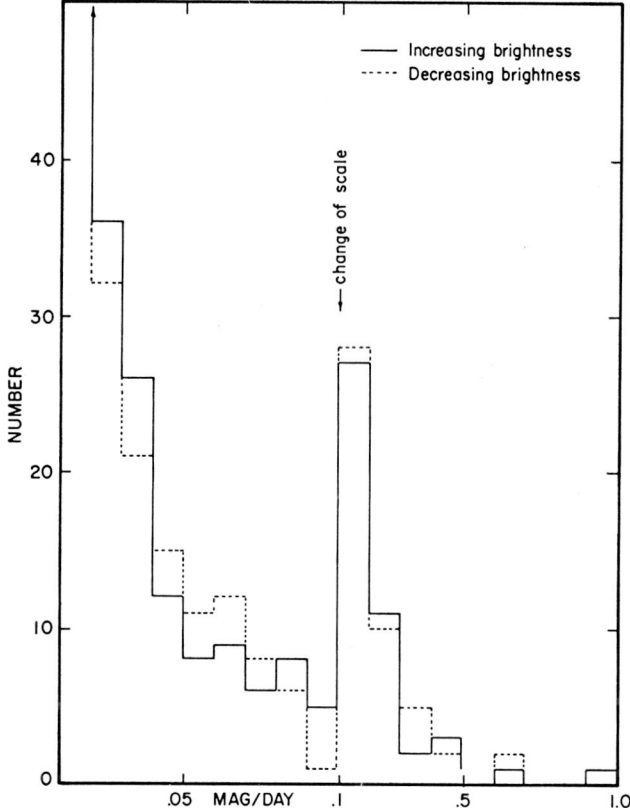

Fig. 2. Comparison of rates of increase and rates of decrease of QSO brightness.

interesting result that within the limits of the present measurements the QSOs display the same average frequency of rates of decline as rates of rise over all the ranges accessible to our tests. While many quasar models can undoubtedly be adjusted to fit such behavior, it is an expected aspect of at least two: a large rotating mass (galactic pulsar), or a gravitational lens effect.

4. Secular Trends

Five of the QSOs individually showed secular trends persisting for more than fifty years for which a linear least-squares fit indicated a 4σ or greater confidence level:

2135-14 and 3C 232 declining at rates respectively of 0.58 and 0.80 mag. per century; Ton 256, 0405-12, and 0237-23 brightening by 1.58, 1.04, and 0.64 mag. per century respectively. These numbers would be consistent with major changes in the behavior of about 25% of QSO cores over time scales of 10^3 yr; alternatively the time-scale of fluctuation about some average brightness may simply be longer than we have yet been able to observe.

One may also question whether the quasar phenomenon in general leads to an average detectable systematic increase or decrease in brightness over time-scales of the order of a century; such a systematic trend might be detected in the mean behavior of a large number of objects. The average slope for the entire group of QSOs corresponds to a brightening at about 0.12 mag. per century (slightly dependent on weighting factors used). Since a few high or low observations, especially in the early years when good plates were rare, can substantially affect this result, and since even a single case of a relatively steep slope can strongly bias the average, we do not believe that these results establish any general trend in the optical brightness of QSOs.

5. Periodic Behavior

An improved form of periodogram analysis by Dr. Terence J. Deeming of the University of Texas Astronomy Department was used to analyze the original data without

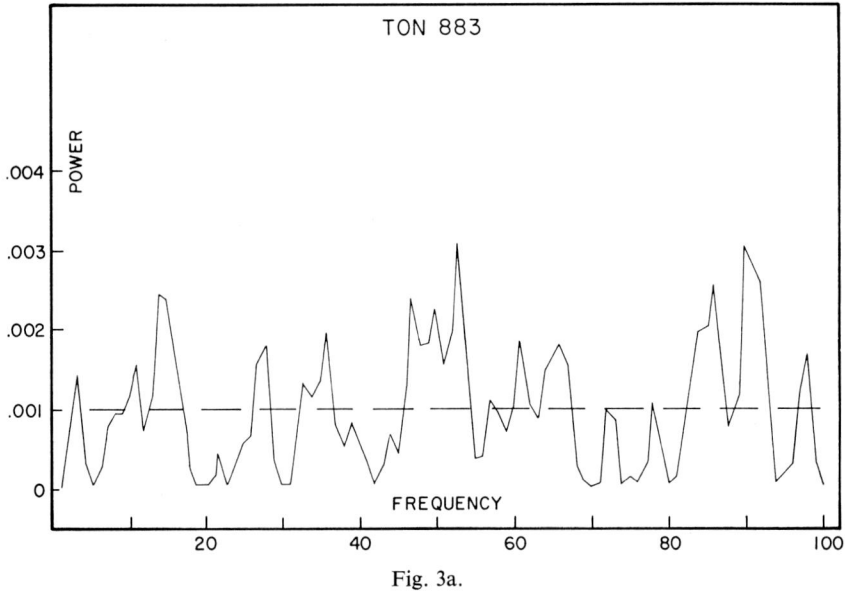

Fig. 3a.

smoothing to uniform intervals of time. Periodograms were formed for each of the light curves; as expected, random fluctuations are shown by the data for the stars inadvertently included in the program (e.g. Figure 3a). In three of the QSOs (3C 273, 2135-14, 0405-12; Figures 3b, 3c, 3d) about half of the spectral energy is concentrated

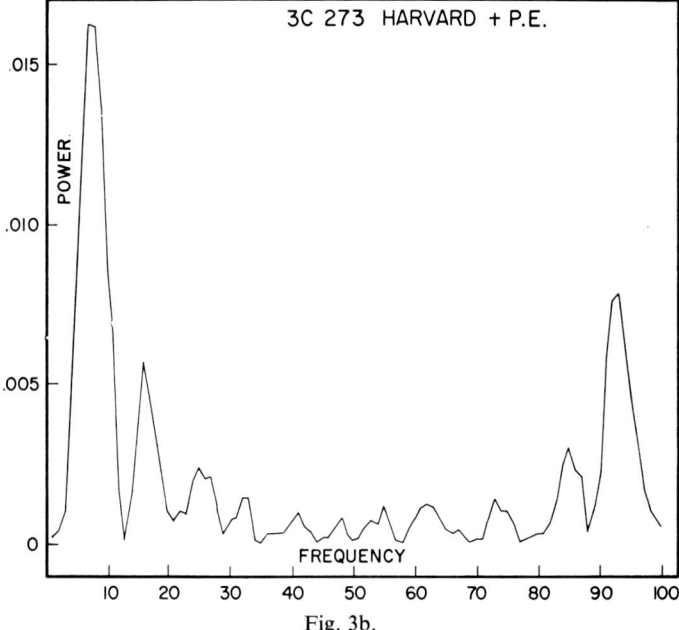

Fig. 3b.

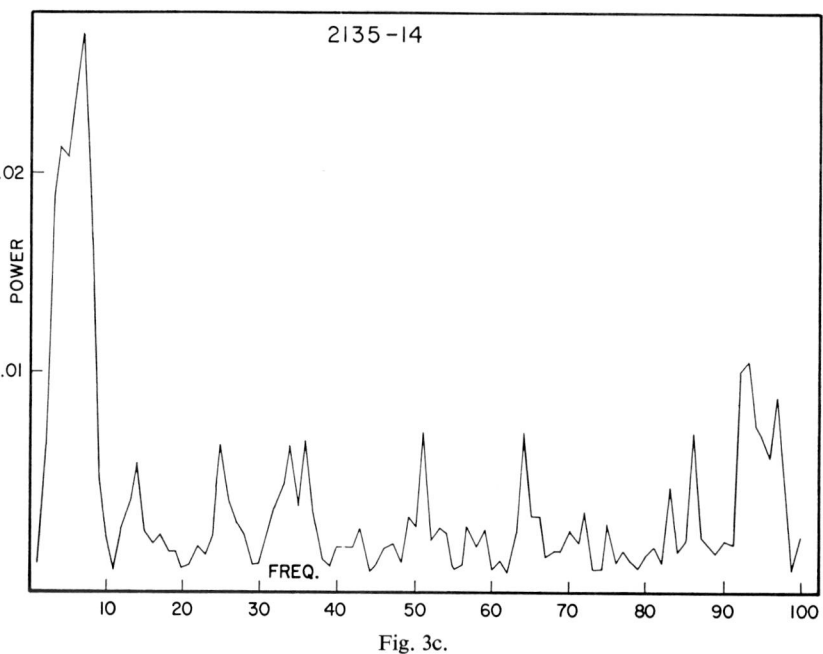

Fig. 3c.

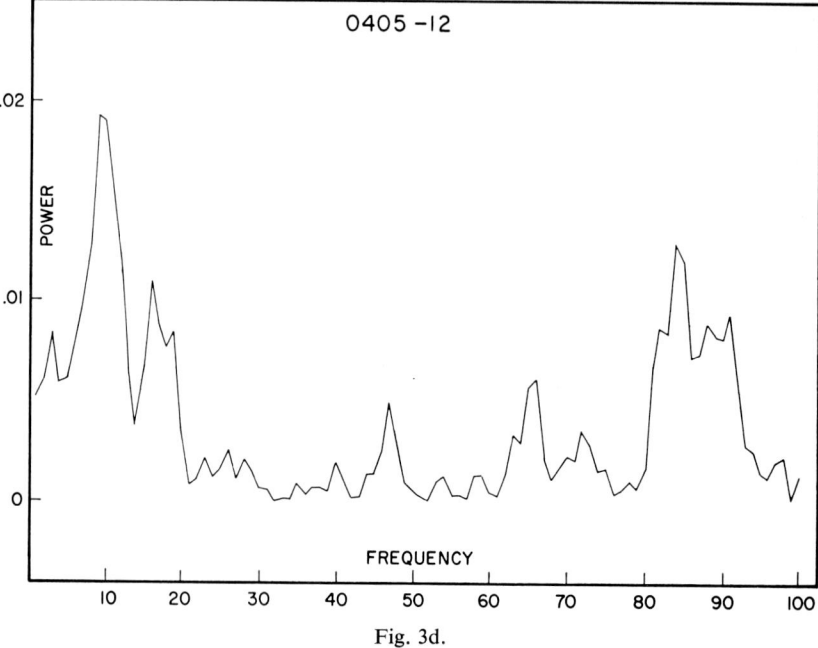

Fig. 3d.

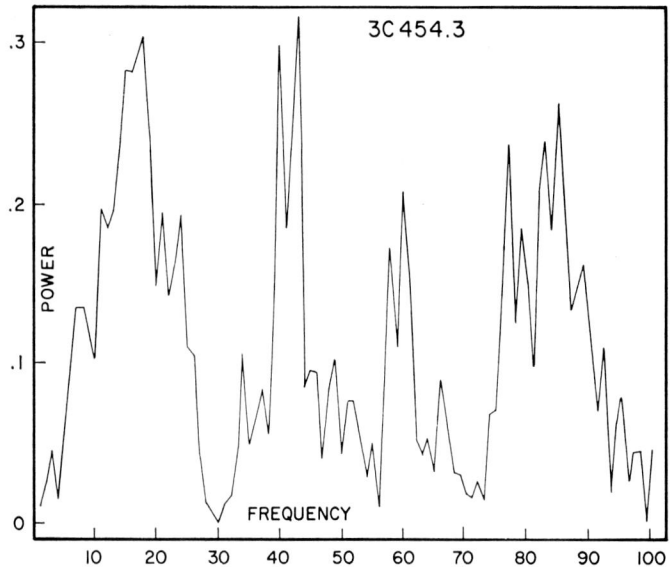

Fig. 3a–e. Typical power spectra. The abscissa frequency unit is 0.01 yr^{-1} (cycles per century). Each diagram thus covers periods from 100 yr to 1 yr. Long-term trends were first removed by fitting a straight line through the data. (a) A non-variable star Ton 883. (b), (c), (d) QSOs of the 3C 273 type. (e) A violent QSO, 3C 454.3.

in the interval around 11 to 13 yr. The figures also show that in each of these QSOs there is appreciable power at about 1.2 yr. 3C 273, for which the data are much the most extensive, appears to be the most nearly periodic object in the entire group of QSOs. The majority of the periodograms are noisy (e.g. 3C 454.3, Figure 3e), although with some suggestion of harmonics. In summary, no simple clearcut periods greater than one year have been found, but nearly all the QSOs show several regions of frequency in which there are substantial concentrations of variational power.

6. Amplitude-Luminosity Relation

Variability offers a test, informally suggested by J. Wheeler, whether individual quasars are fundamentally multiple in nature (e.g. collections of large numbers of randomly flaring objects such as supernovae), or are essentially single coherent objects. The criterion follows from a simple correlation of amplitude of variation vs absolute magnitude. On a cosmological interpretation of the redshifts, the absolute magnitudes

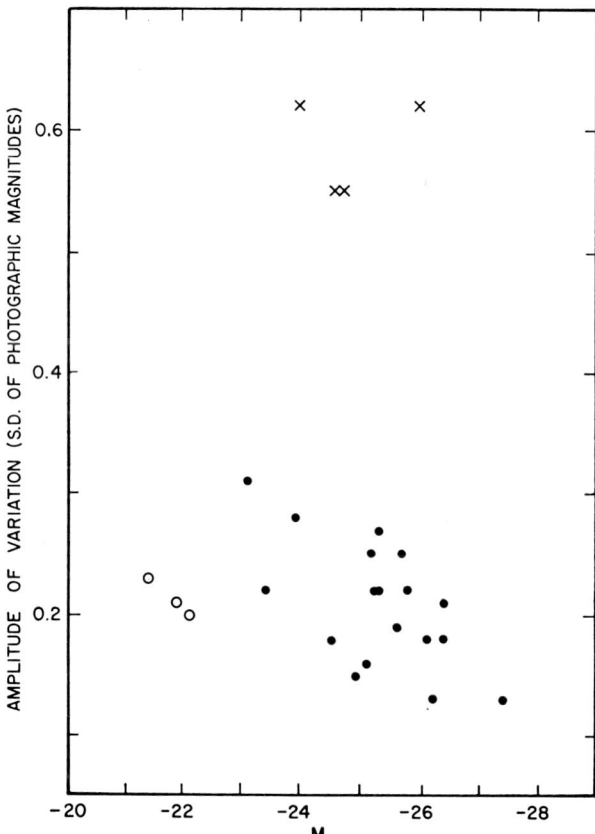

Fig. 4. Amplitude-luminosity relation for the objects in this study. Crosses denote the 'violent' QSOs, 3C 279, 345, 446 and 454.3. Open circles denote low redshift, low-luminosity objects probably of Seyfert or N-galaxy nature. Filled circles denote the remaining more ordinary QSOs.

of the QSOs follow from their apparent magnitudes and from an assumed Hubble constant. If the total luminosity of each QSO represents the sum of a number of separate randomly-occurring events of a relatively similar character throughout the universe (e.g. the explosion of very large stars as supernovae), then the fluctuations in integrated brightness must correlate inversely with the absolute luminosity. This follows since the fluctuation in brightness must be proportional to the inverse square root of the total number of (similar) sources contributing to the brightness at any time.

Here again the results are suggestive (Figure 4). The amplitude-luminosity relation of the QSOs shows them separated into at least two and perhaps three discrete groups: the four violent variables, 3C 279, 3C 345, 3C 446 and 3C 454.3, are in a class by themselves at the top. The three objects (3C 120, Ton 730 and Ton 1542) indicated in Figure 4 by open circles, most resemble Seyfert galaxies. Their small redshifts imply luminosities four magnitudes fainter than the regular QSOs. These last, indicated by filled circles, form a group which, albeit with much scatter, does suggest the slope predicted by Wheeler's argument for multiple-source models, although the fit would be improved on the assumption that the more luminous QSOs also tend to be composed of brighter individual events.

Reference

Angione, R. J.: 1970, Dissertation, Astronomy Department, University of Texas at Austin (June, 1970); in preparation for publication.

CAN THE OPTICAL FLUCTUATIONS OF 3C 273 BE RANDOM?*

JAMES TERRELL and KENNETH H. OLSEN

Los Alamos Scientific Laboratory, University of California, Los Alamos, N.M., U.S.A.

Abstract. The fluctuations in optical brightness of the quasi-stellar object 3C 273 have been investigated to determine whether the suggested periodicity of ~ 10 yr is supported by the observational data extending over 80 yr. New methods of obtaining information from the power spectrum have been used, and moments and trends have also been investigated. No statistically-supportable evidence has been found in the power spectrum for such a non-random variation. The moments and trends are consistent with random fluctuations. If the observed fluctuations are of shot-noise character, due to random outbursts of light, the individual pulses must occur at the rate of 15 ± 5 per year and have an average effective length of 3.2 ± 1 yr. These conclusions were verified by computer-generation and power-spectrum analysis of such random signals. Thus any periodic variation in the brightness of 3C 273, if present, is completely obscured by random fluctuations. The power spectrum, moments, and trend are all consistent with random but long-lasting outbursts of light.

Although 3C 273 was not discovered to be a quasi-stellar object until 1963, we fortunately have an 80-yr record of its optical brightness. It is bright enough at thirteenth magnitude to show up on old sky survey plates extending back to 1887. Harlan Smith and Dorrit Hoffleit (1963), examining these plates, found that 3C 273 has been fluctuating in brightness by considerable amounts.

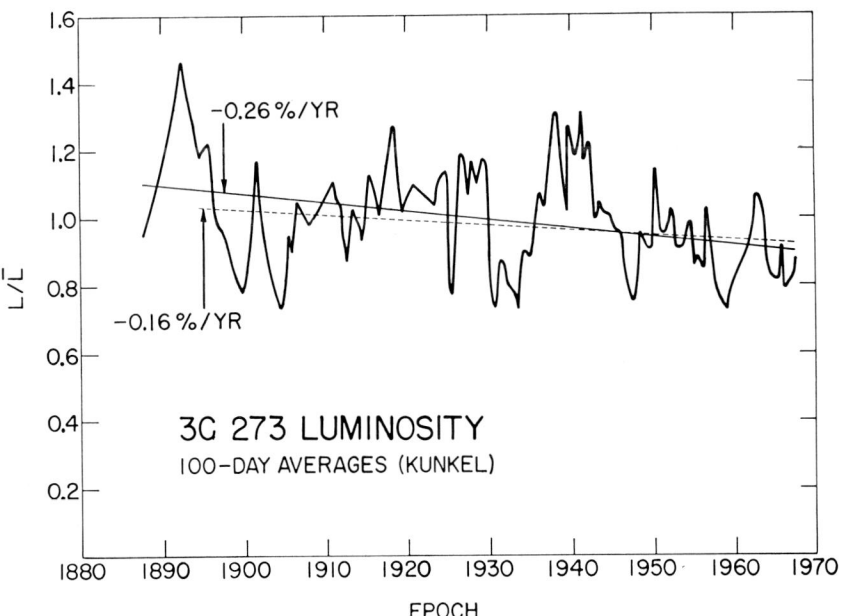

Fig. 1. Luminosity data for 3C 273, relative to the average luminosity, $\bar{L}$, based on photographic magnitudes averaged over 100-day intervals by Kunkel (1967). Least squares linear trends are shown, both for the full data (1887–1967) and for the truncated data (1894–1967).

* Work performed under the auspices of the U.S. Atomic Energy Commission.

As seen in Figure 1, its brightness has continually varied by 10%, 20%, or more, occasionally doubling in brightness in a few days, sometimes dimming just as suddenly. There is also a trend toward lower luminosity, amounting to a decrease of 26% per century. If the first 2500 days of data are omitted as of little reliability, being more interpolation than data, this trend is reduced to 16% per century. The data shown here are as averaged by Kunkel (1967) over 100-day intervals.

Smith (1965) found indications of a 13-yr period, more or less sinusoidal, in this record. Gudzenko *et al.* (1967) analyzed the data and found a 9- or 10-yr period, with 99% certainty (or, more recently, 97%). Manwell and Simon (1966) found no evidence of periodicity and stated that the fluctuations could be reproduced by the superposition of about 25 very short outbursts of light per year. Kunkel found possible evidence of a fundamental 13-yr cycle, plus several harmonics.

Thus the question whether the light from 3C 273 has a fundamental fluctuation period, or is due to the overlap of random unrelated outbursts, has been a controversial one, and the result obviously has a bearing on models of quasars. Ken Olsen and I undertook to find the most definite answer possible, using the techniques of moment and trend analysis and power spectra.

Figure 2 shows the result of a straightforward determination of the power spectrum of these light fluctuations. The power spectrum is essentially the square of the coefficients of a Fourier series representation of the data. It represents the fluctuation power as a function of frequency – extending in this case from zero to $\frac{1}{2}$ cycle per year.

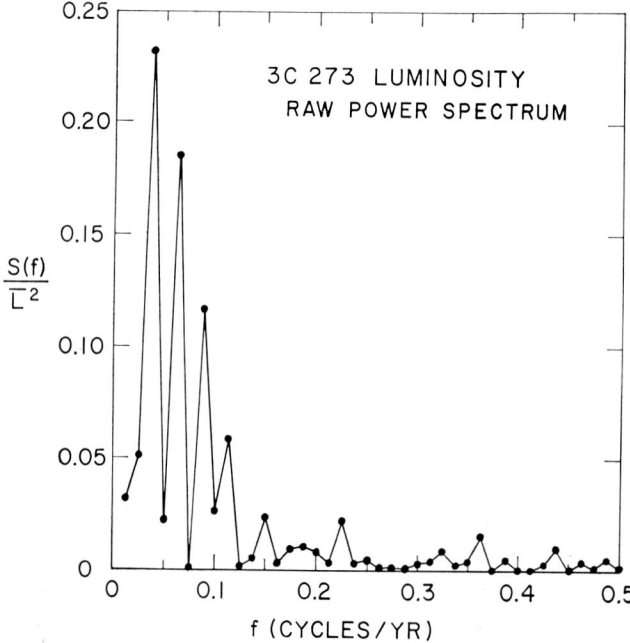

Fig. 2. Power spectrum for 3C 273 luminosity, 1887–1967, based on Kunkel's averages, not detrended. The spectrum is raw, i.e., not smoothed to reduce random fluctuations.

In the raw, unsmoothed form seen here the points fluctuate considerably, being uncertain by a standard deviation of ±100%. Thus these many narrow peaks are not proof of many resonances, but of the lack of smoothing.

Figure 3 shows the result when a standard form of smoothing, due to Parzen, is applied to the power spectrum. The spectrum is much smoother. There still appear

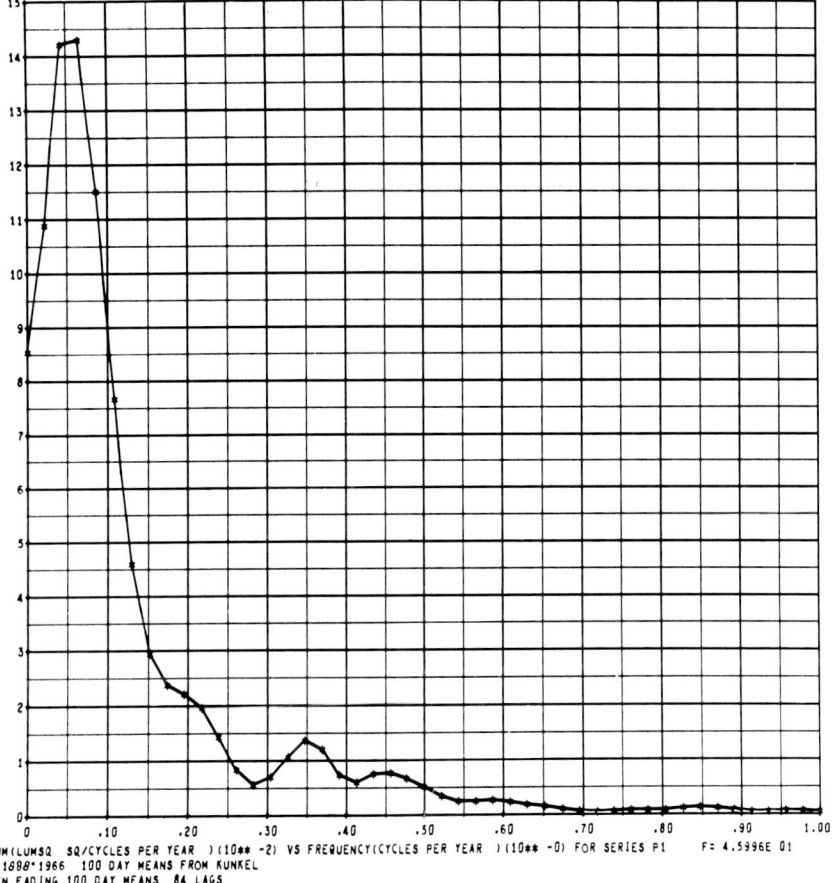

Fig. 3. The power spectrum of 3C 273 smoothed by the standard method of Parzen.

to be a number of weak resonances – but each peak has a standard deviation of ±40%. There is thus no real evidence for resonances or harmonics in the spectrum, except perhaps the peak near a frequency of 0.07 – about a 14-yr period. However, the part of the power spectrum nearest to zero has unfortunately been reduced in height by the standard methods of analysis, which involve removing the average, removing the trend, and Parzen smoothing. Removal of the average reduces the zero-frequency point to zero, and smoothing then averages this value into the lowest frequency points.

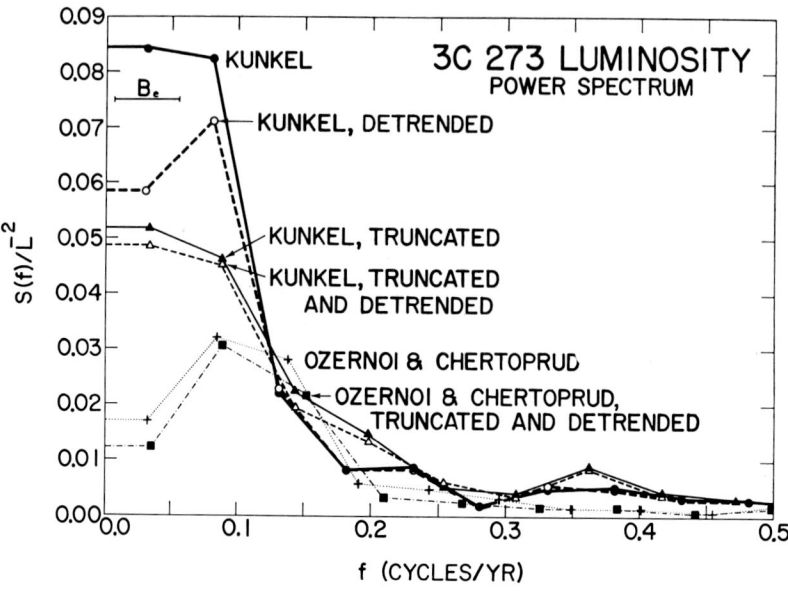

Fig. 4. Power spectra for 3C 273 luminosity for various forms of the data. The points shown have been smoothed over four Fourier harmonics, giving the effective bandwidth B_e shown (for Kunkel's full data), and standard deviations of $\pm 50\%$ for individual estimates.

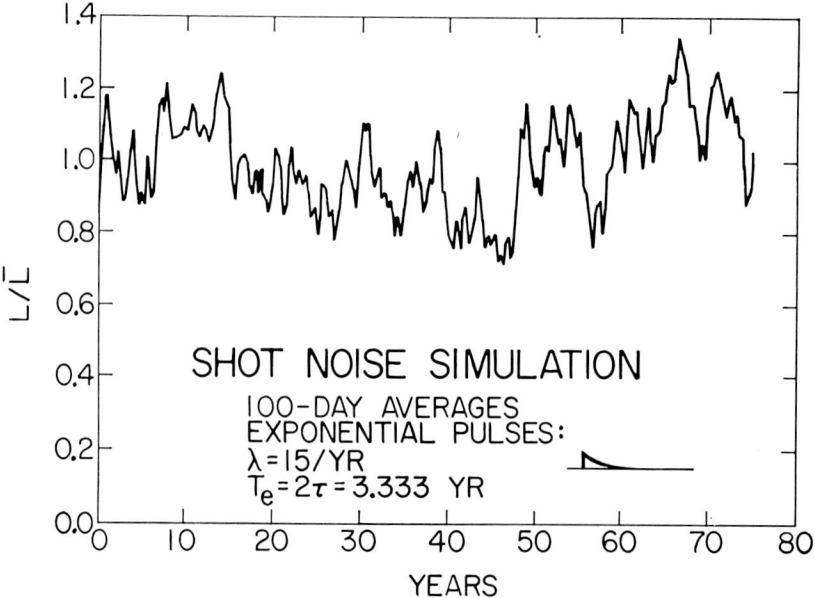

Fig. 5. Typical shot noise generated by computer, consisting of random exponentially-decaying pulses, as shown, with rate and pulse length T_e similar to 3C 273 parameters.

Detrending removes more power from the lowest frequencies, and the end result is misleading evidence of a low-frequency peak.

Figure 4 shows the results we obtained using a different method of smoothing, which does not give such a misleading result by reducing the power at low frequencies. The uppermost curve is the power spectrum obtained from Kunkel's data without detrending. The power spectrum points are each averaged over four Fourier coefficients – not including the one for $f=0$ – and have a standard deviation of $\pm 50\%$. They are also essentially independent. Notice that when the trend is removed from the data before analysis the low frequency power is thereby decreased, giving a false indication of a resonance. When the first 2500 days of data, representing mostly interpolation, are truncated, the result shown here is a power spectrum of somewhat less low-frequency power but otherwise similar. The data as averaged by Ozernoy and Chertoprud (1966) give even less low-frequency power, and a possible resonance – with considerably less than 50% certainty. It should be noted that their data are

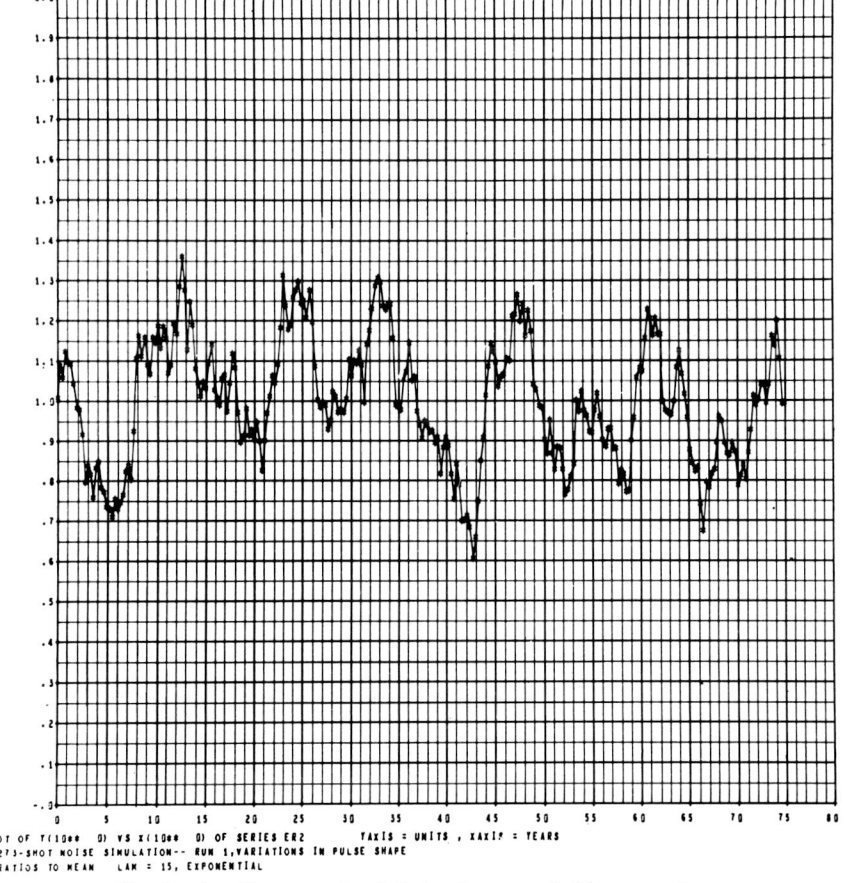

Fig. 6. Another example of shot noise generated by computer.

published in detrended form, and thus with reduced low-frequency power. Truncation of their data does not have much effect.

If the fluctuations of 3C 273 are due only to random outbursts of light – a situation known as shot noise – it can be shown that the maximum power *must* occur at $f = 0$.

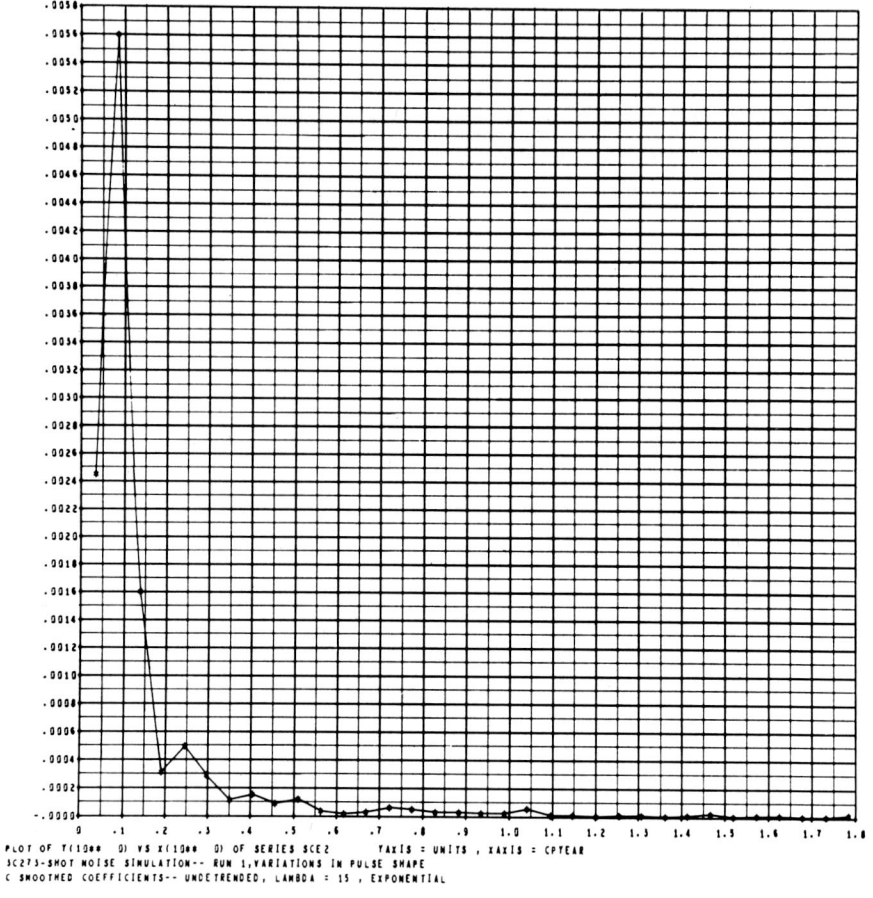

Fig. 7. The power spectrum for the shot noise run of Figure 6.

The value of the two-sided power spectrum at this point, when normalized as here, is equal to $1/\lambda$, where λ is the rate of occurrence of the pulses. We see that if these fluctuations are random shot noise, the rate of occurrence from Kunkel's data is $\lambda = 15 \pm 5$ per year.

The pulse length is then given essentially by the width of the zero-frequency peak, including both positive and negative frequencies, and a more precise analysis gives an effective pulse length $T_e = 3.2 \pm 1$ yr. Thus Kunkel's data can be produced by about 15 outbursts per year of 3.2-yr long pulses, which leads to the average overlap of 48 pulses at a time. The area under the power-spectrum curve is equal to the variance, which, of course, corresponds to an average of about 48 overlapping random pulses, or a standard deviation of ±14%. Thus, if these fluctuations of 3C 273 are due to random pulses, the pulse parameters are given directly and simply by the power spectrum. The moments and trend of the data have also been found to be consistent with random shot noise (Terrell and Olsen, 1970).

Figure 5 shows random shot noise fluctuations which we generated in a computer, with pulse rate and length similar to those determined from the 3C 273 spectrum, in order to verify our results. The pulses in this case were taken to be of exponentially-decaying shape, with an effective pulse length of 3.333 yr. Figure 6 shows the generated shot noise in another run. This particular run is the one (out of five) which most resembled an oscillation, with a period of perhaps 10 yr. Figure 7 shows the power spectrum for the shot-noise run just shown – and it does seem to have a peak corresponding to a 10- or 15-yr period, but not a statistically defensible one.

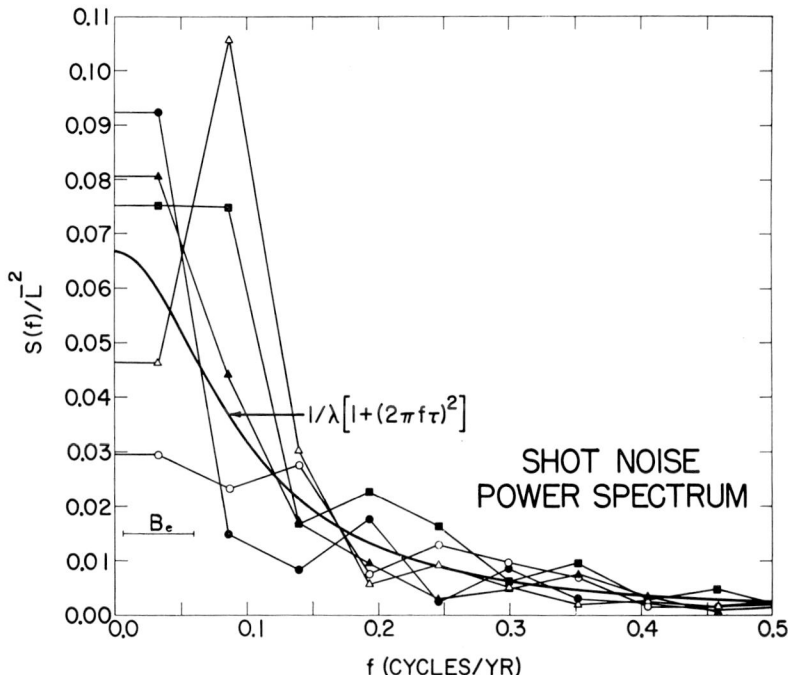

Fig. 8. Power spectra for five 75-yr runs of generated shot noise, not detrended, with parameters similar to those given by 3C 273 data. Also shown is the expected form of the power spectrum for the exponentially decaying pulses used. The points have been smoothed over four Fourier harmonics to decrease statistical standard deviations to ±50%.

Figure 8 shows our power spectrum results for all five runs with the 3C 273 parameters. Although the power spectra show the expected fluctuations of ±50%, the results fall close to the expected theoretical curve – for exponentially decaying pulses in these examples – and give on the average the expected value of zero-frequency power and peak width. We also generated and analyzed some shot noise cases with rectangular pulses with rather similar results.

Thus, to summarize, we find that the luminosity fluctuations of 3C 273 give no definite evidence of a 10- or 13-yr period, but are quite consistent with random outbursts of 3.2-yr equivalent length and at a rate of about 15 per year. Unfortunately, this still does not tell us whether 3C 273 is necessarily many independent sources or a single fluctuating light source, since randomness can never be proved – only non-randomness. However, if any periodic variation is present in the brightness of 3C 273, it is completely obscured by random fluctuations. The power spectrum, moments, and trend are all consistent with random outbursts of $\sim$3-yr pulses of light.

References

Gudzenko, L. I., Ozernoy, L. M., and Chertoprud, V. E.: 1967, *Nature* **215**, 605.
Kunkel, W. E.: 1967, *Astron. J.* **72**, 1341.
Manwell, T. and Simon, M.: 1966, *Nature* **212**, 1224.
Ozernoy, L. M. and Chertoprud, V. E.: 1966, *Astron. Zh.* **43**, 20.
Smith, Harlan J.: 1965, *Quasi-Stellar Sources and Gravitational Collapse*, (I. Robinson, A. Schild, and E. L. Schücking, eds.), University of Chicago, 221.
Smith, Harlan J. and Hoffleit, Dorrit: 1963, *Nature* **198**, 650.
Terrell, J. and Olsen, K. H.: 1970, *Astrophys. J.* **161**, 399.

Discussion of Papers Read by Smith and Terrell

Lasker: I would like to make some comments on the power spectra which we have just seen. Firstly, it is often useful to present power spectra in a log P–log f plane. Then one frequently notes that the data over many spectral elements can be represented by a straight line. This simplicity is of possible physical importance. Secondly, it is noteworthy that Dr Smith's power spectra frequently have two peaks. This is very different from what we find for variable stars in an admittedly very different frequency range; the power spectra of ordinary irregular variable stars is generally monotonic.

Smith: QSOs are probably quite different from variable stars.

Terrell: We found the power spectrum of 3C 273 to be of the form f^{-2} at high frequencies. This corresponds, in the case of random pulses, to a finite discontinuity in the pulse amplitude.

We generated a number of sets of random shot noise data, some with exponentially-decaying pulses and some with rectangular pulses. There was no detectable difference in the appearance of these two types of time series, nor in the trends.

Noerdlinger: Have you separated the long – and short – period amplitudes to examine their correlation with luminosity, or do you consider this worth doing when more data are available?

Smith: We have not done this, but it should be worthwhile to try.

J. Barnothy: I know that many look at the gravitational lens as if it were a sledge hammer which destroys the possibility of beautiful speculations regarding the nature of these new and interesting astronomical objects, but to deny the existence of gravitational lenses would be the equivalent of denying the existence of Newton's gravitational law. On the other hand, if the majority of QSOs were gravitational lens images of, say nuclei of Seyfert galaxies, then the variation in the brightness of the image becomes an inherent characteristic of QSOs. The power spectrum of these brightness variations does not depend on the number of secondary light sources within the object, it depends merely on the speed with which the optical axis scans the object disk. A simple computation as well as a trial scan over a random source field shows that the pattern and power spectrum of the variations of 3C 273 can be faithfully reproduced at a scanning speed of 25000 km s^{-1} in the frame of the object. Considering the rotation of the galaxy, that of the supergalactic system and the galactic and supergalactic coordinates of 3C 273, such an extremely high scanning-speed would follow if the lens were nearer than about 12 Mpc. As 3C 273 is in the direction of the Virgo cluster, which is at a distance of 11.4 Mpc, this suggests that the lens is a compact galaxy (of the type Zw 0930–5527 or Zw 1117–5141) in the Virgo cluster. The nuclei of the two compact galaxies mentioned are small enough that from a distance of 12 Mpc their apparent diameter would be less than one arc sec. The faint wisp-like nebulosity seen near the radio source of A of 3C 273 could then be the irregular nebulosity often accompanying compact galaxies.

Ozernoy: Gudzenko, Chertoprud and I analyzed the paper by Terrell and Olsen and found that their conclusions are invalid. The reason is that Terrell and Olsen reach their conclusions on the basis of the behaviour of sampling power spectrum in the low-frequency region where it is not representative. For example, it can be seen from Figure 8 of their paper, that the sampling variability coefficient becomes near to one. The second point is the following. Even if we shut our eyes to the inconsistency of Terrell's and Olsen's parameters of the pulses with natural physical requirements upon pulses, they are inconsistent with the statistical properties of the 3C 273 light curve. Namely, we have shown using analytic techniques (instead of numerical calculations by Terrell and Olsen) that the sampling variability coefficient of the envelope of 3C 273 brightness, v, must exceed 0.34 with 98% confidence. Meanwhile, for 3C 273 it has been found earlier that $v = 0.34$. Of course, choosing pulses of complex form with a great number of parameters, it would be possible to approximate the statistical characteristics of the observed process as well as we like. However, this would rather resemble the description of the motion of the Sun and the planets around the Earth by epicycles and deferents, performed as is well known by Ptolemy with very great precision.

Terrell: Apparently Dr Ozernoy has misunderstood our results. We were not presenting physical arguments, but merely investigating whether there is any reliable evidence for nonrandom fluctuations in 3C 273. The reasons for the usual unreliability of low-frequency power spectra were, in fact, pointed out in our paper. Using new methods which avoided these pitfalls, we found that there was then no remaining evidence of periodicity.

The complex methods used by Ozernoy and his co-workers depend critically on the assumed existence of a fundamental dynamical period of ~ 9 yr, plus perturbations (and would thus seem to have more connection with epicycles than ours). We find that random shot noise generated by computer often gives the impression of such a period, which we deduce to be ~ 10 yr (by methods similar to Ozernoy's) for *random* data similar to the 3C 273 data used by Ozernoy and Chertoprud. (*Astron. Zh.* **43**, 20, 1966).

Furthermore, it seems premature to draw such firm conclusions as Ozernoy's from the deviation of one parameter ($v = 0.34$) by 2 standard deviations from the value (0.52) calculated for random fluctuations. As a matter of fact, we find this parameter, the variability coefficient of the envelope for a 9-yr period, to be given by the data of Kunkel as $v = 0.56$, which is quite closely the expected value for random pulses. It is surprising that Ozernoy and his co-workers (*Astron. Zh.* **46**, 1317, 1969) did not discover this in their re-evaluation of their results based on Kunkel's data.

A STUDY OF THE CONTINUA OF THE NUCLEI OF GALAXIES*

D. ALLOIN, Y. ANDRILLAT and S. SOUFFRIN

Institut d'Astrophysique, Paris, France and Faculté des Sciences, Montpellier, France

Abstract. Many people have attacked the problem of synthesizing the stellar population of the galaxies. We have performed such a synthesis, by using only the intensities of absorption lines in the nuclei of galaxies. It is then possible to obtain a synthesized continuous spectrum even if there are several possible solutions for the stellar composition, and the computed continuum does not vary much with the assumed model of stellar population.

The method is to fit the equivalent widths of the absorption lines using different stellar compositions. In order to avoid any instrumental effect, we have observed different lines and bands from 3500 to 8500 Å in stars of various well-known stellar types and luminosity classes, under the same conditions as the galaxies.

We have applied this method to some 'ordinary' nuclei and to some Seyfert ones. As an example, we show the results for M81 and for NGC 1068.

M81

The first result concerns the stellar population: as was already noticed by Spinrad and Wood, we find an important contribution by M dwarfs (40 to 50% of the visible continuum) and also by K stars (30%). It is, therefore, likely that this nucleus has a very strong infrared spectrum.

The results for the continuum spectrum are shown in Figure 1.

The observed continuum does not show any reddening, when compared to the computed one. The lack of data concerning the continuum of M stars beyond 7000 Å

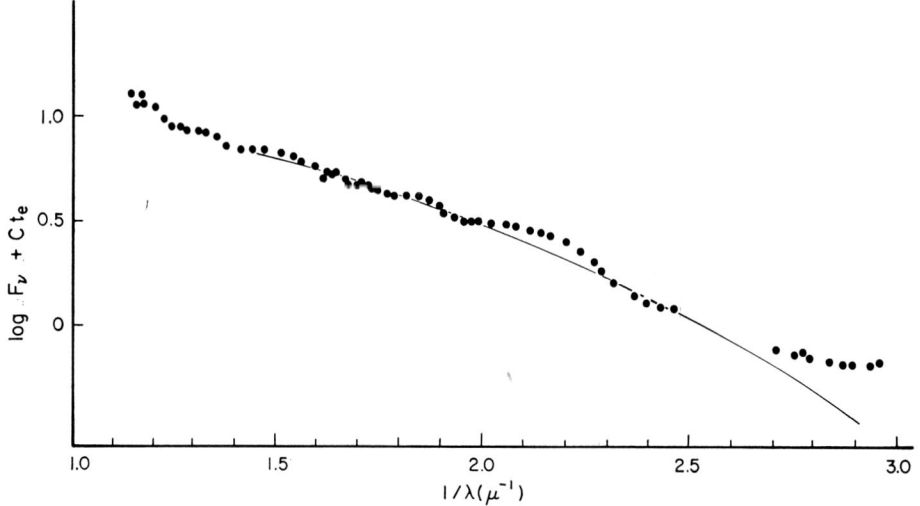

Fig. 1. The continuum energy distribution in the optical spectrum of the nucleus of M81.

* The observations were performed at the Haute-Provence Observatory (CNRS).

prevents us from computing the synthetic continuum in the near infrared. However, we can conclude from a study of the CaII 8542 line that there is no important non-stellar contribution at wavelength 8500 Å. Finally, the figure shows the presence of a non-stellar continuum in the ultraviolet range, which can be identified with the thermal emission of the gas.

The conclusions of this study are as follows:

there is no reddening other than that due to the tilt of the plane of the galaxy;
there is no infrared non-thermal continuum;
the thermal Balmer continuum comes from the gas.

NGC 1068

In the Seyfert nuclei, we have taken into account the possibility of a non-stellar component in the visible range by adding 'fictitious' stars contributing to the continuum with a given spectral distribution, but not contributing to the lines. For NGC 1068 the spectral distribution of the non-thermal continuum has been deduced from published polarization measurements by Visvanathan and Oke, and the spectral distribution of the thermal emission has been corrected for the same amount of reddening that Wampler has found for the emission lines.

One can then obtain a computed continuum fitting the observed one, with a very cold stellar population (K and M stars). The stellar continuum does not show any reddening; hence, the reddening of the emission lines should come from a small region of the nucleus, probably filamentary clouds ejected from the nucleus and containing dust, as Hoyle and Wickramasinghe have suggested.

Two papers relevant to the subject should appear shortly in *Astron. Astrophys.*

Discussion

de Vaucouleurs: How many absorption lines did you measure for equivalent width, and how many stellar spectral types did you use in the population fit?

Mme Souffrin: So far we have measured seven lines: Ca II $\lambda 3933$, Ca II + H I $\lambda 3968$, Ca I $\lambda 4227$, G-band, Mg I triplet, Na D lines, and Ca II $\lambda 8542$. We have used about five spectral types in the population fit.

King: Spinrad has also observed M81, at many line and continuum wavelengths, and also concludes that the population of the center includes many K and M dwarfs. Minkowski and I have included M81 in our velocity dispersion and photometry program, and preliminary results indicate a high M/L.

RADIO EMISSION FROM COMPACT OBJECTS

K. I. KELLERMANN

National Radio Astronomy Observatory, Green Bank, W. Va., U.S.A.*

Abstract. Compact radio stars are associated with both galaxies and quasi-stellar objects; and there appears to be no way to distinguish between the radio galaxies and the quasi-stellar radio sources from their radio properties alone.

The compact radio sources are opaque at the longer radio wavelengths and have spectra that are either peaked or complex. They have a complex brightness distribution and often contain components less than 0.001 arc sec in size.

Many of the compact sources show large intensity variations and in NGC 1275 there is evidence for a change in the angular size during one year.

1. Introduction

Radio emission is observed from all of the various types of compact extragalactic objects. These include not only quasi stellar objects, but also the nuclei of compact, Seyfert, and N type galaxies.

There are, at present, insufficient systematic data to determine accurately the radio luminosity function of any of the compact objects or the probability that any particular object can be detected as a radio source above some given flux limit.

A. SEYFERT GALAXIES

Because two of the earliest identified radio sources (3C 71 and 3C 84) are identified with Seyfert galaxies (NGC 1068 and NGC 1275 respectively), it has been widely thought that the Seyfert phenomenon is associated with intense radio emission. Although at least one other galaxy, 3C 120, is identified with a compact radio source and exhibits the Seyfert phenomenon, there is no evidence that Seyfert galaxies are particularly likely to be strong radio sources. NGC 1068 (3C 71), for example, has an absolute radio luminosity near 10^{40} erg s^{-1} or only slightly more than a normal spiral galaxy.

Attempts have been made to observe most of the other known Seyfert galaxies (Heeschen and Wade, 1964; de Jong, 1967; Pauliny-Toth and Kellermann, 1968a). Although most of the known Seyfert galaxies are weak radio sources, their radio luminosity in general does not significantly exceed that of normal spiral galaxies. The fraction of Seyfert galaxies which are strong radio sources in fact does not appear to differ significantly from that of giant elliptical galaxies (e.g. Rogstad and Ekers, 1969).

B. COMPACT GALAXIES

Similarly, the compact galaxies of the type described by Zwicky (1964) and Sargent (1970) do not in general appear to be particularly prominent as radio emitters. Only

---

* Operated by Associated Universities, Inc. under contract with the National Science Foundation.

two of these, Zw 1727+50 (Pauliny-Toth and Kellermann, 1968a) and III Zw 2 (Kellermann and Pauliny-Toth, unpublished), have been detected as radio sources. Both have relatively flat spectra, suggesting that the radio sources are compact.

C. N GALAXIES

The probability of radio emission from N-type galaxies is difficult to discuss, since most of the galaxies of this type were first recognized in the identification of known radio sources.

D. QUASI-STELLAR OBJECTS

In this paper we shall use the term quasi-stellar radio source, or QSS, to refer to a radio source associated with a quasi-stellar object (QSO) independent of the size or nature of the radio source. More than 150 such QSS are known. Most of these are the result of identification of a catalogued radio source with a quasi-stellar object, based essentially on the good agreement in the position of the radio source and QSO.

Other QSOs have been isolated as a result of multicolor photometry (e.g. Sandage and Luyten, 1967; Braccesi *et al.*, 1970). These do not appear to differ in any obvious way from the QSS except that they are not in existing radio catalogues and are usually referred to as 'radio quiet' QSOs. Several groups have attempted to observe radio emission from these objects and a small number have been detected (Braccesi, 1967; Pauliny-Toth and Kellermann, 1968a; Lang and Terzian, 1969; Grueff, 1970). Although the statistics are still poor, it appears that 10 or 15% of the optically discovered QSOs may be detected as radio sources. The higher fraction of detection found by Lang and Terzian (1969) appears to be incorrect as more recent observations made with higher resolutions have shown these to be mainly due to chance coincidence (Grueff, 1970).

All of the known QSS have significant red shifts, generally of the order of unity. Thus, if the red shifts are cosmological, even the QSS with very low measured flux density have large absolute radio luminosities, typically in the range of 10^{44-45} erg s^{-1}, which is comparable to the strongest radio galaxies such as Cygnus A and 3C 295. Since the QSOs are thought to be at great distances, the absence of detectable radio emission usually implies only a modest limit to the radio luminosity of about 10^{43} erg s^{-1}. Because their space density is low, there are very few nearby QSOs and there is therefore little chance of detecting intrinsically weak QSS if they exist. Figure 1 shows the distribution of absolute monochromatic radio luminosity at 6 cm for identified galaxies (Figure 1(a)) and for identified QSS (Figure 1(b)). Figure 1(c) shows a similar histogram for those objects initially thought to be radio quiet.

It has often been suggested that the radio-quiet objects may show significant emission at very short wavelength. There is no evidence that this is the case. Most of the 'radio-quiet' objects which have been subsequently detected have normal spectra and were missed in the early surveys simply because their flux densities are small.

The radio luminosity function of identified QSS in the 3CR catalogue has been investigated by Schmidt (1970) who finds that the space density of QSS is approxi-

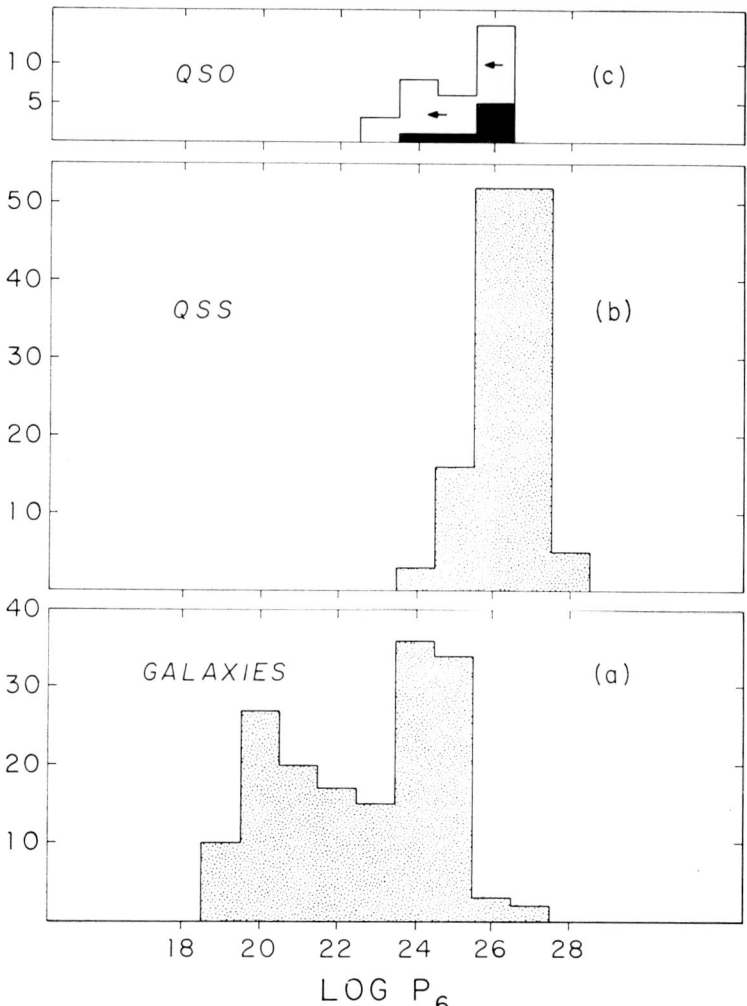

Fig. 1. Histograms showing the distribution of absolute monochromatic luminosity (P_6) at 5 GHz (6 cm) for (a) galaxies and (b) QSS which have measured redshifts. The upper histogram, (c), also shows the distribution of upper limits for radio quiet QSOs. The filled in part of (c) indicates radio quiet QSOs which have subsequently been detected or recognized as radio sources. The absolute luminosities are calculated assuming the redshifts are cosmological, $H = 100$ km s^{-1} Mpc^{-1}, and $q_0 = 1$. These histograms are in no way intended to represent a luminosity distribution as the sources are taken from an inhomogeneous sample. They are intended to show only the range of luminosity covered by the various objects.

mately inversely proportional to absolute luminosity. Only about 1 QSO in 300 is strong enough to be in the 3C catalogue (e.g. Schmidt, 1969). Thus, if Schmidt's luminosity function can be extrapolated by several orders of magnitude, most QSOs should be stronger than a few hundredths of a flux unit at 178 MHz or a few milliflux units at centimeter wavelengths. This is not inconsistent with the fraction of QSOs which have been detected at intermediate flux density levels. However, systematic surveys of radio quiet QSOs down to a very low flux density level are clearly important the better to establish their radio luminosity function. On the basis of existing data, it appears that the probability of finding radio emission from compact objects such as QSOs, Seyfert, compact, or N galaxies is comparable to that of giant elliptical galaxies.

E. CLASSIFICATION OF THE RADIO EMISSION FROM COMPACT OBJECTS

It is perhaps surprising to note that of the identified radio sources which have been studied in some detail, there is no clear difference in the radio emission from galaxies and from QSOs or other compact objects. Both classes of objects have similar radio frequency spectra, similar radio structure and polarization, and assuming that the redshifts are cosmological, a similar range of linear dimensions.

A rough separation of all extragalactic sources can be made into two groups:

(1) The compact radio sources which have angular dimensions much less than 1 arc sec and are opaque over at least part of the radio spectrum; and

(2) the extended radio sources which typically are greater than 1 arc sec in angular extent and are optically thin throughout the observable spectrum.

It must be emphasized, however, that this division is based on the observed radio properties, and in no way separates radio galaxies from QSS. Small radio sources are found not only in QSS but in galaxies as well (e.g. Heeschen, 1970) and QSS often have dimensions comparable with the classical extended radio galaxies (e.g. Hogg, 1969).

It is true that the majority of the *identified* radio galaxies are extended and have optically thin radio spectra, while the majority of the *identified* QSS are small and often have components which are optically thick (Bolton, 1966; Pauliny-Toth and Kellermann, 1968b). The unidentified sources generally have angular dimensions and spectra similar to those of the identified radio galaxies, and it has been thought that the unidentified sources must therefore be radio galaxies beyond the 48-in. plate limit (e.g. Bolton, 1966). An alternative interpretation, however, is that there is no relation at all between radio properties and optical identification, but that in the process of identification, the association with a stellar object is only accepted if the position agreement is very good, and this will discriminate against the identification of extended radio sources with stellar objects. More complete samples of identified sources are required before definite conclusions can be made. However, the identification of faint optical objects with extended radio sources will always be uncertain, particularly if the centroid of radio emission does not coincide with the optical object. It is interesting to note, as shown in Figure 2(a) and (b), that if the QSS redshifts are presumed to be cosmological, then in the 3CR catalogue where the identifications are most complete, there is no systematic difference between the linear dimensions of galaxies and QSOs. As has been pointed out many times, this gives great weight to the interpretation of QSO redshifts as being cosmological.

Because the radio properties of galaxies and the more compact objects appear to be indistinguishable, the emphasis in this review will be on the compact radio sources independent of their optical identification.

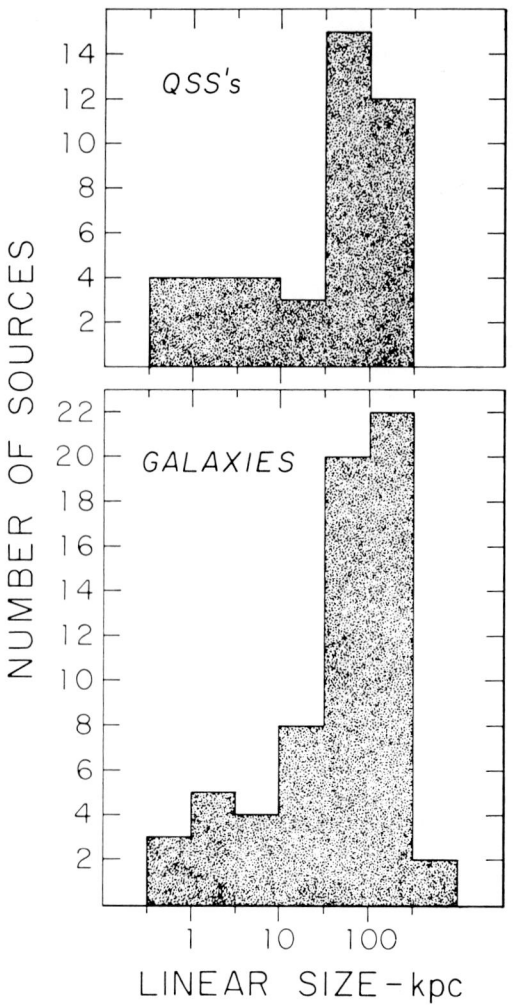

Fig. 2. Histograms showing the distribution of linear dimensions of QSS and radio galaxies in the 3CR catalogue which have measured redshifts. The sizes are computed from published and unpublished (Miley, 1970) measurements of angular sizes and assume that the measured redshifts are cosmological, $H = 100$ km s^{-1} Mpc^{-1}, and $q_0 = 1$.

2. Radio Spectra

A. CLASSIFICATION OF SPECTRA

Many extragalactic sources have been observed over a very wide range of wavelengths, typically from 3 cm to 10 m, but in some cases extending from 3 mm to 30 m. This range of 10^4 in wavelength may be compared with the range of only 2:1 covered by conventional optical spectra. It is not surprising therefore that a wide variety of spectral shapes is found. Nevertheless, no distinction can be made between QSS and radio galaxies either from a statistical comparison of the spectra (e.g. Kellermann *et al.*,

1969) or on the basis of the detailed radio spectra. This is illustrated in Figures 3, 4 and 5, which show the spectra of a number of galaxies and QSS for which there are data over a wide range of wavelength. The spectra are conveniently divided into three groups:

(i) Sources where the flux density decreases monotonically over the entire range of observed frequencies. These sources are optically thin at all wavelengths (Figure 3);

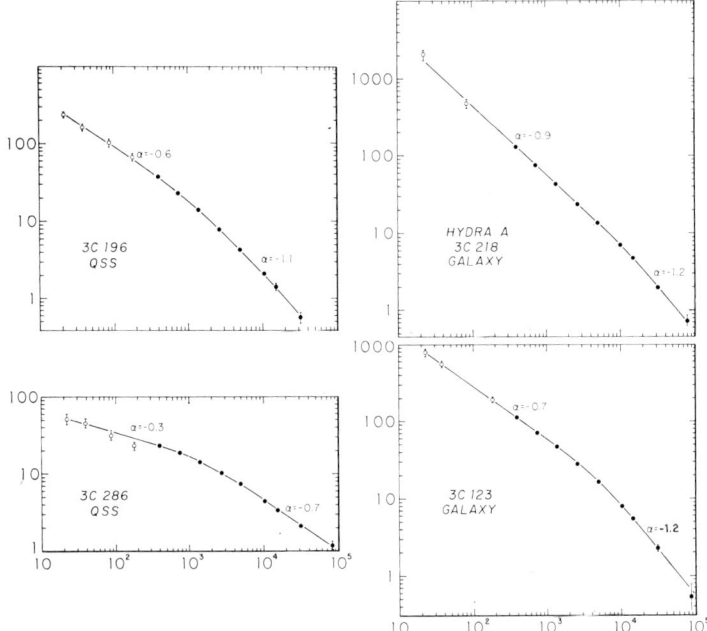

Fig. 3. The radio spectra of several galaxies and QSS which are optically thin over the entire range of observed frequencies. The data between 3 mm and 75 cm are taken from published and unpublished measurements made at NRAO by Pauliny-Toth and the author and are shown with closed circles. Data at longer wavelengths come from published observations made at several observatories and are shown as open circles.

(ii) Sources where the flux density decreases with frequency at high frequencies but has a sharp cutoff at low frequencies which is probably due to self-absorption (Figure 4);

(iii) Sources which have complex spectra with one or more relative minima and maxima. These are thought to be composed of several distinct components which become opaque at different wavelengths (Figure 5).

B. OPTICALLY THIN SOURCES

The spectral index, in the optically thin part of the spectrum, is directly related to the electron energy distribution through the well-known relation $\alpha = (1-\gamma)/2$ where α is the spectral index defined by $S \propto \nu^\alpha$, and γ is the energy index defined by $N(E) \propto E^{-\gamma}$, where S is the flux density, ν the frequency, and $N(E)$ the number of electrons

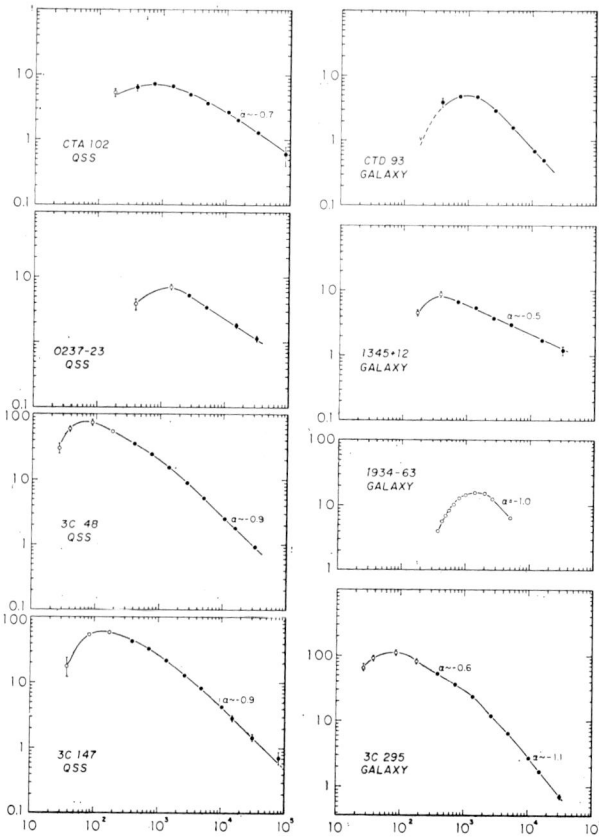

Fig. 4. The radio spectra of several galaxies and QSS which become optically thick at long wavelengths. The source of data is the same as in Figure 3.

with energy, E. At wavelengths where the sources are optically thin, the spectra are typically power law or dual power law with indices usually in the range $\alpha = -0.7$ to -1.2. In the case of the dual power law spectra, the curvature in the spectrum generally extends over one decade or more of wavelength. At significantly longer or shorter wavelengths the sources usually have well defined power spectra with a difference in index close to 0.5 which is the value expected when relativistic particles are continuously supplied and the rate of injection is balanced by the losses due to radiation (e.g. Kardashev, 1962).

In none of the sources with optically thin spectra is there sign of a high frequency cutoff due to a cutoff in the electron energy distribution. An upper limit to the electron energy must exist due either to the acceleration mechanism or to energy losses by synchrotron radiation or inverse Compton losses. In particular, as pointed out by Rees and Setti (1968) and emphasized by van der Laan and Perola (1969), the inverse Compton scattering by photons of the 3K background will limit the lifetime of the high energy particles, especially at large redshifts where the effective background

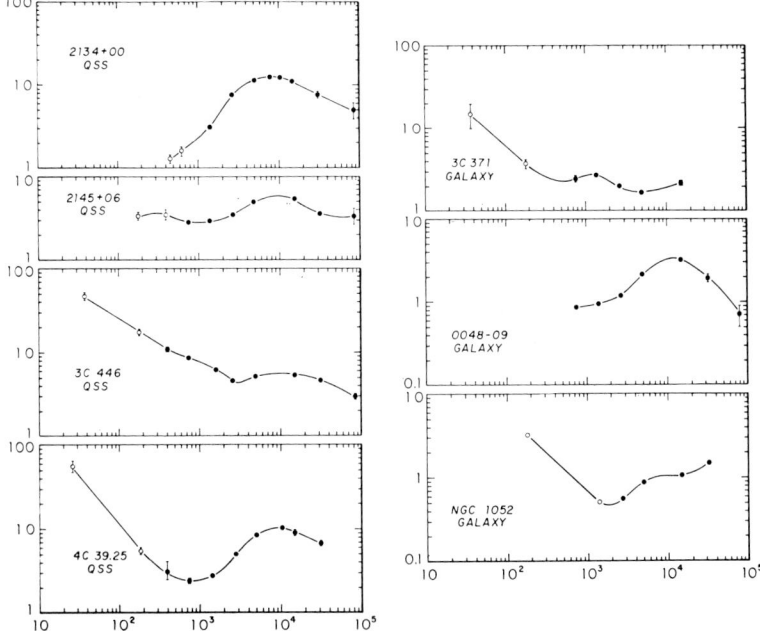

Fig. 5. The radio spectra of several galaxies and QSS which have complex spectra. The source of data is the same as in Figure 3, except for NGC 1052 (Heeschen, 1970).

temperature is increased by a factor of $(1+z)$ and the Compton lifetime is decreased by $(1+z)^4$.

The available data show no evidence of a cutoff in any of several hundred sources out to 10 GHz, or in the smaller number which have been studied at 15 or 30 GHz. In a few cases data exist near 100 GHz still with no evidence of the expected cutoff (Kellermann and Pauliny-Toth, 1971).

C. OPTICALLY THICK SOURCES

Many of the more compact sources are opaque over at least part of the radio spectrum. Contrary to several published remarks that such sources are rare or unusual (e.g. Shimmins *et al.*, 1968, MacLeod and Andrew, 1968), they are in fact quite numerous, with approximately one-half of all sources observed in centimeter wavelength surveys being of this type (Kellermann *et al.*, 1968; Brandie, 1970).

Because the compact sources are partially opaque, their spectral indices are flat or even inverted so that they are relatively inconspicuous at meter wavelengths where all of the earlier radio surveys have been made. Most of the known opaque sources have therefore been initially found in the limited surveys made at shorter decimeter or centimeter wavelengths, and it is becoming increasingly clear that compact opaque radio sources having complex radio spectra are quite common and comprise a significant fraction of all radio sources that can be detected with current instruments.

The number of identifications of these sources is limited, but they are with both galaxies and QSOs.

Many of the compact sources have now been studied at millimeter wavelengths where, although few sources remain opaque even at 3.5 mm, most are becoming optically thin and show a turnover in their spectra near 1 cm (Kellermann and Pauliny-Toth, 1971). For the same reason that these sources are undetected in long wavelength surveys, sources which may be opaque at short millimeter wavelengths might only be detected in surveys made at wavelengths less than 1 cm. Presently available sensitivities do not, however, yet permit reasonably efficient surveys to be made at wavelengths less than a few centimeters.

For the extended radio sources, the spectral index is always steeper than -0.5, corresponding to an index $\gamma > 2$. In nearly every case where flatter spectra are observed, accurate flux measurements or long-baseline interferometry indicate that the flat spectrum is due to the superposition of several optically thick components, and not to a flat electron energy distribution. The only exception to this is in the optically thin part of the spectrum of the very young variable components; these appear to have indices between 0 and -0.25 or, $1 < \gamma < 1.5$.

In a recent paper based on observations at Ohio State, Andrew and Kraus (1970) have remarked that some of the sources with flat spectra appear to have smooth power law spectra, and they suggest that these are optically thin and do indeed indicate a flat electron energy distribution. More accurate flux density measurements (Jauncey et al., 1970a) of many of these sources, however, show fine structure in the spectra suggesting the presence of several separate opaque components. In addition, some of these sources have been observed with high resolution interferometers and show the expected complex structure (e.g. Kellermann et al., 1970).

D. RELATION BETWEEN RADIO, INFRARED OPTICAL, AND X-RAY EMISSION

It is by no means clear what relation there is, if any, between observed radio emission from compact objects and the radiation observed at infrared optical, or X-ray wavelengths. It is perhaps significant that the three extragalactic X-ray sources, M87, 3C 273 and Centaurus A (e.g. Bowyer et al., 1970) all contain a very small diameter radio source*, suggesting that the X-ray emission may be due to inverse Compton scattering in the compact radio source (e.g. Burbidge, 1970). Since the ratio of inverse Compton emission to synchrotron emission depends only on the peak radio brightness temperature, detectable X-ray emission may also be expected from other radio sources, with equally great brightness temperature and comparable radio flux density such as 3C 279, 3C 345, and 3C 454.3.

A number of compact objects are particularly strong in the far infrared (Kleinmann

---

* Observations of the small radio source in M87 have been published (e.g. Cohen et al., 1969). The small source in Centaurus A was found in unpublished measurements made with the NRAO three element interferometer. The small source contains a few percent of the total flux density at 3.7 cm; it appears to be unresolved at a resolution of one arc sec. The spectrum between 3.7 and 11 cm indicates that it is opaque at centimeter wavelengths and thus most likely $\theta \lesssim 0.''001$.

and Low, 1970a). In none of these objects is there any evidence of an increase in the radio flux density at short millimeter wavelengths toward the very large infrared flux density. This is illustrated in Figure 6 which shows the combined radio and infrared spectrum of the galaxies NGC 1068 (3C 71) and NGC 3034 (M82, 3C 231). Schorn

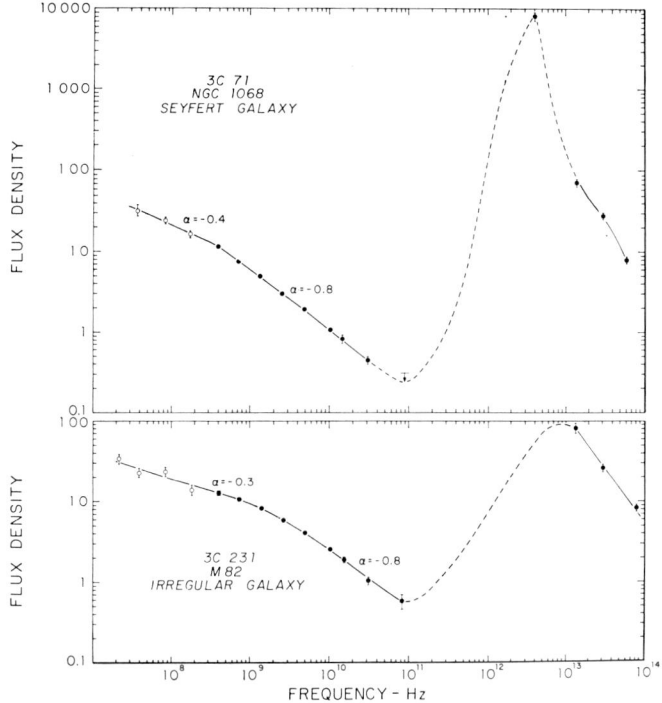

Fig. 6. Spectrum of NGC 1068 (3C 71) and NGC 3034 (3C 231, M82) at radio and infrared wavelengths. The source of data at radio wavelengths is the same as in Figure 3, and the infrared data are taken from Kleinmann and Low (1970) and Low (1970).

et al. (1968) have measured a flux density of about 10 flux units* for NGC 1068 at 3.4 mm. This has not been confirmed by more recent measurements made at NRAO at the same wavelength and with considerably greater sensitivity, or by measurements made near the same time but at the longer wavelength of 2 cm (Kellermann and Pauliny-Toth, 1971).

The NRAO limit of a few tenths of a flux unit combined with the high flux density of about 10000 flux units measured by Low (1970) near 100μ requires a spectral index in this wavelength range near the limit of +2.5 of an opaque synchrotron source. It has also been suggested that the low frequency cutoff of the infrared emission may be due to free-free scattering in a dense electron core (Weymann, 1970).

NGC 3034 (M82) is of particular interest since the infrared and radio sources coincide and have similar dimensions of approximately 20×10 arc sec (Kleinmann

* 1 flux unit = 10^{-26} W m^{-2} Hz^{-1}.

and Low, 1970b) suggesting that the emission at radio and infrared wavelengths may be in some way related. The radio spectrum, however, at centimeter and millimeter wavelengths is a power law with an index of -0.8 and shows no evidence of enhanced emission even at 3.5 mm. Future work in the difficult range 100μ to 1 mm will clearly be important in establishing the relation between radio and infrared emission in M82 and other galaxies.

3. Radio Intensity Variations

Many of the compact radio sources show pronounced variations in intensity on time scales extending from a few weeks in case of the unusual object, BL Lac (Andrew *et al.*, 1969; Olsen, 1969) to a few years. Since the initial discovery of intensity variations by Dent (1965), a number of sources have been regularly monitored at several observatories at wavelengths from 3 mm to 40 cm. The data show no simple pattern to the observed variations; in particular, there is no evidence for any periodic phenomena. Rather, the variations appear to be in the form of random outbursts which often appear first at short wavelengths and then at a later time at longer wavelengths and with reduced amplitude.

There is no obvious difference in the pattern of intensity variations observed in radio galaxies or in QSS. The typical change in radiated power in the variable radio galaxies such as NGC 1275 and 3C 120 is of the order of 10^{41-43} erg s^{-1} in one year. Because the QSS are generally considered to be at very great distances, their change in radiated power appears to be as great as 10^{45} erg s^{-1} in one year, although the observed changes in apparent intensity are comparable in amplitude and frequency with those of the radio galaxies. Assuming they are at cosmological distances, the change during a few months in radiated power in some of the QSS is comparable with the entire radio luminosity of the strongest radio galaxies such as Cygnus A.

A. THEORIES OF INTENSITY VARIATIONS

There are many reasons why the compact radio sources may be expected to vary with time. These include changes in the rate of production or acceleration of relativistic particles, the loss of energy of the particles due to synchrotron emission, inverse Compton scattering, or adiabatic expansion, and changes in the magnetic field strength or in the size of the source. Probably all of these factors contribute in varying degrees to the variations that are observed in the different sources, and at different epochs in individual sources.

B. EXPANDING SOURCE MODEL

Soon after the initial discovery of intensity variations it was realized (Kellermann and Pauliny-Toth, 1966; Moffet, 1966; Pauliny-Toth and Kellermann, 1966) that the observed variations could be approximately described by a simple model first discussed by Shklovsky (1965). This model assumes that the variations occur as the result of the expansion of a cloud of relativistic particles which is initially opaque out to short wavelengths, but which, due to expansion, becomes optically thin at successively longer wavelengths. If, as supposed by Shklovsky, the magnetic flux is conserved, the magnetic field will decrease and so the peak flux density will decrease with time and increasing wavelength.

The immediately obvious applicability of the expanding source model to explain the observed intensity variations led van der Laan (1966) to reformulate in detail the behavior expected from the simple case where all of the particles are generated instantaneously in an infinitesimal volume of space, where the magnetic field and electron cloud is homogeneous and isotropic, where the expansion occurs at a constant rate or with constant deceleration, and where the magnetic flux is conserved.

At least several events observed in the galaxy 3C 120 (Pauliny-Toth and Kellermann, 1968) and in 3C 273 (Dent, 1968) have followed remarkably closely the quantitative form expected from the simple model. In general, however, most sources show repeated outbursts or a continuous but variable acceleration of relativistic particles, so that individual events are not sufficiently isolated either in frequency or time to permit a detailed analysis. But to the extent that comparison between observation and theory is possible, the variations are reasonably well explained by the 'expanding source model'. The fact that there are sources which deviate from the simplest form of the model described by Pauliny-Toth and Kellermann (1966) and by van der Laan (1966) is not surprising. What is surprising is that the observed variations in most sources do follow remarkably well the theory outlined by Shklovsky (1965) at a time when to most others the very possibility of observable intensity variations in extragalactic radio sources seemed remote.

C. INITIAL CONDITIONS IN VARIABLE SOURCES

In the simple model which assumes the instantaneous production of particles, the source is initially opaque at infinitesimally small wavelengths. In a real source, however, particle production must occur over a finite period of time, and throughout a finite volume of space. Thus below some critical wavelength, λ_c, the source must be always optically thin and the observed intensity variations will reflect the rate of acceleration of relativistic electrons as well as changes in the magnetic field strength and electron energy distribution. For $\lambda < \lambda_c$, there is no expected delay in the time when the peak amplitude is reached at different wavelengths, and the maximum flux density reached is less than expected from extrapolation of the simple model. The data at millimeter wavelengths (Schorn et al., 1968; Kellermann and Pauliny-Toth, unpublished; Simon, 1969) suggest that $\lambda_c \gtrsim 1$ cm. The extensive data obtained by the Canadian observers at 2.8 and 4.6 cm (Locke et al., 1969, 1970) indicate a number of sources for which $\lambda_c \gtrsim 4.6$ cm.

In all of the variable sources, the peak flux density observed at wavelengths where $\lambda < \lambda_c$ is very nearly the same indicating a spectral index in the range $-0.25 \lesssim \alpha \lesssim 0$ or $1 \lesssim \gamma \lesssim 1.5$. A similar value of γ is found in those cases where detailed analysis of the rate at which the peak amplitude changes with wavelength or time has been possible. There is, therefore, growing evidence, that in both galaxies and QSS, the initial electron energy distribution is very flat with an index in the range $1 < \gamma < 1.5$. Since in the older, more extended sources $2.5 < \gamma < 3.5$ electron energy losses must sometime steepen the spectra.

Synchrotron radiation losses will produce a cutoff or a bend in the spectrum at a frequency given by

$$v_0 = t^{-2} B^{-3} \text{ GHz},$$

where t is the age of the source in years and B the magnetic field in gauss. Since some sources clearly show no steepening of the spectrum even at 3.5 mm (85 GHz) for a year or more following an outburst, the initial magnetic field cannot greatly exceed 1 G.

D. MOTION OF VARIABLE COMPONENTS

An interesting variation on the expanding source model has been suggested by Ozernoy and Sazonov (1969) who propose that the observed multiple outbursts are the result of several expanding sources which are 'flying apart' at relativistic velocity. In their model the delay between different observed 'events' is due to the different propagation time from each cloud rather than to independent outbursts. However, contrary to observation, this model predicts that the time between observed outbursts should increase with increasing wavelength, since the radial distance between the separate clouds will increase with time. Figures 7 and 8 show, however, that at least for the sources 3C 120 and 3C 273, where good data exist, the time between different events remains essentially constant. This, of course, does not imply that relativistic motions do not play some role in determining the detailed nature of the light curves of some sources. High resolution interferometer observations made over a period of time and a range of wavelengths are required to determine whether or not the different outbursts are spatially separated and, if so, whether or not there are significant proper motions of the individual components.

E. POLARIZATION VARIATIONS

Another sensitive test of models of intensity variations is the observation of polarization variations. It is well known that in a source of synchrotron emission from a uniform magnetic field and power law distribution of relativistic particles, the fractional polarization, P, for optical depth $\tau \ll 1$ is

$$P = (3\gamma + 3)/(3\gamma + 7) \sim 0.60 \qquad (\gamma = 1)$$

with the electric vector perpendicular to the direction of the magnetic field, and for $\tau \gg 1$

$$P = 3/(6\gamma + 13) \sim 0.16 \qquad (\gamma = 1)$$

and is parallel to the magnetic field (Le Roux, 1961). Thus, in an expanding source, when $\tau \sim 1$, a large increase in the degree of polarization accompanied by a 90° rotation is expected (Kellermann and Pauliny-Toth, 1968). In practice, the situation may not be so simple, since, when $\tau \gg 1$, we observe radiation from only a thin layer

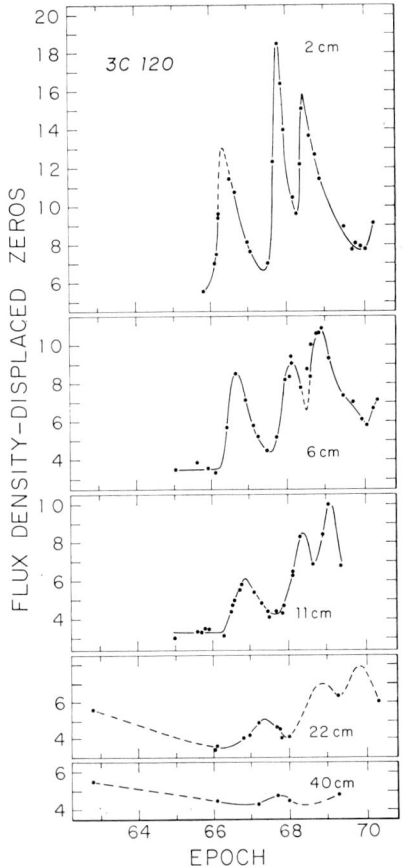

Fig. 7. Intensity variations observed in the radio galaxy 3C 120 at 2, 6, 11, 22, and 40 cm. The data are taken from measurements made by Pauliny-Toth and the author. The vertical axis gives the flux density at each wavelength and the horizontal axis the epoch.

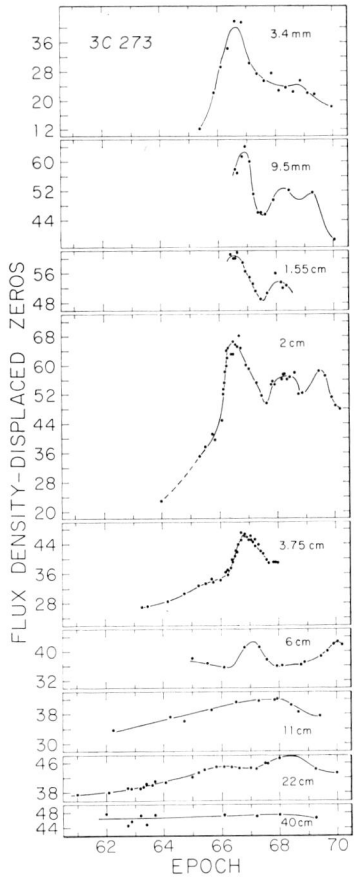

Fig. 8. Intensity variations observed in the QSS 3C 273 at 3.5 and 9.5 mm 1.5, 2, 6, 11, 22 and 40 cm. The axes and the source of the data are the same as in Figure 7 plus Schorn et al. (3.5 mm), and Hobbs et al. (9.5 mm, 1.5 cm).

of the source where the magnetic field may be expected to be more uniform than when averaged over the whole volume that is involved when $\tau \ll 1$.

Measurement of polarization variations is very difficult, since the polarization is typically only a few percent. Observations of polarization variations by Hobbs et al. (1968), Hobbs et al. (1969), Olsen (1969), McCullough and Waak (1969), and Aller (1970a, b) generally are consistent with the expanding cloud model, but the uncertainties are large and more accurate observations are still required.

F. RELATION BETWEEN RADIO AND OPTICAL VARIATIONS

Many compact objects also show significant variations at optical or infrared wavelengths as well. It is not clear, however, if there is any relation between the variations observed at optical and radio wavelengths. There is no evidence for any correlation but there are in fact very little overlapping data. One of the best studied sources is 3C 345, for which the observed radio and optical variations are shown in Figure 9. No relation is apparent, but it is clear that considerably more data are required before any conclusions can be made.

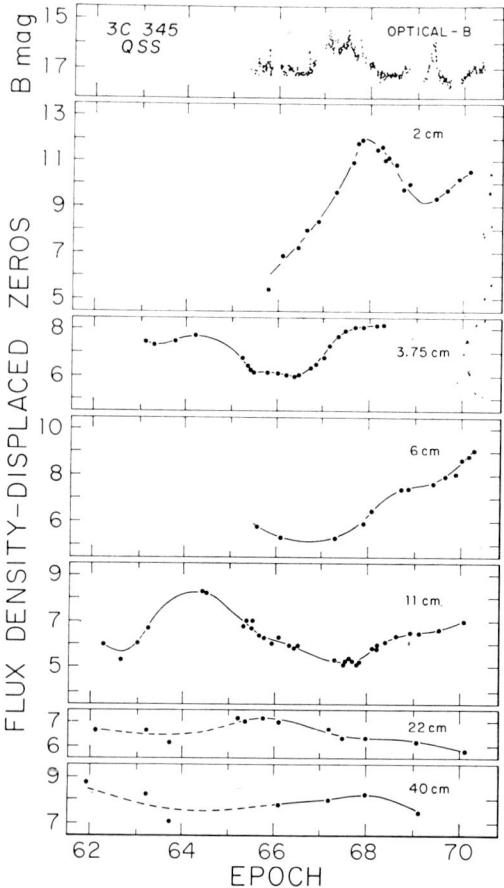

Fig. 9. Radio and optical intensity variations observed in the QSS 3C 345. The axes and radio data are the same as in Figure 7. The optical data are from Kinman (1970).

In general, those sources which have been most active at radio frequencies are those which have been described as also showing large optical variations (e.g. Penston and Cannon, 1968). But there are exceptions such as 3C 273 which has been one of the most active radio sources during the past five years but has had only very small optical variations. On the other hand NGC 4151, which has demonstrated large optical and infrared variations (e.g. Fitch *et al.*, 1967), is a normal weak radio galaxy with no evidence for variable radio emission.

4. The Structure of Compact Radio Sources

It has been realized for some time that radio sources which are opaque at short wavelengths, or which show significant intensity variations, must be very small (e.g. Slish, 1963; Williams, 1963; Shklovsky, 1965). The expected small size has now been directly confirmed using tape recording and radio link interferometers and it appears that of the stronger sources at decimeter wavelengths about 15% have significant structure on a scale of 0.001 arc sec or less (Kellermann *et al.*, 1970). This includes not only QSS but a number of radio galaxies as well.

It is difficult, however, to determine in detail the small scale structure of the sources, since the available interferometer observations are insufficient to allow a full synthesis of the brightness distribution. Although extensive measurements have been made over the wavelength range 6 to 75 cm (e.g. Clark *et al.*, 1968; Kellermann *et al.*, 1968; Clarke *et al.*, 1969; Jauncey *et al.*, 1970; Kellermann *et al.*, 1971) only a crude picture of the small scale structure exists.

These data, together with the measurements of interplanetary scintillations (e.g. Little and Hewish, 1968; Harris and Hardebeck, 1970) indicate a continuous scale of dimensions from 1 to less than 0.001 arc sec. Often sources show up to three or more components which may differ by up to 100:1 in angular dimensions. It is not clearly established whether or not the individual components are spatially separated or are concentric. There is some evidence from the long-baseline interferometer observations made by the Canadian group at 75 cm of component separations in the range 0.01 to 0.1 arc sec, but this is not firmly established (Clarke *et al.*, 1969).

It is clearly very important to measure in greater detail the small-scale structure, particularly to determine the minimum size of the characteristic multiple separated components, and to see whether in fact this type of structure is found in the smallest sources as would be predicted from the theory of Ozernoy and Sazonov (1969).

The core-halo structure described above is found in both radio galaxies and QSS which have flat or complex radio spectra. The smallest components show cutoffs at the shortest wavelengths, as is expected if the cutoffs are due to synchrotron self-absorption. The smallest angular dimensions which are observed are of the order of 0.0004 arc sec or less (Kellermann *et al.*, 1971). The smallest linear dimension which has been directly observed is the source in the nucleus of M87 which contains only about 1% of the total flux density of the galaxy at 13 cm and is about 0.1 pc or less in extent (Cohen *et al.*, 1969). Although no other such small components have been directly observed, it may be inferred from the frequent presence of optically thick radio sources in nearby elliptical galaxies that small intrinsically weak radio sources are not infrequently found in the nuclei of elliptical galaxies (Heeschen, 1970).

Two of the best studied sources are the radio galaxy 3C 84 (NGC 1275) and the QSS 3C 273. Both show the characteristic hierarchal core-halo structure. Figures 10 and 11 show the spectra of the individual components based on published and unpublished long-baseline interferometry measurements made between 6 and 75 cm, conventional interferometry (Ryle and Windram, 1968), observations of lunar occul-

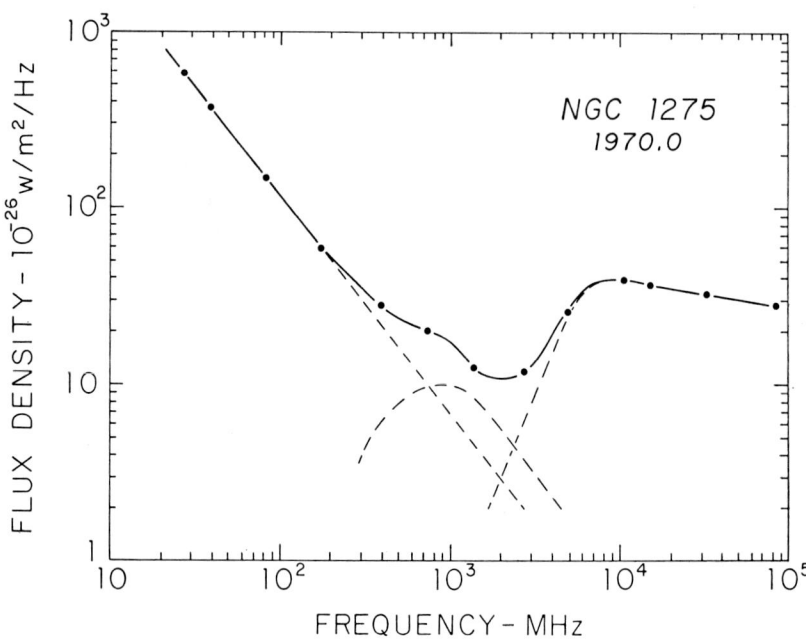

Fig. 10. Radio spectrum of the galaxy NGC 1275 (3C 84) showing the separation into components.

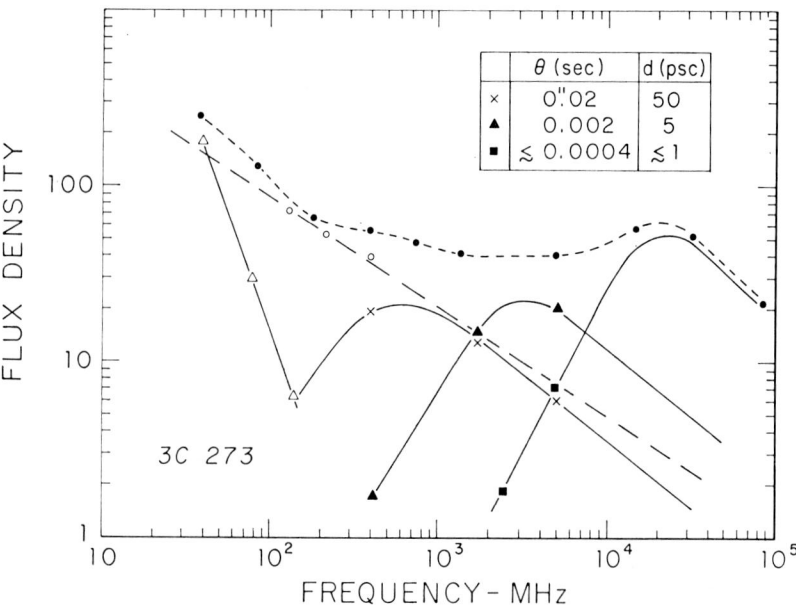

Fig. 11. Radio spectrum of the QSS 3C 273 showing the separation into components.

tations (e.g. Hazard, 1965), and interplanetary scintillations (e.g. Bell and Hewish, 1969). The smallest component in 3C 273 is less than 0.0004 arc sec corresponding to less than 1 pc at a distance of 500 Mpc. This is significantly smaller than the probable dimensions of the optical emission-line region (Bahcall and Kozlovsky, 1969).

The small component in 3C 84 is completely resolved at the longest baselines and appears to be circularly symmetric with dimensions about 0.002 arc sec or 2 light years.

5. Energetics

For those sources where there are accurate measurements of the self-absorption cut-off frequency, v_c, the peak flux density, S_p, and angular size, θ, it is possible to determine directly the magnetic field strength, B, from the expression

$$B \sim 2.4 \times 10^{-5} S_p^{-2} \theta^4 v_c^5 (1+z)^{-1} \text{ G}.$$

It must be emphasized that the magnetic field determined in this way does not require knowledge of the distance to the source and is independent of any assumptions of equipartition. The accuracy with which the field strength is determined, however, is poor since it depends on large powers of the measured quantities θ and v_c.

A. EVOLUTION OF COMPACT SOURCES

Figure 12 shows a plot of self-absorption cutoff frequency, v_c vs surface brightness, S/θ^2, for radio source components for which there are good observational data. Most sources appear to have magnetic field strengths in the range $10^{-4 \pm 1}$ G and a peak brightness temperature between 10^{11} and 10^{12} K. This appears to be the limiting brightness temperature of a source of incoherent synchrotron emission from relativistic particles because of inverse Compton cooling (Kellermann and Pauliny-Toth, 1969).

This type of cutoff frequency-surface brightness diagram is a useful way of displaying the evolution of compact radio sources. Individual components are thought to be born in the upper right hand part of the diagram. If the expansion occurs with conservation of magnetic flux, then the peak brightness temperature remains approximately constant (Kellermann and Pauliny-Toth, 1969) and the source evolves toward the lower left part of the diagram.

B. ENERGY REQUIREMENTS

If the distance of the source is known, the energy contained in relativistic electrons, E_e, and in the magnetic field, E_m, may be calculated without any assumption of equipartition between the energy in particles and in the magnetic field. The uncertainties, however, are very great, and depend on approximately the 10th power of the angular size and cutoff frequency (Kellermann and Pauliny-Toth, 1969).

Several investigators have noticed that some sources with small measured size do not show the expected low frequency cut-off and have estimated $10^{-7} \lesssim B \lesssim 10^{-6}$ G and $E_e/E_m \gg 1$ (e.g. Bridle, 1967). More recent measurements, however, indicate that low frequency emission does not come from the small components, which do in fact

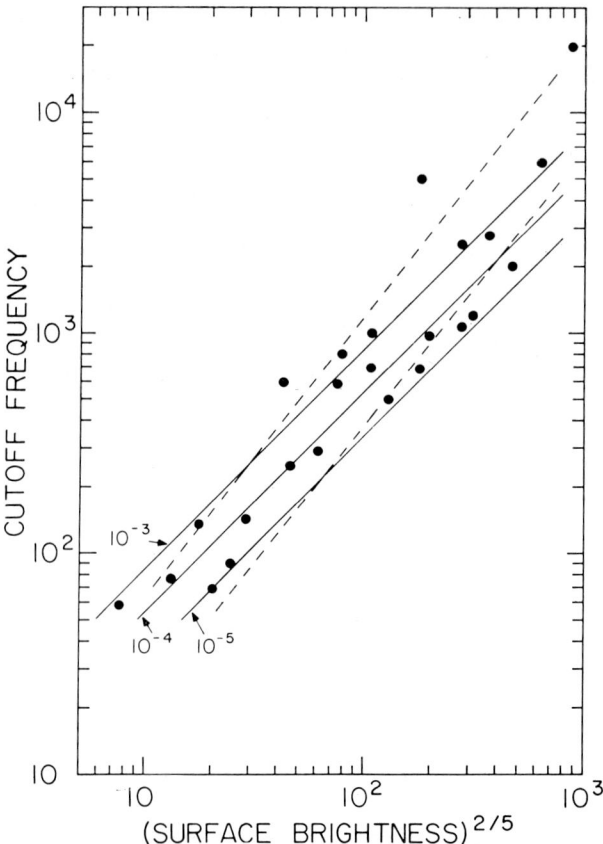

Fig. 12. Relation between observed cutoff frequency and surface brightness. The surface brightness is measured in units of 'flux units arc sec^{-2}'. The solid lines are lines of constant magnetic field strength and the upper and lower dashed lines represent lines of constant peak brightness temperature of 10^{11} and 10^{12} K.

become optically thick at a frequency corresponding to magnetic fields near 10^{-4} G and values of E_e/E_m which do not differ significantly from unity. We must bear in mind, however, that the uncertainty in the value of E_e/E_m is at least a few orders of magnitude.

The total energy in the form of relativistic particles and magnetic field in the small components of radio galaxies such as NGC 1275 determined in this way is about 10^{52} ergs. If the compact QSS are at cosmological distances, their energies are considerably greater and are about $10^{56\pm2}$ ergs. These values are both very much less than the minimum energy of 10^{58-61} ergs estimated for the extended components based on equipartition arguments (e.g. Macdonald et al., 1968). It is therefore clear that it is not possible for a single compact radio source to evolve into an extended source, unless particles are continuously accelerated.

There is, of course, direct evidence of repeated particle accelerations from the observation of intensity variations, and it is clearly important to ask "What is the energy required in a single outburst?". This is difficult to answer, however, since generally it has not been possible unambiguously to resolve the variable components.

A possible procedure in this case is to estimate the maximum linear size from the time scale of the variations, and the limits placed by the light travel-time across the source. Knowing the distance, it is possible to estimate the angular size and thus B, E_e, and E_m. Application of this procedure generally results in relatively weak magnetic fields and the requirement that up to 10^{58} ergs of relativistic particles (Pauliny-Toth and Kellermann, 1966; Moffet, 1966) are repeatedly released in individual outbursts in QSS such as 3C 273 in a time of a few months or less. Not only does this appear to be an unreasonable energy requirement, but because of the high relativistic particle density, inverse Compton radiation causes a very rapid loss of energy (e.g. Hoyle et al., 1966).

C. EXPANSION VELOCITY

Rees (1967), and Rees and Simon (1968) have pointed out that if the cloud of relativistic particles is expanding with relativistic velocity, then because of the finite speed of light the apparent rate of increase of angular size may be very much greater than given by the simple light travel-time argument. E_e can then be reduced by several orders of magnitude for $v/c \gtrsim 0.9$. Van der Laan (1970) has emphasized, however, that there is a limit to the amount that the energy requirements can be reduced since E_m will increase about as rapidly as E_e decreases. The minimum total energy occurs for values of v/c such that $E_m \sim E_e$. In the case of the 1966 outburst in 3C 273 this occurs for $v/c \sim 0.95$ which gives $E_e \sim E_m \sim 10^{54}$ ergs, or about 4 orders of magnitude less than given by the simple light travel-time argument.

Future long-baseline interferometer measurements of variable sources made over a sufficient length of time to determine the expansion velocity will be important in determining whether or not super-relativistic expansion velocities actually do occur. At the present time there are only a few relevant observations, and their interpretation is ambiguous.

The variable component of NGC 1275, which has been steadily increasing in intensity for about 10 years (Kellermann and Pauliny-Toth, 1968), is well resolved at 6 cm and has a diameter about 0.002 arc sec corresponding to a radius of 1 light year. If all the particles were produced when the intensity increase began in about 1960, and if the expansion has been at constant velocity then $v \sim 0.1\ c$. However, the intensity variations show considerable structure (e.g. Hobbs et al., 1968) suggesting recent recurrent activity in this source, so that the 0.002 arc sec component may be considerably younger than 10 yr and thus $v/c > 0.1$.

The only case where direct experimental evidence of a relativistic expansion velocity has been reported is from long-baseline interferometer measurements made at 13 cm (Gubbay et al., 1969) which showed no change in fringe amplitude of 3C 273 during a time when the total flux density decreased by more than 4 flux units. The authors conclude that the variable component was completely resolved by their interferometer and was therefore greater than 0.002 arc sec in diameter in November 1967. They assume that the intensity variations observed at 13 cm between November 1967 and June 1969 are due to the same event which produced the large variations which were observed in 1966 at centimeter wavelength. This event apparently originated in late 1965 (Kellermann and Pauliny-Toth, 1968; Dent, 1968), and therefore was about 2 yr old in late 1967 when the radius of the variable source was greater than 6 light years, and so an apparent expansion velocity of at least 3 times the velocity of light was derived. Figure 8 indicates, however, that the intensity increase observed at the nearby wavelength of 11 cm in 1967 was not of the form expected from the 1965 outburst, but was due to a more slowly varying component, in no way related to the 1965 outburst so that the age estimate of 2 yr is probably incorrect.*

Nevertheless, these measurements do appear to establish the simultaneous presence of a variable component which was resolved by the interferometer and at the same time a smaller unresolved component, which did not vary significantly. In the analysis of intensity variations, it is usually tempting to assume that the most rapid variations occur in the smallest components; the observations of Gubbay *et al.* suggest that this is not always true.

6. Summary

There appears to be no significant difference in the radio emission observed from compact objects such as QSOs or from galaxies, so that it is not possible to distinguish from the radio measurements alone, a galaxy from a QSS. But perhaps more important is the growing realization that the same phenomena are occurring both in galaxies and QSOs.

There is no evidence that the classical theory of incoherent synchrotron emission from ultra-relativistic electrons does not adequately explain all of the observed radio phenomena. In particular there is no evidence for any coherent or collective process as had been frequently suggested (e.g. Ginzburg and Ozernoy, 1966; Papadopoulos and Lerche, 1969).

It is of course widely accepted that the radio emission from extended radio sources is due to incoherent synchrotron emission. The good agreement between observed angular dimensions and those calculated from the spectral cutoff frequency, the limiting observed brightness temperature of 10^{12} K, and the success of the expanding source model in explaining the observed intensity and polarization variations all give weight to the same interpretation of the radio emission from the compact sources.

The energy contained in these compact components is of the order of 10^{52} ergs in the nuclei of galaxies, and considerably more in the QSS if they are at cosmological distances. The variable source observations indicate that energies of this order are released in repeated events lasting from a few weeks to a few years and in volumes of space having diameters of the order of one light year or less across. The observations

---

* Evidence for a highly relativistic expansion in 3C 279 is given by Moffet *et al.* in this volume, p. 228.

also determine the initial energy index, $\gamma \sim 1.5$ and initial magnetic field $B \lesssim 1$ G.

Further observations of intensity and polarization variations, particularly at short wavelengths corresponding to early epochs after particle generation will help determine more accurately the rate at which particles gain and lose energy and the initial electron energy distribution and magnetic field strength. The direct determination of the size of the variable components using very long baseline interferometry will measure directly the rate of change of angular size and thus the manner in which the magnetic field strength varies with time. By such studies of the behavior of the very young compact radio sources, we may hope better to understand the source of energy and the way it is converted into relativistic particles, and the relation between the quasi-stellar objects and the nuclei of galaxies where these energetic events appear to occur.

Acknowledgements

I wish to thank my colleagues at NRAO for fruitful discussions and for their helpful comments on the manuscript.

References

Aller, H. D.: 1970a, *Astrophys. J.* **161**, 1.
Aller, H. D.: 1970b, *Astrophys. J.* **161**, 19.
Andrew, B. H., MacLeod, J. M., Locke, J. L., Medd, W. J., and Purton, C. R.: 1969, *Nature* **223**, 598.
Andrew, B. H. and Kraus, J. D.: 1970, *Astrophys. J. Letters* **159**, L45.
Bahcall, J. N. and Kozlovsky, B. Z.: 1969, *Astrophys. J.* **155**, 1077.
Bell, S. J. and Hewish, A.: 1969, *Astrophys. Letters* **4**, 211.
Bolton, J. G.: 1966, *Nature* **211**, 917.
Bowyer, C. S., Lampton, M., Mack, J., and de Mendonea, F.: 1970, *Astrophys. J. Letters* **161**, L1.
Braccesi, A.: 1967, *Nuovo Cim.* **49**, Series X, 151.
Braccesi, A., Formiggini, L., and Gandolfi, E.: 1970, *Astron. Astrophys.* **5**, 264.
Brandie, G. W.: 1970, *Nature* **225**, 352.
Bridle, A. H.: 1967, *Observatory* **87**, 263.
Burbidge, G. R.: 1970, *Astrophys. J. Letters* **159**, L105.
Clark, B. G., Kellermann, K. I., Bare, C. C., Cohen, M. H., and Jauncey, D. L.: 1968, *Astrophys. J.* **153**, 705.
Clarke, R. W., Broten, N. W., Legg, T. H., Locke, J. L., and Yen, J. L.: 1969, *Monthly Notices Roy. Astron. Soc.* **146**, 381.
Cohen, M. H., Moffet, A. T., Shaffer, D. B., Clark, B. G., Kellermann, K. I., Jauncey, D. L., and Gulkis, S.: 1969, *Astrophys. J. Letters* **158**, L83.
De Jong, M.: 1967, *Astrophys. J.* **150**, 1.
Dent, W. A.: 1965, *Science* **148**, 1458.
Dent, W. A.: 1968, *Astrophys. J. Letters* **153**, L29.
Fitch, W. S., Pacholczyk, A. G., and Weymann, R. J.: 1969, *Astrophys. J. Letters* **150**, L67.
Ginzburg, V. L. and Ozernoy, L. M.: 1966, *Astrophys. J.* **144**, 599.
Grueff, G.: 1970, *Astrophys. J. Letters* **160**, L41.
Gubbay, J., Legg, A. J., Robertson, D. S., Moffet, A. T., Ekers, R. D., and Seidel, B.: 1969, *Nature* **224**, 1094.
Harris, D. E. and Hardebeck, E. G.: 1970, *Astrophys. J. Suppl.* **19**, 115.
Hazard, C.: 1965, *Quasi-Stellar Sources and Gravitational Collapse*, Robinson, Schild, Schücking (eds.), University of Chicago Press, p. 135.
Heeschen, D. S.: 1970, *Astrophys. Letters* **6**, 49.
Heeschen, D. S. and Wade, C. M.: 1964, *Astrophys. J.* **69**, 277.

Hobbs, R. W., Hollinger, J. P., and Marandino, G. E.: 1968, *Astrophys. J. Letters* **154**, L49.
Hobbs, R. W., Corbett, H. H., and Santini, N. J.: 1969, *Astrophys. J. Letters* **156**, L15.
Hogg, D.: 1969, *Astrophys. J.* **155**, 1099.
Hoyle, F., Burbidge, G., and Sargent, W. L.: 1966, *Nature* **209**, 751.
Jauncey, D. L., Niell, A. E., and Condon, J. J.: 1970a, *Astrophys. J. Letters*.
Jauncey, D. L., Bare, C. C., Clark, B. G., Kellermann, K. I.. and Cohen, M. H.: 1970b, *Astrophys. J.* **160**, 337.
Kardashev, N. S.: 1962, *Soviet Astron.* **6**, 317.
Kellermann, K. I. and Pauliny-Toth, I. I. K.: 1966, in M. Arakeljan (ed.), 'Instability Phenomena in Galaxies', *IAU Symp.* **29** (in Russian).
Kellermann, K. I. and Pauliny-Toth, I. I. K.: 1966, *Nature* **212**, 781.
Kellermann, K. I. and Pauliny-Toth, I. I. K.: 1968, *Ann. Rev. Astron. Astrophys.* **6**, 417.
Kellermann, K. I. and Pauliny-Toth, I. I. K.: 1969, *Astrophys. J. Letters* **155**, L71.
Kellermann, K. I. and Pauliny-Toth, I. I. K.: 1971, *Astrophys. Letters* **8**, 153.
Kellermann, K. I., Clark, B. G., Bare, C. C., Rydbeck, O., Ellder, J., Hansson, B., Kollberg, E., Höglund, B., Cohen, M. H., and Jauncey, D.: 1968, *Astrophys. J. Letters* **153**, L209.
Kellermann, K. I., Pauliny-Toth, I. I. K., and Davis, M. M.: 1968, *Astrophys. Letters* **2**, 105.
Kellermann, K. I., Clark, B. G., Jauncey, D. L., Cohen, M. H., Schaffer, D. B., Moffet, A. T., and Gulkis, S.: 1970, *Astrophys. J.* **161**, 803.
Kellermann, K. I., Jauncey, D. L., Cohen, M. H., Shaffer, D. B., Clark, B. G., Broderick, J. J., Rönnäng, B., Rydbeck, D. E. H., Matveyenko, L., Moiseyev, I., Vitkevitch, V. V., Cooper, B. F. C., and Batchelor, R.: 1971, *Astrophys. J.*, in press.
Kinman, T. D.: 1970, private communication.
Kleinmann, D. E. and Low, F.: 1970a, *Astrophys. J. Letters* **159**, L165.
Kleinmann, D. E. and Low, F.: 1970b, *Astrophys. J. Letters* **161**, L203.
Lang, K. R. and Terzian, Y.: 1969, *Astrophys. J. Letters* **158**, L11.
Le Roux, E.: 1961, *Ann. Astrophys.* **24**, 71.
Little, L. and Hewish, A.: 1968, *Monthly Notices Roy. Astron. Soc.* **138**, 393.
Locke, J. L., Andrew, B. H., and Medd, W. J.: 1969, *Astrophys. J. Letters* **157**, L81.
Locke, J. L., Andrew, B. H., and Medd, W. J.: 1970, paper presented at General Assembly, IAU, Brighton, England.
Low, F.: 1970, *Astrophys. J. Letters* **159**, L173.
Macdonald, G. H., Kenderdine, S., and Neville, A. C.: 1968, *Monthly Notices Roy. Astron. Soc.* **138**, 259.
MacLeod, V. M. and Andrew, B. H.: 1968, *Astrophys. Letters* **1**, 243.
McCullough, T. P. and Waak, J. A.: 1969, *Astrophys. J.* **158**, 849.
Miley, G.: 1970, private communication.
Moffet, A.: 1966, in M. Arakeljan (ed.), 'Instability Phenomena in Galaxies', *IAU Symp.* **29** (in Russian).
Olsen, E. T.: 1969, *Nature* **224**, 1008.
Ozernoy, L. M. and Sazonov, V. N.: 1969, *Astrophys. Space Sci.* **3**, 395.
Papadopoulos, K. and Lerche, I.: 1969, *Astrophys. J.* **158**, 981.
Pauliny-Toth, I. I. K. and Kellermann, K. I.: 1966, *Astrophys. J.* **146**, 643.
Pauliny-Toth, I. I. K. and Kellermann, K. I.: 1968a, *Astron. J.* **73**, 953.
Pauliny-Toth, I. I. K. and Kellermann, K. I.: 1968b, *Astrophys. J. Letters* **152**, L169.
Penston, M. V. and Cannon, R. D.: 1970, *Roy. Obs. Bull.* No. 159.
Rees, M.: 1967, *Monthly Notices Roy. Astron. Soc.* **135**, 345.
Rees, M. and Setti, G.: 1968, *Nature* **219**, 127.
Rees, M. and Simon, M.: 1968, *Astrophys. J. Letters* **152**, L145.
Rogstad, D. and Ekers, R.: *Astrophys. J.* **157**, 481.
Ryle, M. and Windram, M. D.: 1968, *Monthly Notices Roy. Astron. Soc.* **138**, 1.
Sandage, A. and Luyten, W. J.: 1967, *Astrophys. J.* **148**, 767.
Sargent, W.: 1970, *Astrophys. J.* **160**, 405.
Schmidt, M.: 1969, *Ann. Rev. Astron. Astrophys.* **7**, 527.
Schmidt, M.: 1970, paper presented at the *Semaine d'Etude* on the Nuclei of Galaxies, Vatican City, April 1970.
Schorn, R. A., Epstein, E. E., Oliver, J. P., Sater, S. L., and Wilson, W. J.: 1968, *Astrophys. J. Letters* **151**, L27.

Shimmins, A. J., Searle, L., Andrew, B. H., and Brandie, G. W.: 1968, *Astrophys. Letters* **1**, 167.
Shklovsky, I.: 1965, *Soviet Astron.* **9**, 22.
Simon, M.: 1969, *Astrophys. J. Letters* **158**, 865.
Slish, V.: 1963, *Nature* **199**, 682.
van der Laan, H.: 1966, *Nature* **211**, 1131.
van der Laan, H. and Perola, G. C.: 1969, *Astron. Astrophys.* **3**, 468.
van der Laan, H.: 1970, paper presented at the *Semaine d'Etude* on the Nuclei of Galaxies, Vatican City, April 1970.
Weymann, R.: 1970, *Astrophys. J. Letters* **161**, L21.
Williams, P. J. S.: 1963, *Nature* **200**, 56.
Zwicky, F.: 1964, *Astrophys. J.* **140**, 1467.

Discussion

Scheuer: Which sources have the smallest diameters you quoted, of about 3×10^{-4} arc sec?

Kellermann: 3C 273, 3C 279, 3C 454.3 and PKS 2145+06 have components smaller than 3×10^{-4} arc sec. 4C 39.25 and 3C 345 also have very small components, but the limits are about 30% greater.

Scheuer: You mentioned a compact source in Centaurus A. Is this the well-known double, or something smaller discovered recently?

Kellermann: Observations with the NRAO 3-element interferometer at 3.6 cm and 11.3 cm indicate a component which is less than about 1 arc sec, is optically thick at centimeter wavelengths, and is a few flux units. There have not yet been any very long baseline observations of Centaurus A, but the large optical depth implies that the angular size is probably about 10^{-3} arc sec.

Felten: You said something to the effect that the occurrence of flat radio spectra among the 'normal' radio galaxies does not imply flat electron spectra, but rather that the electron spectra are all like the galactic cosmic rays; you attributed these flat radio spectra to spatial distribution. Would you add a little more about this?

Kellermann: I said that the flat spectra were due to partial self-absorption and varying optical depth through the source and did not reflect a flat electron energy distribution.

Felten: But you did say that the electron spectra occurring in outbursts of *variable* sources are flatter, $n(E) \sim E^{-1}$. So are you saying that sources can be divided cleanly into these two groups?

Kellermann: Yes.

Noerdlinger: Did your energy estimates include an allowance for relativistic protons?

Kellermann: No, only electrons, that is what we need to explain the observed emission. The amount of energy in protons or the kinetic energy of expansion is another question.

Miss Harris: What is the reality of the low-frequency spectrum component shown in your slide of 3C 273? If real, do you attach any significance to its steep spectrum?

Kellermann: It comes from occultation measurements at Arecibo. I see no reason to question it. The spectrum is very similar to that of the small low-frequency component in the Crab Nebula.

Shakeshaft: The low-frequency component in 3C 273 has been confirmed by Hewish and Bell by means of studies of interplanetary scintillation of source at a frequency of 81.5 MHz.

THE ANGULAR STRUCTURES OF SOME COMPACT SOURCES

H. P. PALMER

Nuffield Radio Astronomy Laboratories, Jodrell Bank, Cheshire, Great Britain

From 1965 to 1967 radio link interferometers were operated between Jodrell Bank and Malvern, at wavelengths of 21, 11, and 6 cm. Preliminary results were published by Adgie *et al.* (1965), Barber *et al.* (1966), Palmer *et al.* (1967) and Miley *et al.* (1967). The full analysis of these data has now been completed. Donaldson *et al.* (1969) report that six sources were unresolved, and they set size limits in the range 0.1 to 0.01 arc sec. Seven sources were found to contain unresolved components, and similar but less complete data were obtained on 33 other sources. Four sources (3C 119, 138, 147, and 237) were shown to have double or more complex structures. (Donaldson and Smith, 1970; Donaldson *et al.*, 1970). The cores of 3C 274 and four other sources were shown to have elongations in the range 0.3 to 0.03 arc sec. One of these is 3C 287, and when the data on this source are combined with those of Clark *et al.* (1968) at 1660 MHz it appears that this source is a double of separation 0.08 arc sec in position angle 025°. This differs from the interpretation given by Clark *et al.* Some new data were also obtained on the well known double sources 3C 405 and 1938–15.

Critchley and Palmer (1971) will report shortly an investigation of the angular structures at 408 MHz of 105 radio sources catalogued at Parkes in the declination range −13° to −17°. Most of these sources have fluxes in the range 1–6 flux units. The angular structures found have been compared with the information available for a group of more intense sources from the 3CR catalogue. No general trend towards smaller angular sizes can be seen amongst the less intense sources. A comparison of the axial ratios of the double sources in the two groups suggests, however, that widely separated double sources having compact components are two or three times more abundant amongst the group of less intense sources. As one studies sources of decreasing flux density, confusion will eventually produce this effect, but in this data the compact components are 30–100 arc sec apart and they radiate more than 0.5 f.u. Chance coincidences could produce only about one in ten of the wide-spaced double sources found in this sample.

Several explanations have been suggested for this apparent increase in the proportion of sources with high axial ratios. It may arise for one or both of two observational reasons. The angular structures of the two groups were measured with instruments of very different resolving power, and the Parkes sources were selected at 408 MHz, while the 3CR sample was catalogued at 178 MHz. If the effect is confirmed for two groups of sources selected in the same way, and observed with the same instrument, it suggests that, if these faint double sources have red-shifts greater than 1, and hence separations greater than 1.8 Mpc, these small components must have remained confined for longer than 6×10^6 yr. However, these sources have the same angular scale as the more intense sources, so their red-shifts may be nearer 0.3.

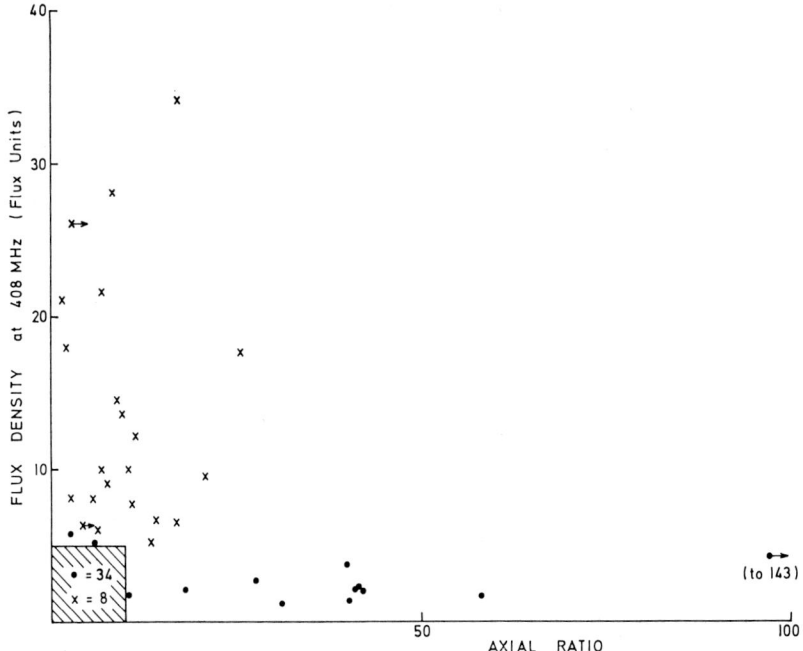

Fig. 1. The axial ratio (separation of components/component size) plotted against flux at 408 MHz, for the 77 double sources found in groups which are complete to 3, and to 1 flux unit respectively. (42 sources fall in the shaded area near the origin.) • The double sources found in a complete sample of 105 Parkes sources ($\delta - 13°$ to $-17°$); × The doubles found in a complete sample of 120 3C sources ($\delta > 35°$ $b'' > 10°$).

The components of classical double radio sources may contain 'cores' which continue to emit after the extended regions have decayed.

References

Adgie, R. L., Gent, H., Slee, O. B., Frost, A. D., Palmer, H. P., and Rowson, B.: 1965, *Nature* **208**, 275.
Barber, D., Donaldson, W., Miley, G. K., and Smith, H.: 1966, *Nature* **209**, 753.
Clark, B. G., Kellermann, K. I., Bare, C. C., Cohen, M. H., and Jauncey, D. L.: 1968, *Astrophys. J.* **153**, 705.
Critchley, J. and Palmer, H. P.: 1971, *Monthly Notices Roy. Astron. Soc.*, in preparation.
Donaldson, W., Miley, G. K., Palmer, H. P., and Smith, H.: 1969, *Monthly Notices Roy. Astron. Soc.* **146**, 213.
Donaldson, W., and Smith, H.: 1970, *Monthly Notices Roy. Astron. Soc.*, in preparation.
Donaldson, W. and Miley, G. K.: 1970, *Monthly Notices Roy. Astron. Soc.*, in preparation.
Miley, G. K., Rickett, B. J., and Gent, H.: 1967, *Nature* **216**, 974.
Palmer, H. P., Rowson, B., Anderson, B., Donaldson, W., Miley, G. K., Gent, H., Adgie, R. L. Slee, O. B., and Crowther, J. H.: 1967, *Nature* **213**, 789.

THE RADIO STRUCTURE OF QUASARS

G. K. MILEY and G. H. MACDONALD*

*National Radio Astronomy Observatory,** Green Bank, West Va., U.S.A.*

Abstract. The statistics on the angular structures of quasars have been more than doubled. Quasars are discussed from both morphological and statistical viewpoints and the angular diameter-redshift relation has been confirmed.

1. Introduction

Measurements of the radio structures of quasars are important for several reasons. Firstly, any understanding of the astrophysical processes in quasars must take into account both the morphology and the physical dimensions of the radio emitting regions. Secondly, if established, a relation between angular size and redshift would place constraints on the nature of the redshifts. Thirdly, if the redshifts are cosmological in origin, the statistics of quasar angular sizes might be a valuable new tool for investigating the geometry of the Universe.

Recently we investigated the structures of seventy-nine quasars using the 2695 MHz three-element interferometer at Green Bank with resolutions of up to 3″. All of these quasars had measured redshifts but structures that were previously unknown. Combining our data with those of other workers gives structural information on 128 quasars at the same observing frequency and with similar resolution.

2. The Morphology

A summary of the results is displayed in Table I. At 2695 MHz about half of all quasars have structures >5″, more than 20% are larger than 30″, and some are as large as 150″. We hope that this will dispel the widely prevalent illusion that the radio dimensions of all quasars are very small.

TABLE I

Interferometer observations of 128 quasars

	Number
Unresolved (<2″ near PA 70°, <7″ in any PA)	55
Slightly resolved (3″–10″)	10
Well resolved (>5″)	63

* Present address: University of Kent, Canterbury, England.
** Operated by Associated Universities, Inc., under contract with the National Science Foundation.

The structures of the well resolved sources can be classified as follows:

'D1': Two well-separated emitting regions, neither of which coincides with the optical QSO (7 sources).

'D2': Two well-separated emitting regions, one of which coincides with the optical QSO (2 sources).

'C': Triple or more complex (7 sources).

Most of the remaining sources appeared double but were not sufficiently resolved to establish whether the two emitting regions were separated or whether there was a bridge or third component between them.

D1 quasars (like 3C 47) and C quasars have quite similar radio components and relatively steep radio spectra which do not vary with time. On the other hand, a D2 quasar (like 3C 273) has one small active component coincident with the optical QSO, having an anomalous radio spectrum and a second larger component with a much steeper radio spectrum. It is likely that some of the sources which we find to be unresolved at 2695 MHz will be found to be D2 quasars when observed at lower frequencies, where the steep spectrum components contribute significantly to the total flux density of the source.

3. The Angular Size-Redshift Relation

We have now enough measurements of quasar angular structures to start playing

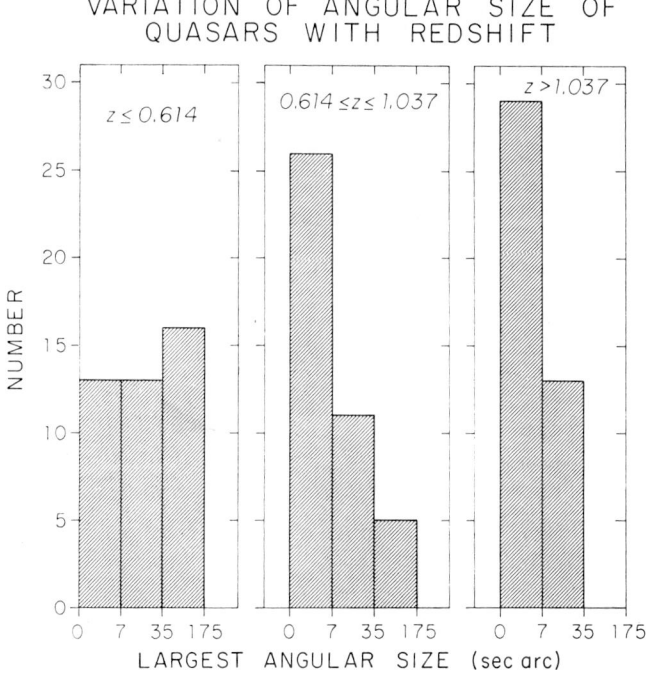

Fig. 1. Distributions of the largest angular size (LAS) of quasars for different ranges of redshifts.

the 'correlation game'. Just over two years ago Miley measured single fringe visibility points for seventy-six quasars. Despite the limited data he showed that for quasars with steep spectra the mean visibility increased with redshift, implying that the angular sizes of their radio components probably decreased with increasing redshift. Subsequently, Legg, and independently Miley and Macdonald, showed that the separation of known double quasars also appeared to decrease with increasing redshift. We have now more than doubled the available data and are therefore in a position to re-examine this correlation.

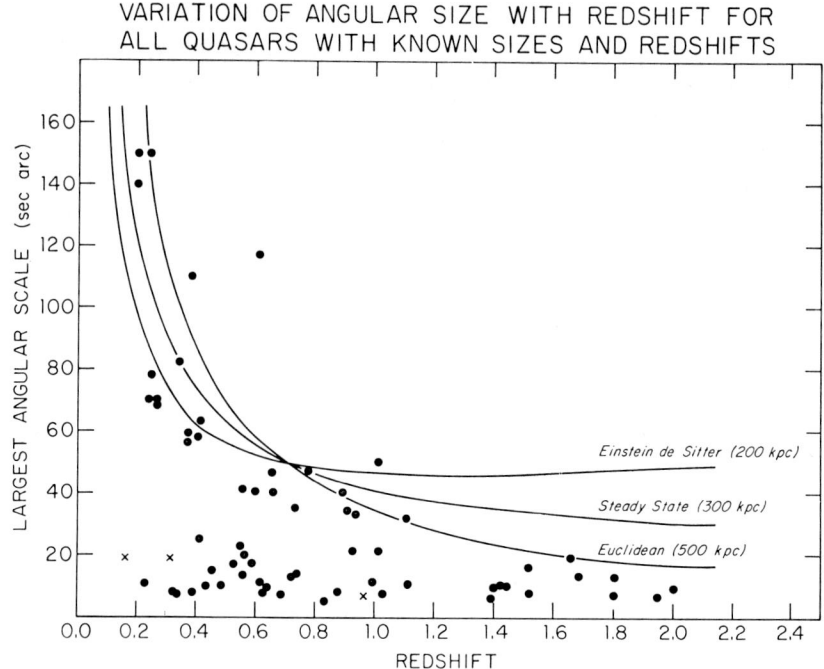

Fig. 2. Variations of LAS with redshift, z, for all quasars with known redshift and with angular size $\lesssim 7''$.

We have first divided our sample into three different redshift ranges with equal numbers in each and have drawn histograms of the three angular size distributions. These are shown in Figure 1 from which the decrease of largest angular size (LAS) with redshift is strikingly evident. Figure 2 is a plot of LAS against redshift for all quasars with known redshift and angular size. The curves show the expected behaviour for sources of constant physical size in different model Universes. The observed angular size will always lie below these envelopes because of projection effects. In Figure 2 we have been arbitrary in excluding the unresolved quasars. We can do this in a more satisfactory way by using the radio spectrum of the source as a suitable criterion. Figure 3 shows the same result for all quasars (including the unresolved ones) which have spectral indices steeper than -0.5 and which definitely have no low frequency cutoff in radio emission between 20 and 100 MHz. As you can see, not many unresolved sources remain here from our original fifty-five.

Although more statistics are needed, there is an indication in Figures 2 and 3 that the angular size falls off with redshift faster than one might expect for most cosmological models and in fact the best agreement occurs for a Euclidean universe. However, Rees and Sciama, van der Laan and Christiansen have all pointed out that inverse Compton losses due to the background radiation should increase as $(1+z)^4$ and would therefore 'snuff-out' physically large quasars at high redshifts.

Also, in our diagrams we have used structures obtained at the same *observing* wavelength (11.1 cm). A more useful comparison would be between structures measured

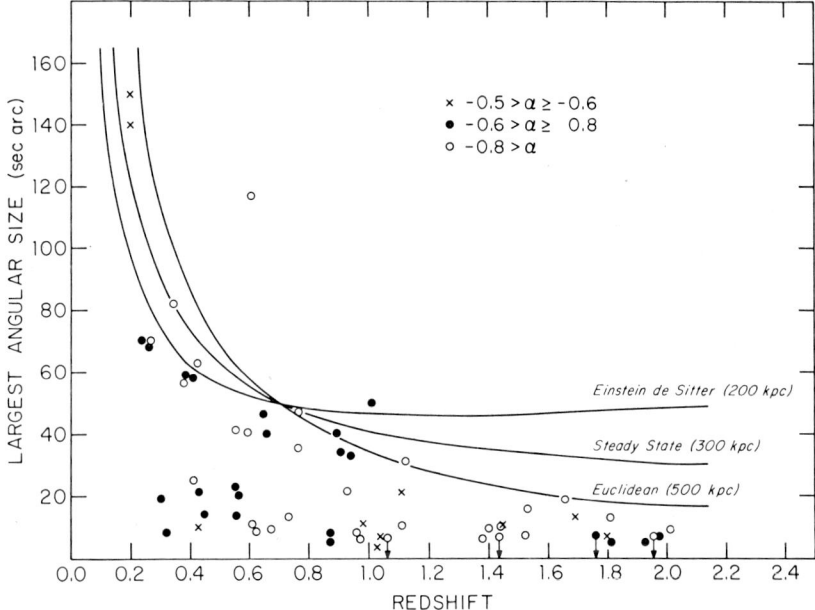

Fig. 3. Variations of LAS with redshift, z, for quasars with known redshift and angular size which have their spectral indices steeper than -0.5 and have no low-frequency cutoff.

at the same *emitted* wavelength. For the high redshift quasars the corresponding emitted wavelength is as short as 3 cm. As yet we have no knowledge of the structures of the small redshift quasars at 3 cm, but there is little evidence to expect that at this wavelength they should have very different structures from those observed at 11 cm.

Figure 4 is a combination of data from quasars and radio galaxies plotted logarithmically. It shows very clearly the continuity in overall angular size. Since this continuity must be explained by any theory of quasar redshifts, it is improbable that the origin of the redshift of quasars differs appreciably from that of the redshift of radio galaxies.

Finally we would like to mention briefly an apparent anisotropy which we observed when we plotted our angular size data on the sky. The small sources tended to cluster in the southern galactic hemisphere and there was a difference in the angular size distributions between the two hemispheres which is statistically significant at better than the 1% level. This can be explained because firstly, we have seen that angular size decreases with redshift and secondly in our complete sample there are relatively more high redshift quasars in the southern galactic hemisphere. The explanation for

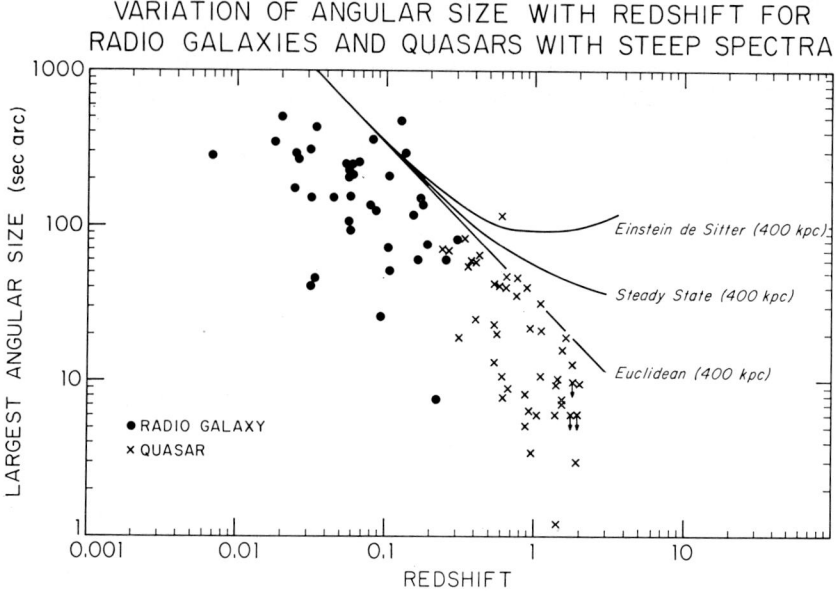

Fig. 4. Variation of LAS with redshift for quasars and galaxies with steep spectra and no low frequency cutoff.

the apparent anisotropy of angular sizes therefore lies with the unequal distribution in the observed redshifts. Complex selection effects occur when comparing optical spectra taken for different redshifts and the reality of this redshift anisotropy is a question for the optical astronomers.

4. Conclusion

We have shown that the largest angular size of quasars with steep spectra decreases with increasing redshift. This is the strongest observed correlation between the radio and optical properties of quasars and indicates that in future the linear diameter may be a more useful 'standard candle' in cosmology than is the luminosity.

Discussion of Papers Read by Palmer and Miley

Longair: Two comments. Firstly, Pooley and I have performed a redshift-angular diameter test for radio sources of large angular size by comparing the numbers of such sources in the 3C and 5C catalogues. We find that there is an excess of such sources in the 5C catalogue. This is, of course, not inconsistent with the present results since our test probably refers to intrinsically weaker radio galaxies.

Secondly, there are difficulties with the hypothesis that inverse Compton scattering of the relict radiation by the relativistic electrons originating in extragalactic radio sources can snuff out radio sources at large redshifts. Normally, this effect will be accompanied by an increase in the radio spectral indices of distant sources at the observing frequency and should be apparent in the spectral index distribution of QSOs at large redshifts. The absence of these changes in mean spectral index suggests that inverse Compton scattering does not affect the properties of distant QSOs. (M. S. Longair: 1970, *Monthly Notices Roy. Astron. Soc.*, in press).

Shakeshaft: Could Dr Palmer please clarify his plot of axial ratio of double sources against flux density? There are dangers in plotting together data from surveys with different resolving power.

Mackay: In connection with the plot of flux density against axial ratio (overall angular size divided by component size) shown by Dr Palmer, I think it unlikely that the lack of bright sources of large axial ratio is a real effect. The majority of sources in the Cambridge work from which Dr Palmer has taken the axial ratios for the bright sources have only lower limits to their axial ratios and so are not shown on his plot. Some of these limits are already very large, and as the Cambridge telescope is very different from that used by Dr Palmer, it seems probable that any effect may be attributed to differences in instrumental limitations.

Palmer: Almost all the comparison group of sources from the 3CR catalogue have also been observed with interferometers of baselines $\geqslant 20000\,\lambda$. I therefore believe that any double sources with separation $\geqslant 30$ arc sec and axial ratios in excess of 20, if present, are already known.

Miley: Because Cygnus A is the strongest known extragalactic radio source, it is possible to investigate its structure in more detail than that of any other such source. As Dr Palmer has mentioned Cygnus A in his talk, I would like to report some recent unpublished observations on the microstructure of this source by myself and Dr Wade. On a large scale Cygnus A has long been known to consist of two very similar components separated by about 2 arc min and located asymmetrically about the optical galaxy. Each component is extended by ~ 30 arc sec and has a small bright core near its outer edge. We have resolved both of these cores with the 2695 MHz radio link interferometer at Green Bank using baselines of up to 35 km or 3×10^5 wavelengths.

The data for the westerly core can be explained by a double model with components extended by ~ 1 arc sec and separated by 4.6 arc sec in a position angle of $130°$ (inclined by $20°$ to the main axis of the source). The data for the easterly core, although more difficult to interpret, show that it consists of at least two components (and not many more) with sizes ~ 0.5 arc sec.

These compact regions that we have found must all have brightness temperatures in excess of 3 million deg, a factor of about ten greater than the background source. Each of them subtends an angle of less than one degree at the parent galaxy. The double nature of at least one of them is of considerable interest since in this respect it would appear that the microstructure of extragalactic sources resembles the large scale structure.

COMPACT RADIO SOURCES IN
THE NUCLEI OF ELLIPTICAL GALAXIES

R. D. EKERS*

California Institute of Technology, Pasadena, Calif., U.S.A.

Abstract. Ten percent of the intrinsically bright elliptical galaxies contain compact radio sources (angular size <3 arc sec) with radio luminosity $\sim 10^{40}$ erg s^{-1}. The presence of a compact source is correlated with the presence of extended radio emission and with the presence of optical emission lines.

In order to investigate whether only giant elliptical galaxies have strong radio sources Rogstad and Ekers (1969) surveyed a sample of elliptical galaxies for radio emission at 10 cm. In addition to showing that being brighter than -20^m is a necessary condition for having a strong radio source, $L_R \geqslant 10^{41}$ erg s^{-1}, we found a number of elliptical galaxies which had weaker radio sources. At the same time Heeschen (1968) showed that NGC 1052 and NGC 4278 contained very small diameter radio sources with spectra of the type usually associated with the compact optically thick radio sources often found in quasi-stellar objects and Seyfert-type galaxies. These two galaxies were also included in our elliptical galaxies with weaker radio sources.

In order to investigate this class of object further I made a new survey at the Owens Valley Radio Observatory at 6 cm with an rms noise level of 0.03×10^{-26} W m^{-2} Hz^{-1} per observation. Twenty percent of the *bright* elliptical galaxies observed were detected. For these I obtained observations with the interferometer at a spacing of 16 000 wavelengths, which permitted division of the sources into two angular size groups. Eight galaxies were found to have radio sources with most of the emission coming from a region <3 arc sec and 12 were found to be >3 arc sec (Table I).

All the more extended sources have power law spectra while the small diameter sources have various spectral forms of the type already mentioned for NGC 1052 and NGC 4278. Heeschen (1970) in an independent survey has obtained similar results.

The radio component in M87 with diameter $<0''.002$ (Cohen *et al.*, 1969) and the optically thick component in Centaurus A, reported at this symposium by Dr K. I. Kellermann, are of similar radio luminosity and are most probably also objects of this type.

More than 4/11 and probably at least 8/12 of the elliptical galaxies containing compact radio sources have extended components, i.e. they are like M87 although the extended components are intrinsically weaker for the other objects. From this it necessarily follows that the total life of the compact source must be comparable with that of the extended source which must be at least 10^6 yr (the light travel time) in some cases. Also it is most likely that there is a continuing causal relationship between the two types of radio emission.

Optical emission lines of Hα, [N II] λ 6583 and others, as well as [O II] λ 3727 (e.g.

---

* Present address: Kapteyn Laboratorium, Groningen, Netherlands.

TABLE I
Elliptical galaxies detected at 6 cm

Extended (> 3 arc sec)			Compact (< 3 arc sec)		
NGC	Flux density 10^{-26} W m^{-2} Hz^{-1}	Other names	NGC	Flux density 10^{-26} W m^{-2} Hz^{-1}	Other names
383	2.1	3C 31	1052[a]	0.72[b]	
741[a]	0.22	PKS 0153+05	1587[d]	0.09	
1399	(0.9)	PKS 0336−35	2911[a]	0.16	
4261	9.0	3C 270	3078	0.14	
4374[a]	2.7	3C 272.1, M84	3998	0.13	
4472[a]	0.10		4278[a]	0.39	
4486[a]	68	Virgo A, M87	4472	0.06[b]	
5128	126	Cent A	4486	0.8[b,c]	Virgo A, M87
6047	⩾ 0.3		4552[a]	0.30	
7385	⩾ 0.14	PKS 2247+11	5077[a]	0.20	
7503	0.52	PKS 2308+07	7385	0.13[b]	
7626[a]	0.24	PKS 2318+07			

[a] Also found by Heeschen (1970).
[b] Both compact and extended components, flux density of compact component given.
[c] 13 cm flux of component < 0.002″ (Cohen et al., 1969).
[d] Or its companion 1′E.

Burbidge and Burbidge, 1965) are seen in 5/8 (62%) of the galaxies with compact radio sources, compared with 37/180 (20%) with any emission lines at all for the other elliptical galaxies.

References

Burbidge, E. M. and Burbidge, G. R.: 1965, *Astrophys. J.* **142**, 634.
Cohen, M. H., Moffet, A. T., Shaffer, D., Clark, B. G., Kellermann, K. I., Jauncey, D. L., and Gulkis, S.: 1969, *Astrophys. J. Letters* **158**, L83.
Heeschen, D. S.: 1968, *Astrophys. J. Letters* **151**, L135.
Heeschen, D. S.: 1970, *Astron. J.* **75**, 523.
Rogstad, D. H. and Ekers, R. D.: 1969, *Astrophys. J.* **157**, 481.

Discussion

Miley: As Dr Ekers says, because M87 is close we can examine the structure of the compact component in very great detail, I would like to report some results which tend to throw doubt on the assumption of circular symmetry which is usually made in the expanding shell model. The compact source in M87 (< 0.05 arc sec) was in fact discovered by the Jodrell-Malvern group and was reported over 3 years ago. A detailed analysis of these results which is about to be published shows that this component is in fact elongated along the optical jet of M87 by ~ 0.3 arc sec, evidence for the injection of electrons from the nucleus of M87 into the radio halo. The recent results of the American VLBI group shows that the compact component probably has the hierarchical structure described by Dr Kellermann, with a sub-component of size ⩽ 0.001 arc sec. The synthesis map of M87 by Hogg *et al.* shows a second small component associated with the optical jet of M87 and the picture that emerges of the micro-structure of the galaxy M87 resembles closely the large scale structure of the quasar 3C 273.

INTENSITY VARIATIONS IN EXTRAGALACTIC RADIO SOURCES AT 13 cm

G. D. NICOLSON

National Institute for Telecommunications Research, Johannesburg, S. Africa

Abstract. Results of a three-year investigation into the variability of 55 Parkes sources at 13 cm are presented. Thirty-six of the sources comprise a complete sample of QSS with fluxes exceeding 2 flux units. The remaining sources include most other known or likely variables stronger than 2 flux units. The relationship between spectra and variability in QSS is investigated and it is confirmed that variables generally have flat low frequency spectra. A possible relationship between redshift and specific types of intensity variations is considered. Limits on the secular change in the intensity of non-variable QSS are set and are generally found to be $\pm 1.5\%$ p.a. at 13 cm. Results for the remaining 19 sources are discussed and some preliminary findings of an extension patrol to include weaker sources in the range 1–2 flux units are given.

OBSERVATIONS OF VARIABLE RADIO SOURCES AT 8.2 mm

V. A. EFANOV, I. G. MOISEEV, H. M. TOVMASJAN,
V. B. SHTEINSHLEGER and V. I. ZAGATIN

Byurakan Astrophysical Observatory, Armenia, U.S.S.R.

Observations of the quasars 3C 273 and 3C 279 at 8 mm were begun in 1967 with the 22-m radio telescope of the Crimean Astrophysical Observatory of the Academy of Sciences of the USSR and have been continued subsequently (V. A. Efanov *et al.*, 1968, 1969, 1970a, 1970b).

Since 1968 observations have been made using a maser at 8.2 mm which improved appreciably the sensitivity of the radio telescope. The same maser was used in observations of the quasars 3C 273 and 3C 345 and of the Seyfert galaxy NGC 1275 (3C 84) in March of 1970. In these observations the feed was installed at the Cassegrain focus of the dish.

The beam of the telescope at half-power points is $1.'7 \times 1.'8$. The radio telescope was calibrated by observations of Jupiter, of which the brightness temperature was taken as 144 K.

Observational results presented in the following table are discussed below.

TABLE I

Flux densities of the observed radio sources

Source	Flux densities (in fl. units)	Number of scans	Date
3C 84	29 ± 1.5	10	17 March 1970
3C 273	45 ± 2.0	7	16 March 1970
3C 345	14 ± 1.0	7	18 March 1970

3C 273

The flux of 3C 273 at 8 mm decreased from October 1968 until March 1969 with a mean rate of 0.4 flux units per year.

3C 84

During the last three years the flux density of 3C 84 was almost constant at millimeter wavelengths. Comparison with measurements at 9.55 mm (Hobbs *et al.*, 1968) and at 4.3 mm (Hobbs *et al.*, 1969) suggests that at this range of wavelengths the spectrum of radio emission of 3C 84 is probably flat.

3C 345

The flux density of 3C 345 at centimeter wavelengths varies rapidly but its behaviour at millimeter wavelengths has not yet been observed.

Future observations of these three, and other variable quasars and galaxies, will be regularly carried out at 8 mm with the 22-m dish of the Crimean Observatory.

References

Efanov, V. A. and Moiseev, I. G.: 1968, *Izv. Krymsk. Astrofiz. Obs.* **40**.
Efanov, V. A., Zagatin, V. I., Moiseev, I. G., Misezhnikov, G. S., and Shteinshleger, V. B.: 1969, *Izv. Krymsk. Astrofiz. Obs.* **41**.
Efanov, V. A., Kisliakov, A. G., Moiseev, I. G., and Naumov, A. I.: 1970a, *Radiophysika* **13**, 219.
Efanov, V. A., Kisliakov, A. G., Naumov, A. I., and Moiseev, I. G.: 1970b, in press.
Hobbs, R. W., Corbett, H. H., and Santini, N. J.: 1968, *Astrophys. J.* **152**, 43.
Hobbs, R. W., Corbett, H. H., and Santini, N. J.: 1969, *Astron. J.* **74**, 824.

VARIABLE RADIO SOURCES: COMPARISON OF OBSERVATIONS WITH THE ADIABATIC SPHERICAL EXPANSION SOURCE MODEL

EUGENE E. EPSTEIN

Aerospace Corporation, Los Angeles, Calif., U.S.A.

Abstract. The adiabatic spherical expansion model of radio source outbursts has been applied to several outbursts in the quasi-stellar sources 3C 273 and 3C 454.3 and the Seyfert galaxies 3C 84 and 3C 120. Considering the simplicity of the model, it represents the data exceedingly well in a number of outbursts. The times of origin of the outbursts, the electron energy distribution indices, and the acceleration of the expansion parameters have been derived. Estimates of the sizes and magnetic fields of the outbursts have been made.

These results are part of a paper to be submitted to the *Astrophys. J.*

Discussion

van der Laan: The expressions which Dr Epstein used to compare data and theory are from the model which is instantaneous, homogeneous, spherically symmetric and non-relativistic. We never expected this model to be conformed to at wavelengths shorter than a few mm. When therefore the 3 mm flux is below the value deduced from longer wavelength data one need not look for a way of quenching the electrons radiating at that wavelength, because it is likely that there the instantaneous approximation is invalid; the generation duration overlaps the decay period computed at the instant of observation. The so-called poor fit at 3 mm is therefore more informative than the good fits mentioned.

As for developing more sophisticated models, there is a lack of input for this work, since the variables conform to the simple model discouragingly well. It is important therefore to continue monitoring sources which do not fit the model, in both intensity and polarization.

HIGH-RESOLUTION OBSERVATIONS OF VARIABLE RADIO SOURCES

A. T. MOFFET

California Institute of Technology, Pasadena, Calif., U.S.A.

and

J. GUBBAY

University of Adelaide, Adelaide, South Australia

and

D. S. ROBERTSON and A. J. LEGG

Weapons Research Establishment, Salisbury, South Australia

Abstract. Preliminary results are reported of a long-term program to test models of variable radio sources by direct observations of diameter changes. Although further confirmation is needed, increase in angular diameter attributed to a centimetric component of 3C 279 found in 1966 implies relativistic expansion with $\alpha \gtrsim 2$.

In order to sort out events taking place in variable sources it is necessary to observe frequently and at many wavelengths, especially at short wavelengths.

This is a progress report on a long-term program designed to test possible models of variable radio sources by direct observation of diameter changes of components of these sources. The observations are made with an independent-local-oscillator tape recorded interferometer using stations of the JPL-NASA Deep Space Network in California and Australia. The wavelength is 13 cm, and the spacing between stations is about 8×10^7 wavelengths, giving a resolution of the order of 0.001 sec of arc. The hardware used is all part of the standard station complement of equipment, so the only logistical problem is that of transporting the California tapes to Adelaide, where all the analysis is done on a CDC 6400 computer.

We have observed five sources in 1967 November, 1968 May and 1969 June. Concurrently with the last of these observations about 80 sources were observed with more sensitive equipment by the National Radio Astronomy Observatory – Caltech group (Kellermann *et al.*, 1970). An improvement in the bandwidth of our recordings increased our sensitivity and permitted us to observe 25 sources in 1969 December and 1970 June. We expect to re-observe this group of sources approximately each six months. The sources were chosen from those which showed the strongest fringes in the 1969 June NRAO-Caltech observations.

Substantial changes in both total flux density and in the correlated flux density over the California-Australia baseline have been noted for many of the sources. When more observations have been made it should be possible to test the application of the expanding source model to many events. The best case for a test so far is in 3C 279. In this source the flux density has increased by 3 flux units between 1967

November and 1969 December (our own observations and Nicolson, 1971). Over this same time the correlated intensity has increased by only 2.5 f.u. with a lower rate of increase than the flux density, particularly in the 1969 June to December interval. The increase in the flux density seems to be related to an expanding component first seen at short centimeter wavelengths in 1966 (Kellermann and Pauliny-Toth, 1968). It appears as if this component is now becoming resolved, implying that its angular diameter has reached about 0.001 sec of arc. If subsequent observations confirm this, it will imply a relativistic expansion with $\gamma \gtrsim 2$ (Rees, 1967).

In an earlier report on this work (Gubbay *et al.*, 1969) we claimed evidence for a relativistic expansion in 3C 273. This claim was based on an identification of a flux density decrease in 3C 273 with an expanding component seen in 1966 at short wavelengths. A more detailed comparison of the data on flux density vs time at several wavelengths shows that this identification is probably not correct, since the expanding component would have a peak flux density at 13 cm of less than one flux unit (Epstein, 1971; more or less this same result was also communicated to ATM by Dr M. Walmsley). This serves to point up the need for continued monitoring of flux densities of variable sources at many wavelengths if we are to be able to sort out the events taking place in these objects. It is clear that repeated interferometer observations should also be made at other wavelengths and at several spacings. In particular, the expanding components will be much more intense at wavelengths of 2 or 3 centimeters, and observations in that range would be very desirable.

Acknowledgements

We are indebted to the staff of the various stations of the Deep Space Network for their assistance in these observations and to NASA and JPL for permission to use their facilities.

References

Epstein, E. E.: 1971, this volume, p. 227.
Gubbay, J., Legg, A. J., Robertson, D. S., Moffet, A. T., Ekers, R. D., and Seidel, B.: 1969, *Nature* **224**, 1094.
Kellermann, K. I. and Pauliny-Toth, I. I. K.: 1968, *Ann. Rev. Astron. Astrophys.* **6**, 417.
Kellermann, K. I., Clark, B. G., Jauncey, D., Cohen, M. H., Shaffer, D., Moffet, A. T., and Gulkis, S.: 1970, *Astrophys. J.* **161**, 803.
Nicolson, G.: 1971, this volume, p. 224.
Rees, M.: 1967, *Monthly Notices Roy. Astron. Soc.* **135**, 345.

POLARIZATION OF QUASARS

R. G. CONWAY

Nuffield Radio Astronomy Laboratories, Jodrell Bank, Cheshire, Great Britain

I report here work done at Jodrell Bank by Mr John Gilbert and myself at a wavelength of 49 cm, on the degree of linear polarization, m_{49}. For most sources the polarization is considerably reduced from its short wavelength value. We presume that most of this depolarization is by Faraday rotation, or rather by Faraday dispersion, between different parts of the source.

The Australian workers have found little if any correlation of short wavelength polarization with other parameters. However, at a wavelength of 49 cm there is good correlation with spectral index, α: flat-spectrum or 'young' quasars show systematically higher polarization than steep spectrum or 'old' quasars.

Because this effect does not show at short wavelengths, it is a difference of depolarization rate. The Faraday dispersion of steep quasars must be mostly >8 rad m^{-2}, whereas the Faraday dispersion of flat quasars is mostly below this figure, either because of a deficit of thermal electrons, or for some other reason.

The reduction of m with increasing wavelength suggests that there ought to be a correlation of m with redshift z, because as z increases the emitted wavelength decreases, and hence m should increase. For $z=2$ the emitted wavelength is 16 cm. We have predicted curves of m vs z both for flat and for steep-spectrum quasars, assuming that all quasars out to $z=2.5$ have the properties of the local ones for which $z<1$.

For each category, the polarization at high z is less than expected by a factor of about 3. Because the effect is present for both categories of quasar it can hardly be an evolutionary effect. It is more likely that distant quasars are covered by some sort of Faraday screen. For example a spiral galaxy would give a rotation measure of about 60 for a line of sight going through a spiral arm, but a much smaller value for a line of sight passing through an interarm gap. Actually our observations require a Faraday dispersion in excess of 80 rad m^{-2}. It is possible that field galaxies could provide this, but rather more likely that the Faraday screen is associated with the quasar.

In either case, one might expect the screen to show absorption lines. Out of 25 quasars with $z>1.25$ we found seven with absorption lines. Those *with* absorption lines have $\bar{m}_{49}=0.56\pm0.21\%$, while those without average $1.38\pm0.29\%$.

Hence it appears that the same clouds cause Faraday polarization and absorption lines. The product $N_e BL$ for such a cloud must be about 10^{-4} cm^{-3} G pc.

I also wish to announce that our most recent measurements made in March of this year, have shown circular polarization in 5 out of 21 quasars, with one further possible case.

We used the interferometer at Jodrell Bank of Mark I and Mark II telescopes, integrating for 8 hr per source. Measurements on 3C 295 were used as calibration.

The apparent value of circular polarization of 3C 147 and 3C 123 agreed precisely with 3C 295 to 0.03%, and we believe that the instrumental correction is right to this accuracy, and that all three sources are unpolarized.

Two sources have polarization which is clearly above noise, at 3.0σ and 3.4σ respectively. Four others have values from 1σ to 2σ which one could not credit on statistics alone, but they do all have the common characteristic of possessing a flat spectrum.

According to simple theory by Legg and Westfold the corresponding value of B is 10^{-3} G for 0.2% polarization. If the field is irregular then $|B| > 10^{-3}$ G.

Discussion

Komesaroff: If the circular polarization is proportional to $f^{-1/2}$ this does not prove the correctness of the 10^{-3} G deduced from the Legg and Westfold paper. Gleeson *et al.* have shown that for a dipole field which would in general yield a much lower circular polarization than the uniform field discussed by the previous authors, the circular polarization may still show an $f^{-1/2}$ dependence. The deduced 10^{-3} G is most probably a lower limit.

Kellermann: At least for CTA 102 the assumption of a power law energy distribution for the relativistic electrons is correct as shown by the short wavelength radio emission where the source is optically thin or has a power law spectrum.

Also for this source the value of the magnetic field derived from the circular polarization measurements is close to that estimated from the measured peak brightness temperature and self-absorption cutoff frequency.

Terzian: It is interesting to see that CTA 102 shows circular polarization. Several years ago CTA 102 was reported to be a periodic variable radio source; however, several series of observations performed with linearly polarized feeds showed no variability. The observations showing the variability were performed with circularly polarized feeds. In your observations did you notice any variability?

Conway: Our integration time was about 8 hr, during which time we noticed no change.

Scheuer: Is there any correlation between the rotation measure and the occurrence of absorption lines in QSOs?

Conway: I shall have to reply privately to this, which I agree is a very interesting question. May I comment that the difficulty in correlating anything against the presence of absorption lines is that the number of reported absorption lines is increasing very rapidly with time just now.

QSQs AND RADIO GALAXIES – THEIR SPECTRA AND TIME VARIATIONS AT RADIO FREQUENCIES

BEVERLEY J. HARRIS*

Mt. Stromlo and Siding Spring Observatories, The Australian National University, Canberra, Australia and The Ohio State University Radio Observatory, Columbus, O., U.S.A.

Abstract. Accurate relative spectra for 300 radio sources from the Parkes catalogue have been measured and a statistical study made of their relation to class of optical identification and to other radio properties. Individual spectra and their time variations have also been investigated.

The results support the contention that radio sources for which no identification appears on the Palomar Sky Survey prints may be galaxies more luminous at radio frequencies than those which are identified. From a study of QSOs, radio galaxies and these blank field objects, it appears (a) that with increasing radio luminosity compact components are more often found, their presence being indicated by synchrotron self-absorption at low and high frequencies, by flat, variable spectra at high frequencies, and by interplanetary scintillations; and (b) that where no compact component contributes to the spectrum at high frequencies, many spectra steepen with increasing frequency, an effect which may be more marked for the more radio luminous objects.

Detailed analyses of the time variations in the compact components of 22 variable sources are generally consistent with the adiabatically expanding, uniform sphere model of Shklovsky, Kellermann, van der Laan and others. The model was modified to include relativistic expansion according to the formulae given by Rees and Simon. The results suggest that these components have evolved within months or years, have linear dimension of 0.1 to 100 pc and magnetic fields of 1 to 10^{-4} G. Some spectra at frequencies above 5000 MHz suggest non-adiabatic expansion which may be the result of continued injection of energy into an expanding region.

1. Introduction

This paper presents the results of a study of the accurate spectra of 300 radio sources from the Parkes catalogue. The relations of these spectra to class of optical identification, radio luminosity, surface brightness, scintillation, polarization and time variability are investigated.

Observations of the accurate spectra were made between frequencies of 500 and 5000 MHz, using the 210-ft telescope and the 210–60 ft interferometer of the CSIRO Division of Radiophysics, at Parkes, Australia. These data have been combined with flux densities measured by other observers to obtain spectra for many of these sources over the frequency range 10 to 10000 MHz. Complete data on the new Parkes flux densities and spectra are presently being prepared for publication (Harris, 1970a). Detailed results of an investigation of these spectra, including their variability with time, are also in preparation (Harris, 1970b, c).

Examples of radio spectra are shown in Figure 1, which illustrates their qualitative classification. The spectra have been classified as having: (a) power law between 100 and 5000 MHz (Figure 1(a)); (b) negative curvature below 100 MHz (Figure 1(b)); (c) negative curvature above 1400 MHz (Figure 1(c)); (d) positive curvature above 1400 MHz (Figure 1(d)).

* Now at Dept. of Astronomy, University of Texas, Austin, Texas.

Polynomial curves have been fitted to all data by the method of least squares. The main objection to this method is that spectra cannot always be represented by simple polynomials. An alternative method, of fitting two linear spectra above and below a 'break' frequency near 600 MHz has been used by Kellermann *et al.* (1969) and Shimmins (1968), to study the statistics of curvature. Few spectra show this break clearly; others are better fitted by a more gradual change in spectral index. This may be expected if physical conditions vary throughout the emitting region, as implied by

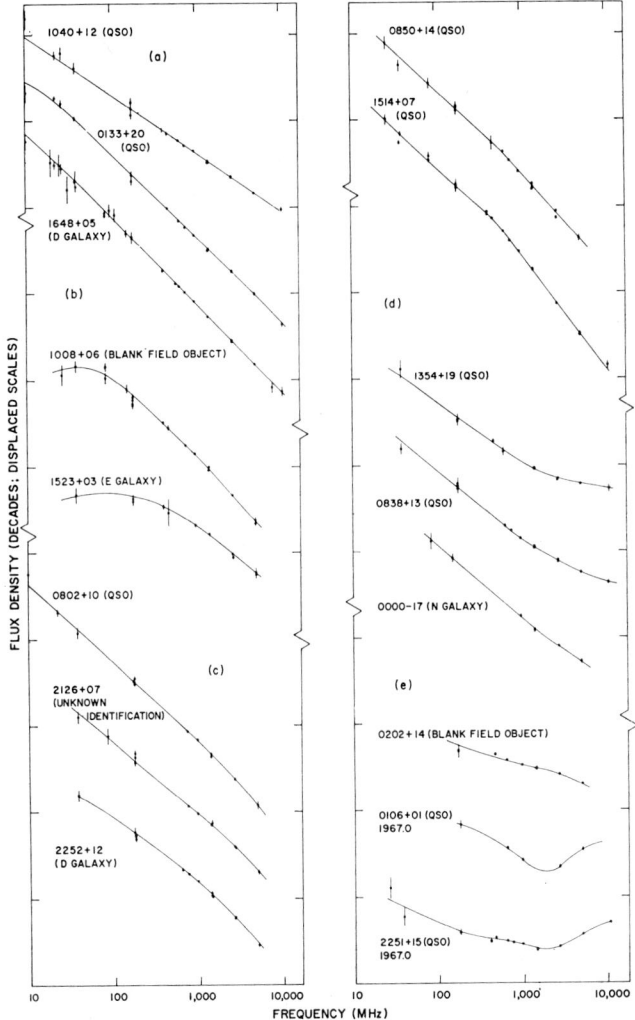

Fig. 1. Examples of spectra from the sample of 300. The following spectral classifications are illustrated: (a) power law spectra, between 100 and 5000 MHz; (b) spectra having negative curvature below 100 MHz; (c) spectra having negative curvature above 1400 MHz (two spectra having well defined 'break' frequencies are shown at upper right); (d) spectra having positive curvature above 1400 MHz; (e) complex spectra. The epoch is given for variable spectra. Note that PKS 1514+07 is identified with a D Galaxy, not a QSO.

observed brightness distributions (for example, Ekers, 1969). Where spectra are more complex than power law, polynomial fitting is the only solution.

Spectral indices α, curvatures, and flux densities were calculated from the fitted curves at observed frequencies f of 400, 1000 and 4000 MHz. These frequencies span the observed frequency range of most spectra, avoiding the extreme ends of the range where the fitted polynomial tends to diverge from the data.

The optical fields of the 300 radio sources have been examined on the Palomar Sky Survey prints by Bolton and others at the CSIRO Division of Radiophysics. The optical fields have been classified as containing:

(a) a galaxy. These have been further classified as S (spiral), E (elliptical), D, db (dumbbell), N, or 'faint' (too faint to be classified);

(b) a quasi-stellar object (QSO). All QSOs have been confirmed, most by their optical spectra, a few by their ultra-violet excess;

(c) no likely optical candidate within two standard deviations of the radio position. Unidentified radio sources whose optical fields are crowded or obscured were rejected from the present sample.

2. The Spectra and Radio Variability of QSOs, Radio Galaxies and Blank Field Objects

A. THE RELATION BETWEEN SPECTRA AND IDENTIFICATION CLASS

The number distribution with respect to the classes of radio spectra for QSOs, S, E, D, db, N, faint galaxies and blank field objects is summarised in Table I, and is discussed below:

(a) Spectra of power law form between 100 and 5000 MHz. There is a significantly smaller percentage of power law spectra in the QSO class than in the galaxy class. There are no power law spectra amongst blank field objects. Although data from studies of sources found in low frequency surveys were used to interpret spectral shape near 100 MHz, these conclusions are also expected to apply to the present sample selected at the higher frequencies of the Parkes 408–1410 MHz survey.

(b) Spectra having negative curvature below 100 MHz. If this negative curvature is attributed to synchrotron self-absorption by regions of different optical depth throughout the radio source angular dimensions of $0''.1$ to $1''$ are implied. The figures therefore show that roughly equal percentages of QSOs and galaxies emit an appreciable fraction of their radiation from regions of these dimensions, and that more than half of the blank field objects contain structure of this order. For many sources these small angular sizes are confirmed by observations of interplanetary scintillation and very long baseline interferometry. Since the low frequency curvature is determined mainly from flux densities of sources found in low frequency surveys, this result is applicable only to sources common to both the low and higher frequency surveys.

(c) Spectra having negative curvature above 1400 MHz. The fraction of sources having steeper spectra at high frequencies increases strikingly from QSOs, through galaxies to blank field objects. As will be shown later this trend is also followed in the degree of negative curvature. Some spectra having different spectral indices above and

TABLE I

Relation between optical identification and spectrum class

Class		QSOs	Galaxy types						All galaxies	Blank field objects
			S	E	D	db	N	faint		
(a) Spectra having power law between 100 and 5000 MHz	Number	9	0	12	12	4	1	4	33	0
	Percent of sample	22	0	28	52	50	14	36	35	0
	Total no. in sample	41	5	42	22	8	7	11	95	28
(b) Spectra having negative curvature, below 100 MHz	Number	16	4	12	5	4	3	5	33	17
	Percent of sample	38	50	27	20	60	40	55	33	50
	Total no. in sample	42	8	44	25	7	7	9	100	32
(c) Spectra having negative curvature, above 1400 MHz	Number	8	2	22	9	1	2	6	42	22
	Percent of sample	15	25	45	33	11	22	55	37	71
	Total no. in sample	53	8	49	27	9	9	11	113	31
(d) Spectra having positive curvature, above 1400 MHz	Number	16	0	2	1	0	3	0	6	1
	Percent of sample	30	0	4	4	4	33	0	5	3
	Total no. in sample	53	8	49	27	9	9	11	113	31

below a 'break' frequency fall in this class; examples are illustrated in Figure 1(c).

(d) Spectra having positive curvature above 1400 MHz. Whereas the low frequency spectra of QSOs and galaxies are indistinguishable, the same is not true above 1400 MHz. For 30% of the QSOs the flux density increases with increasing frequency. This is attributed to components of the source which become optically thick above 1400 MHz. Apart from Seyfert galaxies, which are excluded from these statistics, and N galaxies, this property is rare in the other radio sources. This result for QSOs has been given previously for a similar sample (Shimmins, 1968).

B. THE SPECTRA OF BLANK FIELD OBJECTS

One of the most important results of the present study is the continuous and marked negative curvature characterizing the spectra of nearly all blank field objects. This may be due in part to the redshifting of the high frequency spectra of distant galaxies into the observed frequency range. However, the high frequency observations of galaxy spectra (for example at 8000 MHz by Dent and Haddock, 1966) do not seem to support this explanation.

In general, above 600 MHz, this continuous curvature is more likely a result of aging than synchrotron self-absorption. Below about 600 MHz, synchrotron self-absorption is a probable mechanism.

C. COMPLEX SPECTRA

Examples of these are shown in Figure 1(e). Other examples include PKS 1434+03, 0859−14, 1229−02 and 1603+00. All complex spectra may be attributed to composite spectra of several emitting regions at least one of which is optically thick in the observed frequency range. Most are variable.

D. TIME VARIATION AND OPTICAL IDENTIFICATION

The previously detected time variations occur predominantly in QSOs. Only two galaxies, both of the Seyfert type, were observed to vary. Most of the radio sources were observed at least twice at 2650 and/or 5000 MHz. As the most accurate flux densities are available at these frequencies, and also as past observation and theory show that time variations are more probable at the higher frequencies, the following statistics will employ these data.

Table II gives for each identification class, the number of sources observed at least

TABLE II

Variability as related to flat and steep spectra for QSOs, radio galaxies and blank field objects

Identification class	Type of spectrum[a]	Number having 2 well-spaced observations at 2650 and/or 5000 MHz	Number found to vary	
			definitely	possibly
QSOs	steep	27	0	0
	flat	35	16	7
Galaxies	steep	62	0	1
	flat	7	4	1
Blank fields	steep	13	0	1
	flat	7	0	1
Others[b]	steep	12	0	1
	flat	9	4	1

[a] The division between steep and flat spectra is made at $\alpha_{1000} = 0.5$.
[b] Most of the identifications classed as 'others' could probably be included with the blank field objects; many are sources whose original identifications were rejected because they show no UV excess, or for which a more accurate position became available during the course of observations.

twice at 2650 and/or 5000 MHz, where the interval between observations was longer than $2\frac{1}{2}$ months. This number is compared with the number of sources in which time variations were detected, either definitely or possibly. The statistics for sources with steep spectra, and those having flat or positively curved spectra at high frequencies, are compared.

These results show that variability was detected in about 60% of the 'flat' spectrum sources, regardless of identification. This figure is certainly a lower limit to the true number of variable sources as the observations were not ideally spaced in time; also as many of the variations are known to be small, others must fall within the errors of measurement. For all identifications and blank field objects less than 2% of steep spectrum sources were found to vary. These findings are in essential agreement with those of Medd et al. (1968) at 10630 MHz, although they found definite variations in PKS 0518+16 (3C 138) and 3C 196, both of which have steep spectra at lower frequencies.

E. COMPARISON OF SPECTRAL INDEX AND CURVATURE BETWEEN QSOS, GALAXIES AND BLANK FIELD OBJECTS

Table III shows mean spectral indices α with their dispersions σ_α.

TABLE III

Mean spectral indices for QSOs, radio galaxies and blank field objects

Class of identification	Approx. no. of objects	$f = 200$ MHz		$f = 1000$ MHz		$f = 4000$ MHz	
		α	σ_α	α	σ_α	α	σ_α
QSOs	59	0.62 ± 0.05	0.37	0.67 ± 0.04	0.29	0.67 ± 0.05	0.34
All galaxies	120	0.65 ± 0.04	0.42	0.72 ± 0.02	0.23	0.85 ± 0.03	0.32
Blank fields	31	0.58 ± 0.05	0.28	0.73 ± 0.04	0.24	0.92 ± 0.04	0.25

Apart from the faint, unclassified galaxies, no significant distinction was found among various kinds of galaxy. Note that the results are not corrected for selection, which will reduce the numbers of steeper spectra at low flux densities particularly if the dispersion in spectral indices is large, as for QSOs. Because the ranges of flux density covered by different classes of object are similar, comparison of their mean spectral indices is still valid, but the above selection effect would tend to mask distinction between these classes.

While the mean spectral indices at low frequencies are very similar for the three classes there is a progressive increase in high frequency index from QSOs to galaxies to blank field sources. The flux densities relative to QSOs differ by 18% for galaxies to 30% for blank field objects at 4000 MHz. Figure 2 illustrates the above result in the form of histograms of spectral indices. Gaussian distributions were assumed in evaluating the dispersions of the spectral indices. This is a close approximation for galaxies and blank field objects. For QSOs the distribution has a tail of flat spectra which accounts for their lower mean spectral index and higher dispersion. At 1000 MHz the peak spectral index is 0.82 ± 0.06, values similar to those found for galaxies and blank field objects.

Kellermann's theory of the evolution of radio spectra for recurring injection of relativistic electrons predicts distinct spectral regions. The region $f_{MHz} > 10^3 B_G^{-3} t_{yr}^{-2}$ where synchrotron losses predominate, is steeper by ~ 0.5 than the region $f_{MHz} < 10^3 B^{-3} t^{-2}$, where these losses are balanced by recurring influx of particles. The maximum possible spectral index difference due to this effect is about 0.5 and occurs over a 10 to 1 frequency range. This is consistent with the data except where synchrotron self-absorption is affecting the lower frequencies; usually the difference is significantly less than 0.5 which would imply a variation in 'break' frequency through-

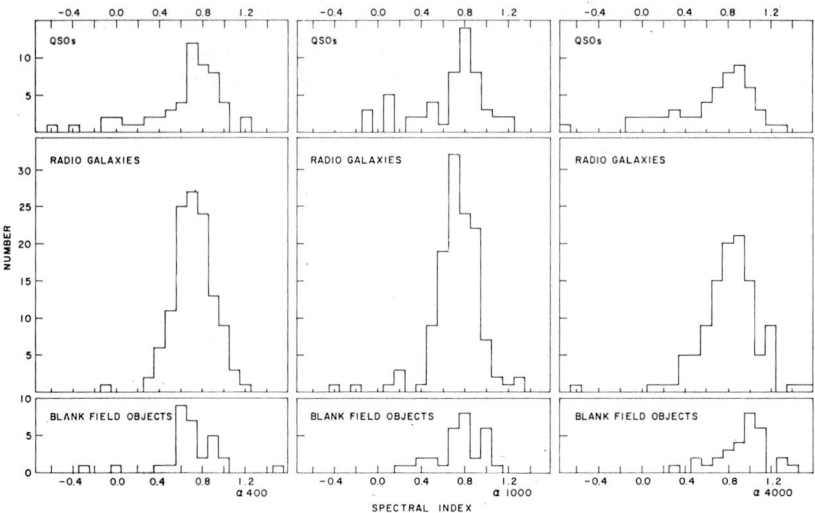

Fig. 2. Histograms of spectral indices α_{400}, α_{1000} and α_{4000} at frequencies of 400, 1000 and 4000 MHz, for QSOs, radio galaxies and blank field objects.

out the emitting region. The linear nature of the spectra of many sources over a wide frequency range implies that their equilibrium electron energy distribution is similar throughout this region.

3. Correlation of Spectra with Other Radio Properties

A. CORRELATION OF SPECTRAL INDEX AND LUMINOSITY

Figure 3(a, upper) shows the spectral index, α_{1000}, for E galaxies plotted against V_c, the photoelectric visual magnitude with correction for galactic absorption and K-effect. Figure 3(a, lower) shows α_{1000} for all identifications plotted against redshift for galaxies. At least for E galaxies, there is a suggestion of steepening of spectra with increasing luminosity (or redshift). It is not clear whether the lower boundary of spectral indices for QSOs may continue this trend to higher luminosities (or redshifts). Observational selection cannot account for the steeepening of galaxy spectra with redshift, since the distribution of redshift is essentially independent of flux density.

To investigate further whether spectra steepen with luminosity rather than with redshift, spectral index has been graphed against the logarithm of luminosity at 400 MHz, $\log P_{400}$, for galaxies at approximately the same distance in Figure 3(b). To increase the numbers of sources and extend the luminosity range, data have been taken from Kellermann (1964) for the strong, extended galaxies Fornax A, Centaurus A, and Virgo A, and also from Hobbs *et al.* (1968) for Cygnus A. The separate graphs for different ranges of V_c suggest the interpretation of a luminosity dependence of spectral index.

Kellermann (1966) has suggested that the more luminous radio sources have greater

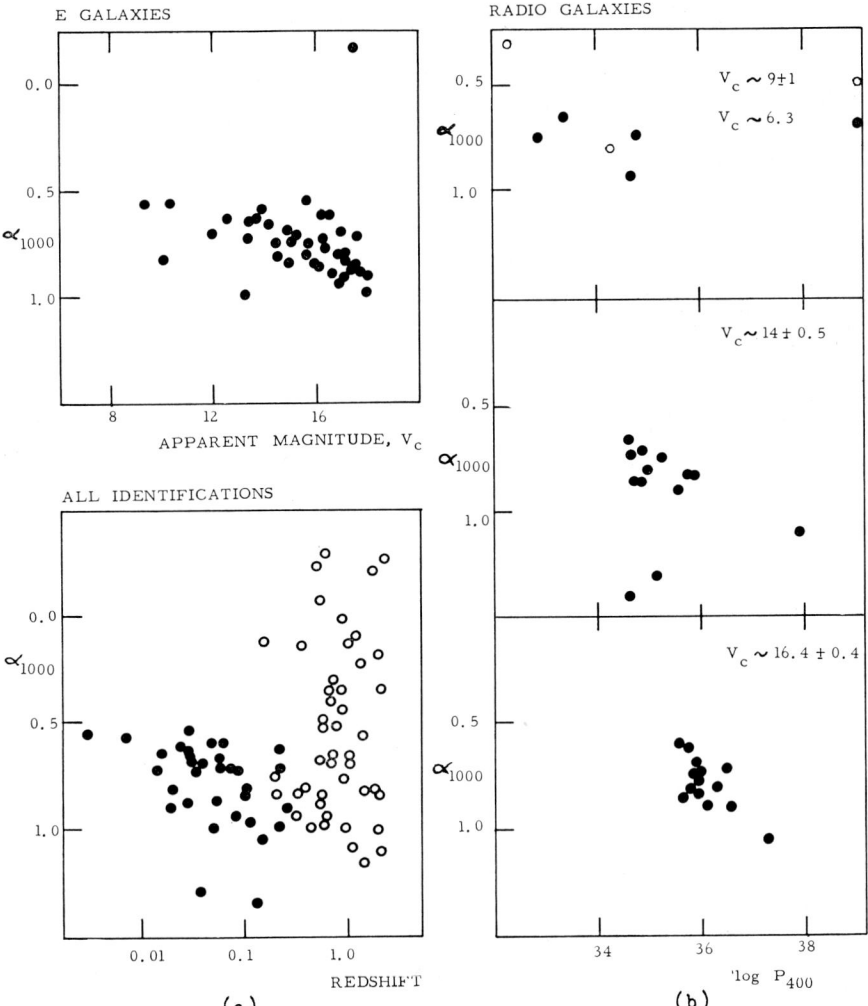

Fig. 3. (a, upper) The spectral index α_{1000} at 1000 MHz for E galaxies plotted against V_c, the photo-electric visual magnitude with correction for galactic absorption and K-effect; and (a, lower) for all identifications vs redshift, (b) spectral index α_{1000} vs $\log P_{400}$ where P_{400} is the luminosity at 400 MHz, for different ranges of V_c as indicated.

emissivities and hence evolve more rapidly. The steeper spectral indices could then be a consequence of this.

B. THE LUMINOSITY – SURFACE BRIGHTNESS DIAGRAM

The logarithm of luminosity at 400 MHz has been graphed against the logarithm of surface brightness (not shown). Surface brightness, or a lower limit, has been calculated using available data on brightness distributions found from long baseline interferometry and interplanetary scintillations. The same general trends are apparent as in the diagram by Heeschen (1966) or Longair and Macdonald (1969). These trends are:

(a) for radio galaxies, increase in luminosity roughly proportional to surface brightness which is well defined toward the upper left of the diagram, but poorly defined to right (high surface brightness);

(b) for QSOs, a horizontal branch of high luminosity which joins the galaxy branch at low values of surface brightness and extends to $B_{400} > 10^{-13}$ W m^{-2} ster^{-1} Hz^{-1}.

Heeschen interprets the diagram in terms of an evolutionary sequence in which an optically thin cloud of relativistic electrons expands, thus decreasing the surface brightness while maintaining constant luminosity. On approaching the branch at low surface brightness the expansion is retarded by magnetic fields and/or the intergalactic medium, so that further evolution occurs with constant or slowly increasing source dimensions. Luminosity decreases as particle energies decay, and the surface brightness decreases proportionally.

The relative positions of QSOs and galaxies in the luminosity-surface brightness diagram indicate a similar evolutionary pattern for both, but with reduced luminosity in the case of radio galaxies.

C. SPECTRAL INDEX AND LINEAR DIMENSIONS

If the above evolutionary scheme is correct, one might expect a steepening of the spectrum with decreasing surface brightness or with increasing linear dimensions as particle energies decay. The steep spectra of the extended halos about some individual radio galaxies are well known (for example, Ekers, 1969; Macdonald et al., 1968). For galaxies a trend of increasing spectral index with surface brightness was found. This may be accounted for by the spectral index – luminosity and luminosity – surface brightness relations. Apart from this the present results show no correlations between spectral index or curvature with surface brightness or overall source dimensions, above a 5% level of significance.

D. SPECTRAL INDEX AND INTERPLANETARY SCINTILLATIONS

The relation between spectral index and linear dimensions may also be studied through interplanetary scintillation. The detection of scintillation indicates that more than 50% of the flux density is radiated from a region of angular dimensions $<0\overset{''}{.}2$ (for example, Gardner and Whiteoak, 1969). Since the observed range of linear size among radio sources is much greater than their range of distances, the presence or absence of scintillation implies the presence or absence of components of compact linear dimensions.

Mean spectral indices for scintillators and non-scintillators are given in Table IV for 400, 1000 and 4000 MHz. QSOs, galaxies and blank field objects are distinguished. Both steep and flat spectrum objects belong to the class of scintillating QSOs, whereas nearly all non-scintillating QSOs have steep galaxy-like spectra. The flat spectra may thus be associated with angular dimensions $<0\overset{''}{.}2$. Because the observed range of flux densities is small compared with the range of linear (or angular) dimensions,

TABLE IV

Scintillation statistics for QSOs, radio galaxies and blank field objects

Identification class	Percentage of scintillators		α_{400}	α_{1000}	α_{4000}
QSOs	70	Scintillators (39)	–	0.55 ± 0.03	0.51 ± 0.03
		Non-scintillators (16)	–	0.78 ± 0.07	0.78 ± 0.07
Radio Galaxies	25	Scintillators (19)	–	0.78 ± 0.03	0.89 ± 0.03
		Non-scintillators (57)	–	0.74 ± 0.02	0.85 ± 0.02
Blank field objects	60	Scintillators (11)	0.43 ± 0.11	0.68 ± 0.07	0.94 ± 0.05
		Non-scintillators (7)	0.57 ± 0.05	0.75 ± 0.05	0.92 ± 0.05

compact components are always associated with high surface brightness for QSOs. For a flux density of 10×10^{-26} W m^{-2} Hz^{-1} at 300 MHz, and angular diameter $< 0''.1$, $T_b > 10^{10}$ K, so the flat spectra are probably a result of synchrotron self-absorption.

For the galaxies, 14% of those brighter than $17^m.0$ scintillate compared with 48% of the fainter galaxies. This can be explained by the decreasing angular size with increasing distance. As shown in Table IV the scintillating galaxies have a larger mean spectral index than the non-scintillating galaxies. This difference is of marginal significance; it is the result of the spectral index – luminosity and luminosity – distance relations.

The results for blank field objects imply: (a) that if they are of similar radio structure to the galaxies, they contain a higher proportion of scintillators by virtue of their greater distances; and (b) that those with the flattest spectra at frequencies near 400 MHz (and hence of highest surface brightness) contain structure on the smallest scale. These flat spectra are accounted for entirely by self-absorption and not by a flat, presumably young, electron energy distribution. At 4000 MHz there is no significant difference in spectral index between the scintillators and non-scintillators.

E. SPECTRA AND POLARIZATION

For QSOs, α_{1000} has been plotted against percentage polarization at 2650 MHz (not shown). There are significantly more flat spectrum objects having low polarization than high polarization (at the 5% level of significance on a chi-squared test). This trend is not evident for 1410 MHz polarization. This result is consistent with the lower average polarization found for sources having compact components at high frequencies. The explanation may be that magnetic fields are highly disordered in young objects. Fields become aligned as the source expands, resulting in a higher degree of integrated polarization, as in the halo components of radio galaxies (for example, Morris and Whiteoak, 1968; Seielstad and Weiler, 1969). This would be in general agreement with the model proposed by Gardner and Whiteoak (1969). The large scatter in percentage polarization for steep spectrum objects may be due in part to different orientations of the magnetic field to the line of sight.

No other correlations were found (at the 5% level of significance) between spectral index or curvature, and percentage polarization at 1410 or 2650 MHz, or depolarization (as measured by the ratio of degrees of polarization at these frequencies). Nor were any significant correlations found between percentage polarization at 2650 and 1410 MHz and luminosity, redshift, or surface brightness although there is some indication that for radio galaxies a high percentage of polarization is associated with low surface brightness, and low luminosity.

4. Time Variations of Spectra

Observations of time variations of spectra in the northern hemisphere to December 1967, have been reviewed by Kellermann and Pauliny-Toth (1968). They have applied the theory of an adiabatically expanding cloud of relativistic electrons to determine ages of the order of a few years for six sources whose flux densities have been well observed as a function of time and frequency. These authors have clearly demonstrated the good fit of the expanding source model to time variations observed in the Seyfert galaxy 3C 120 (PKS 0430+05) for frequencies below 5000 MHz. Locke et al. (1969) and Medd et al. (1968) report that observations at 6600 and 10600 MHz are generally consistent with the theory, although the simple model does not appear to explain the variations of PKS 1510−08, PKS 0763+01 and NRAO 512.

Variable spectra of 22 sources amongst the present sample have been analyzed individually in the light of the expanding source model of Kellermann, van der Laan, Shklovsky and others. However, this model was modified to include relativistic expansion according to the theory of Rees and Simon (1968).

In many cases linear extrapolation of the low frequency spectrum was assumed in order to extract higher frequency, optically thick components. Examples of the time variations of flux density and spectra for three sources are given below.

A. PKS 0430+05 (3C 120, SEYFERT GALAXY)

Comparison of the variations of flux density (Figure 4(a)) illustrates strikingly the time delay between the same event at different frequencies. Figure 4(b) gives the total observed spectrum in 1968.2. Figure 4(c) gives the spectra of the time varying components for several epochs, obtained by subtracting the spectrum of constant component A, assumed to be an extrapolation of the low frequency spectrum. Figure 4(d) shows the variable spectrum at three epochs analyzed into components C, D, E and F. Component B is not well determined and may not even exist. Component F may not be distinct, but may be an indication of continued outbursts at higher frequencies. These components may be determined uniquely, if their spectral indices in the optically thick region are assumed to be -2.5. For components C, D, and E the expansion ages at epoch 1968.2 are 3.0, 3.3, and 0.8 yr, respectively. If E is identified with the component having angular size $<0''.0008$ found from very long baseline interferometry by Kellermann et al. (1968) and Clark et al. (1968) then $B < 2.4 \times 10^{-3}$ G in 1968.1.

The angular diameters of the components of PKS 0336−01, 1127−14, 1226+02 (3C 273), 1253−05 (3C 279) and 2251+15 (3C 454.3) found from time variation theory have been compared with the diameters and limits to the angular diameters

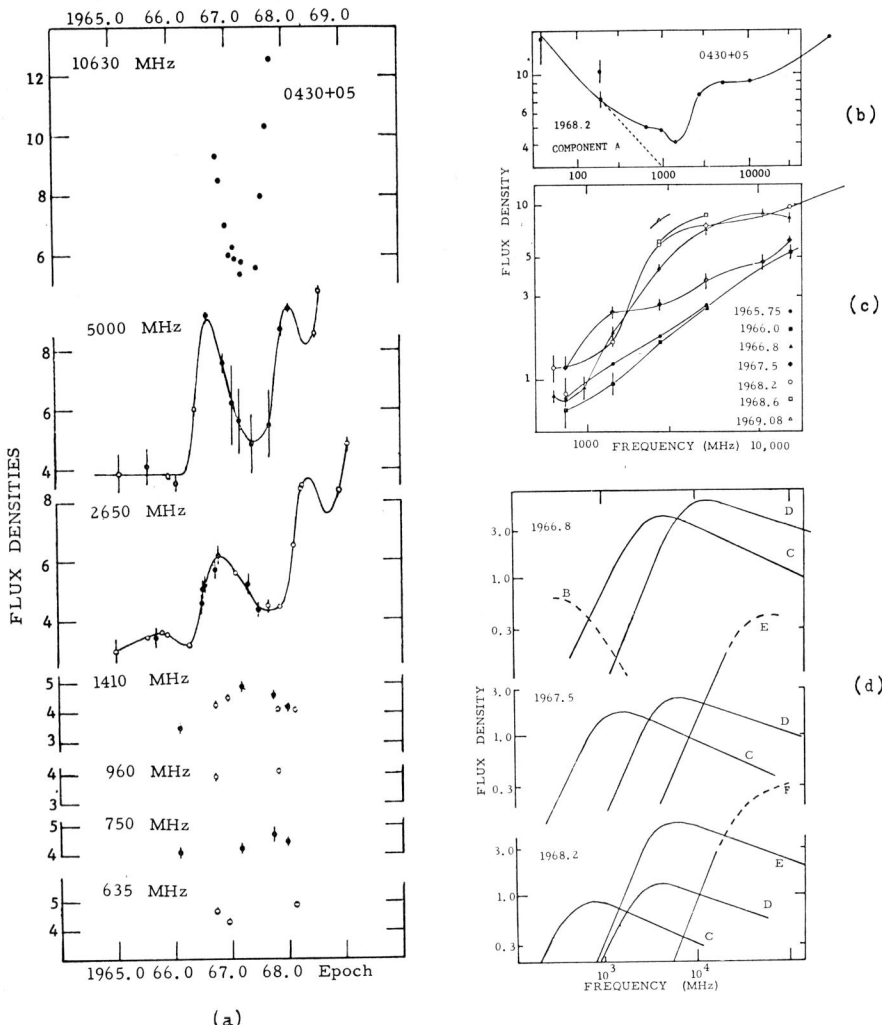

Fig. 4. The variable spectrum of the Seyfert galaxy, PKS 0430+05, showing (a) variations of flux density with epoch, at different frequencies; (b) the total observed spectrum in 1968.2; (c) the spectra of time-varying components at several epochs, found by subtracting the spectrum of component A from the total spectrum (spectrum component A at high frequencies assumed to be an extrapolation of the total spectrum at low frequencies); (d) analysis of the total variable spectrum, at three epochs, into components C, D, E and F. These components may be determined uniquely, if their spectral indices in the optically thick region are assumed to be -2.5. The open circles are data points from the present study. Closed circles are from other published results.

found from VLB interferometry. They show good agreement if magnetic fields are between 1 and 10^{-4} G within these variable components. The same is true for the sources 3C 84 and 3C 345.

The observations for most sources were consistent with the adiabatic, relativistic expansion model. The following are examples which do not appear to be consistent with adiabatic expansion:

B. PKS 0736+01 (QSO)

The Parkes observations at 2700 MHz bear out the very sharp decline observed at 6630 and 10630 MHz by Locke *et al.* (1969) (Figure 5(a)). The spectrum in 1967.74 (Figure 5(b)) shows no detectable high frequency components. In 1968.62 flux densities

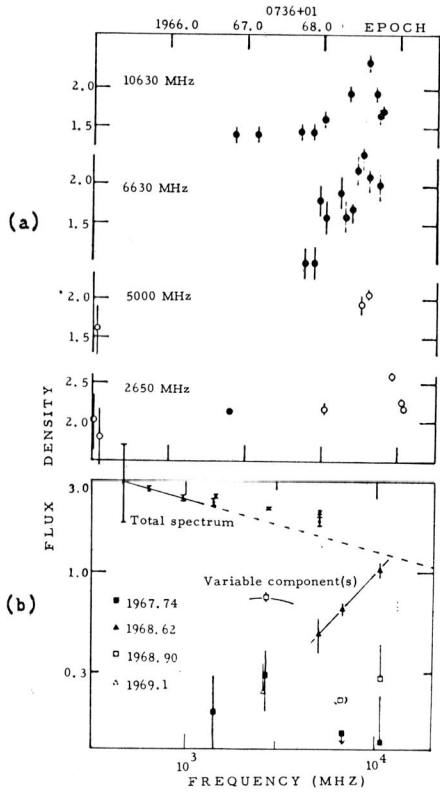

Fig. 5. The variable spectrum of the QSO, PKS 0736+01, showing (a) variations in flux density with epoch, for several frequencies (the open and closed circles have the same meaning as in Figure 4); (b) the total spectrum, and spectrum of the variable component at four epochs. The dashed line is an extrapolation of the low frequency spectrum, which was subtracted from the total spectrum at higher frequencies, to obtain the variable component.

at 5000, 6630 and 10630 MHz reached a maximum almost simultaneously, in conflict with the adiabatic expansion of an optically thick region. At this time, $\alpha = -1.0$ for the high frequency component indicating significant absorption. Although this high frequency component was barely detectable at 6630 and 10630 MHz in 1968.9 the source must have been optically thick at 2700 MHz until this time, after which the flux density rapidly decreased due to adiabatic relativistic expansion. The disappearance of the source at these frequencies by 1969.0 could be explained by relativistic expansion of an optically thin region. Locke et al. (1969) suggest that we have observed the birth of a variable component.

C. PKS 2345−16 (QSO)

Below 5000 MHz the variations are consistent with an expanding optically thick source, becoming optically thin above 2700 MHz at 1967.5 (Figure 6). At 5000 MHz

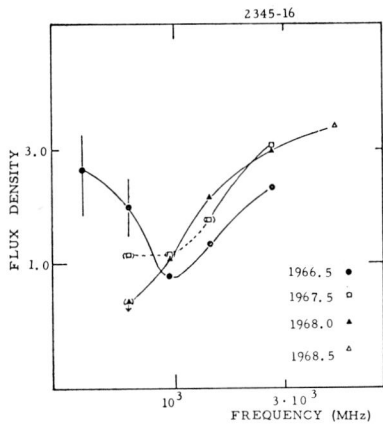

Fig. 6. The spectrum of the QSO, PKS 2345−16, at several epochs.

the flux density variations suggest continued injection of relativistic electrons. The very rapid decrease from 3.66 ± 0.07 to $3.46 \pm 0.07 \times 10^{-26}$ W m^{-2} Hz^{-1} between 1968.54 and 1968.60 (not shown) implies expansion velocities approaching that of light. These events may be compared with those in PKS 0736+01 and 1510−08, also the northern source NRAO 512 (Locke et al., 1969).

The results of the analysis of 22 variable spectra show that, for the observed variable components:

(a) Ages range from a few months to more than 100 yr;
(b) The linear dimensions vary between ∼0.1 pc and 100 pc;
(c) Relativistic expansion velocities are general, even if the QSOs are more local than implied by a cosmological interpretation of their redshifts. If the expansion is relativistic, energy requirements are reduced and possible inverse Compton catastrophes are avoided (Rees and Simon, 1968). Such high velocities appear to be necessary if the expanding synchrotron model is to explain the rapid variations of flux density observed in PKS 2251+15, 0736+01, 1510−08 and 2345−16.
(d) The derived angular sizes are in excellent agreement with those of very long baseline interferometry, if magnetic fields of 1 to 10^{-4} G are accepted. This agreement is independent of a cosmological interpretation of the redshifts.

Typical parameters for these young sources, obtained from the present study, are summarized in Table V. These are compared with the results of Colgate (1968) derived

TABLE V

Typical data derived for variable radio components

	Observed variable components[a]	Colgate's theory	
		Seyfert or N galaxy	QSO
Expansion age, t_0 (years)	1–100		
γ	1.0–1.5		
Expansion velocity (units of c)	0.03–0.9 (10^{-4})		
	0.3–0.996 (1.)		
Linear size (pc)	0.5–20 (10^{-4})	~ 0.01–1	~ 0.01–1
	5–200 (1.)		
Electron density (cm^{-3})	10^4–10^6 (10^{-4})		
	0.1–10 (1.)		
Mass (relative to Sun)	$\sim 10^4$–10^6 (10^{-4})	30–50	30–50
	~ 1–10^2 (1.)		
Rate of formation (yr^{-1})	~ 1	0.1–1	5–10
Internal energy (erg)	$\sim 10^{53}$ (relativistic expansion)	10^{53}–10^{54}	10^{53}–10^{54}
	$\sim 10^{58}$ (non-relativistic expansion)		
Power output (erg s^{-1})	$\sim 10^{46}$ (non-relativistic expansion)	10^{45}–10^{46}	10^{46}–10^{47}

[a] Figure in parentheses is assumed magnetic field, in G.

from his theory of QSOs and related objects. The theory predicts a rapid succession of supernova events resulting from multiple stellar coalescence in a nucleus of 1 to 100 pc dimensions, containing $\sim 10^8$ to 10^{10} stars. This idea is particularly attractive in view of the approximately similar energies of variable components suggested by the observations, and that of a large supernova explosion (for example, Cas A). The observations of supernova remnants in the Galaxy suggest that an expanding shell and/or filamentary model may be more appropriate than the exploding uniform sphere fitted to the present observations. Woltjer (1966) has shown that energy requirements and inverse Compton losses in variable radio sources could then be reduced.

5. Conclusions

The QSOs, radio galaxies, and blank field objects may be distinguished statistically by their high frequency spectra. Above 1000 MHz, more than 30% of QSOs have flat spectra ($\alpha < 0.5$). The remaining QSOs have spectra similar to those of the radio galaxies. More than 30% of N and Seyfert galaxies (the most radio-luminous galaxies) also have flat spectra. The spectra of 15% of QSOs and 40% of radio galaxies show a definite increase in spectral index with increasing frequency. This steepening is even more marked in the case of blank field objects.

Near 100 MHz, 50% of the most luminous galaxies (N and faint galaxies) show synchrotron self-absorption compared with 25% for the remaining galaxies. This figure is 40% for QSOs and more than 50% for blank field objects.

Near 1000 MHz, a weak trend of increasing average spectral index with increasing luminosity is suggested.

The above results suggest that the blank field objects may be galaxies, more luminous at radio frequencies than those which are close enough to appear on the prints of the Palomar Sky Survey. Further evidence for this is provided by the fact that 60% of blank field objects show interplanetary scintillation, compared with 25% for galaxies. This may be attributed to their greater distance and also to the presence of compact components which, as is suggested by the above results, may be a characteristic of the most radio-luminous identified radio sources.

The detailed analyses of individual spectra bear out the above results. Although 'break' frequencies are observed in a few radio galaxies and one QSO, the increase in spectral index appears to be gradual in many other sources. This is particularly well illustrated by the spectra of the blank field objects.

Low percentage polarization at 2650 MHz is associated with QSOs having flat spectra; the QSOs with the steepest spectra have the highest percentage polarization at 2650 MHz. This result confirms that suggested by the observations of others and has been attributed to highly tangled magnetic fields within compact sources, which untangle as the source expands, resulting in a higher degree of integrated polarization.

Where sufficient radio observations are available, variability is suspected or confirmed for nearly all sources with flat spectra. Detailed analyses of variations of 22 individual spectra are consistent with the adiabatically expanding uniform source model of Shklovsky, Kellermann, van der Laan, and others. In the present investigation, in order to explain the rapid time variations it was necessary to introduce the modification suggested by Rees and Simon – namely that the expansion occurs with relativistic velocities.

The results of this study may be summarized as follows:

(a) Expansion ages are of the order of months to tens of years. The present observations were not designed to detect variations on time scales less than one or two months; but others to detect variations with time scales of days to months were also made (Harris, 1970c).

(b) Linear dimensions are between 0.1 and 100 pc.

(c) Comparison of the predicted angular dimensions with those measured by very long baseline interferometry suggests magnetic fields between 1 and 10^{-4} G in variable components.

(d) Relativistic expansion velocities are general, but are less for observed variable components of Seyfert galaxies than for QSOs.

(e) Observations of some variable components at frequencies of 5000 MHz and above suggest non-adiabatic expansion which may be the result of continued injection of energy into an expanding region.

Acknowledgements

The author is grateful to Drs Allan Sandage, J. B. Whiteoak, F. F. Gardner, D. Morris, B. E. Westerlund and J. V. Wall for supplying data in advance of publication. Help with the observations, and useful discussions with staff members of the CSIRO

Division of Radiophysics, Sydney, and the Mt. Stromlo and Siding Spring Observatories of the Australian National University, Canberra, is appreciated. Thanks are due to members of the Ohio State University Radio Observatory for help in preparing this manuscript. The observations on which the present study is based were made at the Australian National Radio Observatory, Parkes, N.S.W., operated by the CSIRO Division of Radiophysics. The receipt of an Australian National University Research Scholarship is acknowledged.

References

Clark, B. G., Kellermann, K. I., Bare, C. C., Cohen, M. H., and Jauncey, D. L.: 1968, *Astrophys. J.* **153**, 705.
Colgate, S. A.: 1968, *Astrophys. J.* **73**, 905.
Dent, W. A. and Haddock, F. T.: 1966, *Astrophys. J.* **144**, 568.
Ekers, R. D.: 1969, *Australian J. Phys. Suppl.* **6**.
Gardner, F. F. and Whiteoak, J. B.: 1969, *Australian J. Phys.* **22**, 107.
Harris, B. J.: 1970a, b, c, in preparation.
Heeschen, D. S.: 1966, *Astrophys. J.* **146**, 517.
Hobbs, R. W., Corbett, W. W., and Santini, N. J.: 1968, *Astrophys. J.* **152**, 43.
Kellermann, K. I.: 1964, *Publ. Owens Valley Radio Obs.* **1**, No. 2.
Kellermann, K. I.: 1966, *Astrophys. J.* **146**, 621.
Kellermann, K. I., Clark, B. G., Bare, C. C., Rydbeck, O., Ellder, J., Hansson, B., Kollberg, E., Hoglund, B., Cohen, M. H., and Jauncey, D. L.: 1968, *Astrophys. J.* **153**, L209.
Kellermann, K. I. and Pauliny-Toth, I. I. K.: 1968, *Ann. Rev. Astron. Astrophys.* **6**, 417.
Kellermann, K. I., Pauliny-Toth, I. I. K., and Williams, P. J. S.: 1969, *Astrophys. J.* **157**, 1.
Locke, J. L., Andrew, B. H., and Medd, W. J.: 1969, *Astrophys. J.* **157**, L81.
Longair, M. S. and Macdonald, G. H.: 1969, *Monthly Notices Roy. Astron. Soc.* **145**, 309.
Macdonald, G. H., Kenderdine, S., and Neville, A. C.: 1968, *Monthly Notices Roy. Astron. Soc.* **138**, 259.
Medd, W. J., Locke, J. L., Andrew, B. H., and van den Bergh, S.: 1968, *Astron. J.* **73**, 293.
Morris, D. and Whiteoak, J. B.: 1968, *Australian J. Phys.* **21**, 475.
Rees, M. J. and Simon, M.: 1968, *Astrophys. J.* **152**, L145.
Seielstad, G. A. and Weiler, K. W.: 1969, *Astrophys. J. Suppl. Ser.* **18**, 85.
Shimmins, A. J.: 1968, *Astrophys. Letters* **2**, 157.
Woltjer, L.: 1966, *Astrophys. J.* **146**, 597.

M87 AND THE X-RAY EMISSION FROM COMPACT OBJECTS

M. S. LONGAIR

Mullard Radio Astronomy Observatory, Cambridge, Great Britain

1. Observations of M87 at 5 GHz

(Abstract of a paper by I. Graham (1970))

M87 (Virgo A, 3C274) has been observed at 5 GHz with the Cambridge One-Mile telescope, with a resolution of 6.5 arc sec. These observations do not include the extensive halo but show that the central region of the source consists of at least three components. The central of the components is unresolved and appears to coincide with the optical nucleus, the outer pair extending from a point about 5 arc sec south of the optical nucleus. These components are unresolved perpendicular to their extension axes. The north preceding component is related to the optical jet but is more extended.

The physical conditions in the radio components and the optical jet are investigated in the light of these new observations. It is concluded that the two outer components probably consist of a number of small sub-components, each confined by its supersonic passage through the gas in the galaxy. It is shown that the nuclear component probably consists of two sub-components, the smaller of these producing the observed X-rays by the inverse Compton mechanism. The structure of the halo is used to derive information about the past activity of the source.

2. The Diffusion of Relativistic Electrons from Infrared Sources and Their X-Ray Emission

(Abstract of a paper by A. S. Webster and M. S. Longair (1971))

A model for X-ray sources in which relativistic electrons originate in compact nuclei which are also powerful sources of infrared emission is analyzed. The distortions of the injection electron spectrum due to inverse Compton losses and the spectrum of the resulting X-rays are determined for the case in which the propagation of the electrons from the nuclei can be described by a simple diffusion model. It is possible to explain the detailed features of the X-ray background spectrum in the range $0.25 < \varepsilon_x < 1000$ keV provided there is little dispersion in the properties of the sources. If Seyfert galaxies are the source of the X-ray background, their infrared spectra should have a maximum about 70 μm, in agreement with observation, and the diffusion coefficients should be 10^{29} cm^2 s^{-1}, similar to that of cosmic ray protons in the Galaxy.

References

Graham, I.: 1970, *Monthly Notices Roy. Astron. Soc.* **149**, 319.
Webster, A. S., and Longair, M. S.: 1971, *Monthly Notices Roy. Astron. Soc.* **151**, 261

Discussion

Kellermann: It is difficult to understand the explanation of X-ray sources by inverse Compton scattering. The ratio of inverse Compton scattered X-rays to synchrotron radio emission depends only on radio brightness temperature. There are many compact radio sources of equally high brightness temperature as the compact component of M87 and with much greater radio synchrotron flux, that do not show X-ray emission.

Longair: Dr Kellermann's remark refers to models in which the infrared emission and the X-rays result from the same relativistic electrons. In our model these are decoupled. We postulate a compact source of infrared photons and relativistic electrons, both being generated by some unknown mechanism.

Hogg: The polarization structure of M87 at 11.1 cm has been measured by Dr Conway and myself, using the NRAO interferometer. The compact source associated with the nucleus is unpolarized, with an upper limit of 0.5%. The two extended components are each polarized. The maximum polarization, 2.5%, is found in the region of the optical jet.

Arp: An interesting detail is that the optical jet in M87 appears to emerge from the nucleus of M87 slightly offset to the north – only a few sec of arc but definitely noticeable on high resolution plates. The radio jets, on the other hand, now appear to center a few sec south of the nucleus.

A matter which now requires further optical investigation is what is the exact optical center of M87? The spot in the center of M87 (about 1.5 arc sec) which radiates strongly in [O II] emission appears to define the center on most photographic plates. It would be worthwhile now to determine independently the exact center of M87 from photographic plates in wavelengths which eliminate this $\lambda 3727$ Å emission.

IDENTIFICATION OPTIQUE ET PHOTOMÉTRIE DE RADIOSOURCES DU CATALOGUE 3CR

GÉRARD WLÉRICK, GÉRARD LELIÈVRE, et PHILIPPE VÉRON

Observatoire de Paris, Meudon, France

Abstract. Using a Lallemand electronic camera, optical identifications for several 3CR radio sources have been made. Photometry has been obtained for objects as faint as $B \simeq 21.25$. It is suggested that some of the identifications may have redshifts in excess of $z = 3$.

1. But

Il serait souhaitable d'identifier toutes les sources du catalogue 3CR pour disposer d'un échantillon *complet* d'objets optiques correspondant aux radiosources les plus brillantes à 178 MHz.

2. Situation au 1er Janvier 1970

Le catalogue 3CR contient environ 250 sources hors de la Voie Lactée ($|b^{II}| > 20°$).

A. IDENTIFICATION

Environ 150 de ces sources sont bien connues: 44 quasars (Schmidt, 1968) et une centaine de radiogalaxies (*cf.* Burbidge à paraître). Une centaine de sources environ ne sont pas encore identifiées. Dans 50 cas, on trouve un 'champ vide' (empty field) sur les cartes du Palomar Sky Survey. Dans les 50 autres cas, on sait seulement qu'il y a un astre très faible 'au voisinage' de la position radio.

B. PHOTOMÉTRIE

La photométrie photoélectrique UBV est un moyen de discriminer entre quasars et radiogalaxies. Elle s'arrête à $V \leqslant 19,0$ et $B \leqslant 19,4$, sauf pour deux sources, 3C 2 ($V = 19,35$) et 3C 280,1 ($V = 19,44$).

3. Méthode

Nous avons utilisé un télescope moyen (Tél. 193 cm, O.H.P.) et la caméra électronique Lallemand, qui permet:
   (i) d'augmenter la magnitude limite (voir figures 1 et 2 du champ de 3C 49);
   (ii) d'effectuer une photométrie photoélectrique quand on dispose d'au moins une étoile étalon (standard star) (Lallemand *et al.*, 1966; Walker et Kron, 1967);
   (iii) d'effectuer en particulier la photométrie d'un astre très faible près d'une étoile brillante.
   L'association télescope-caméra électronique a déjà été décrite (Wlérick, 1969).

4. Observations

Nous avons étudié une quinzaine de sources en choisissant de préférence des astres qui, en ondes radio, ont une faible étendue (Elsmore et Mackay, 1969) et des sources situées dans des champs où il y a déjà une étoile étalon (Sandage, 1970).

5. Résultats

Nous proposons ou confirmons ici une identification pour huit sources: 3C 14, 3C 42, 3C 49, 3C 173, 3C 190, 3C 194, 3C 208,1 et 3C 460 (voir figures 2, 3 et 4).

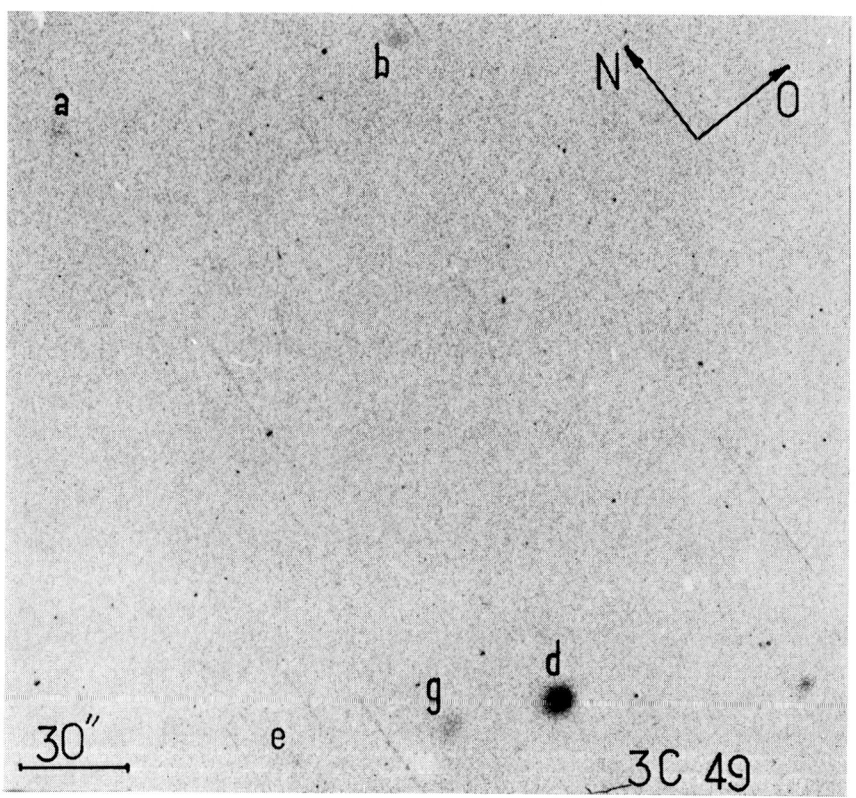

Fig. 1.

A. MESURES DE POSITION

La précision des mesures optiques est, pour chaque coordonnée, de 1″ environ. Les positions sont en excellent accord avec les mesures radio les plus précises, dues à Adgie (1970), Fomalont (1970) ou Elsmore et Mackay (1969). Dans quatre cas (3C 14, 3C 190, 3C 194 et 3C 208,1 composante x), l'écart entre les positions optiques et radio est inférieur à 1″.

B. PHOTOMÉTRIE

3C 190

Nous avons pris un cliché dans la couleur B. Pour les deux étoiles étalons, a et c, nous notons l'accord entre les mesures de Sandage ($\Delta B = 1{,}65$) et nos mesures ($\Delta B = 1{,}73 \pm 0{,}04$). Pour la radio source nous trouvons: $B = 21{,}25 \pm 0{,}20$, la forte erreur provenant du fait que le cliché est sous-exposé.

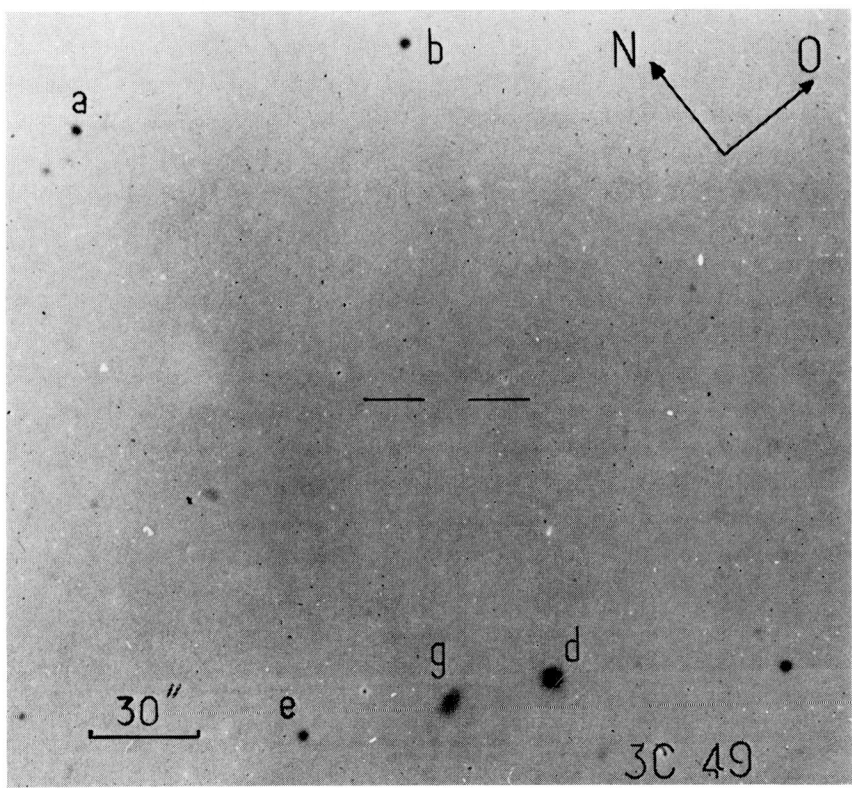

Fig. 2

Figs. 1, 2. Amélioration de magnitude limite par l'emploi de la caméra électronique (3C 49).

3C 173

Nous avons obtenu un cliché bleu en Octobre 1969, un cliché bleu et un cliché jaune en Mars 1970 et la photométrie a conduit aux résultats suivants:

Oct. 69 $B = 20{,}00 \pm 0{,}15$
Mars 70 $\begin{cases} B = 21{,}02 \\ B - V < 0 \end{cases}$

3C 173 est donc un astre variable et bleu. L'erreur $\pm 0{,}15$ est dûe au fait que la radio-

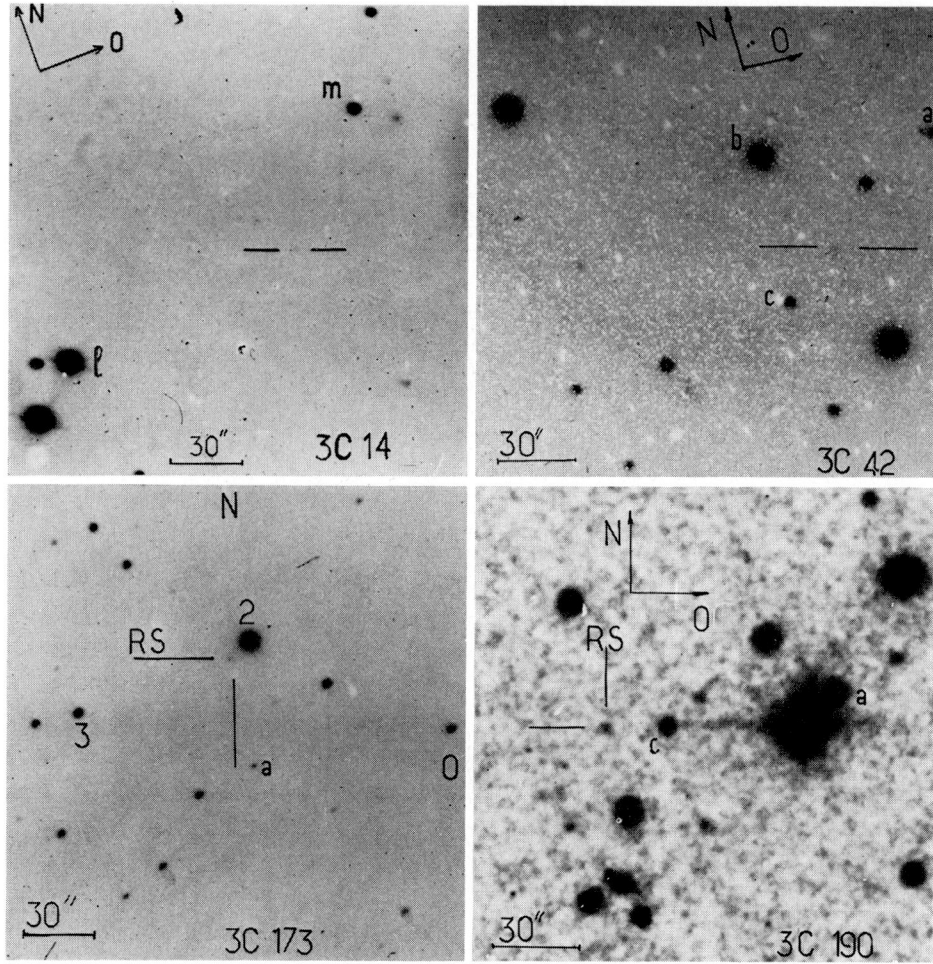

Fig. 3. Identification des sources optiques pour 3C 14, 3C 42, 3C 173 et 3C 190.

source est très voisine (environ 10″) d'une étoile 'brillante' de magnitude visuelle $m_v \approx 12{,}6$ selon Wyndham (1966) (voir la coupe photométrique de la figure 5). Pour quatre astres très faible du champ, nous trouvons que l'erreur quadratique moyenne est seulement B = ±0,08.

Autres sources

Nous ne disposons pas encore d'étoiles étalons dans les champs correspondants. Ces sources sont faibles. Si elles ne sont pas variables, leur magnitude B est probablement comprise entre 20,0 et 22,0, sauf peut-être pour 3C 460.

C. CLASSIFICATION

3C 14 et 3C 42 sont probablement des radiogalaxies. 3C 49, 3C 173 et 3C 190 sont

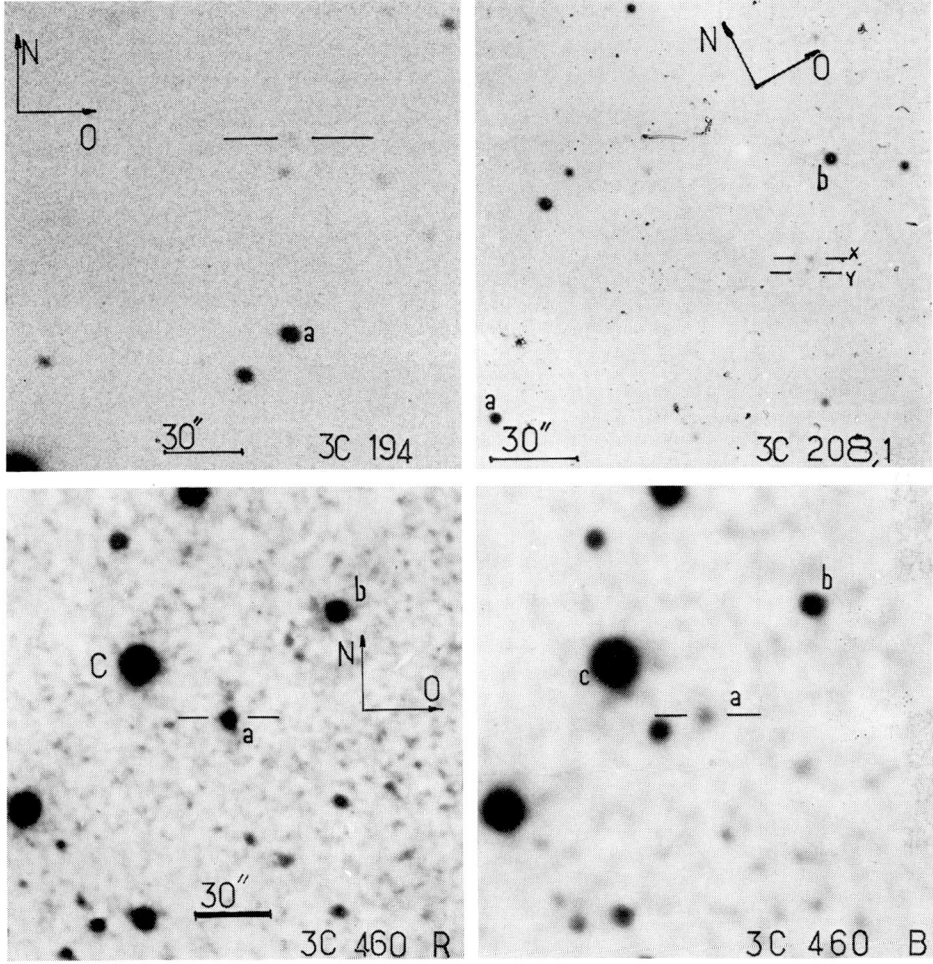

Fig. 4. Identification des sources optiques pour 3C 194, 3C 208,1 et 3C 460.

probablement des quasars: la première et la dernière scintillent fortement et nous avons déjà indiqué que la deuxième est variable et bleue.

On manque de renseignements pour classer 3C 194 et 3C 208,1. 3C 460 semble un cas particulier. L'astre est *très* rouge, comme le montrent les reproductions des clichés bleu et rouge du Palomar Sky Survey (figure 4) et la radiosource scintille fortement.

6. Conclusions

(a) La caméra électronique permet de mesurer les sources du catalogue 3CR, avec un télescope moyen, jusqu'à des magnitudes élevées. On a déjà étendu de 2 unités les magnitudes des sources photométrées et il est possible de mesurer des sources encore plus faibles d'environ 1 à 2 mag.

(b) Il n'y a pas d'indication de discontinuité dans les magnitudes optiques et, pour la très grande majorité des astres de ce Catalogue, on doit avoir $B \leqslant 22,5$.

(c) G. et M. Burbidge (1967) et Schmidt (1968) ont établi des diagrammes 'décalage vers le rouge z – magnitude optique' pour les quasars. Si ces diagrammes peuvent être extrapôlés dans une gamme de 2,5 magnitudes, des valeurs de z nettement supérieures à 3 sont possibles pour les sources optiquement faibles.

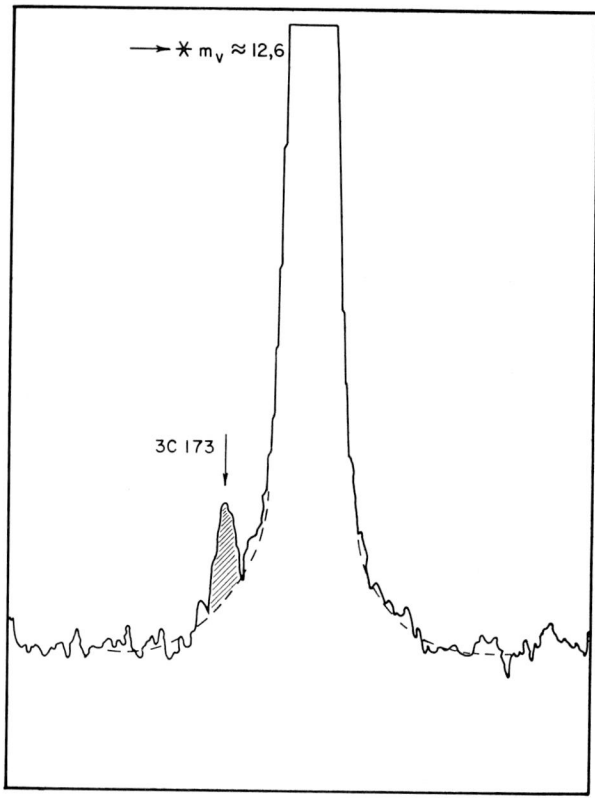

Fig. 5. Coupe photométrique montrant une étoile brillante au voisinage du source 3C 173.

Remerciements

Nous avons bénéficié de concours précieux à Paris, à Meudon, et à l'O.H.P. Nous tenons à remercier J. Delhaye, A. Lallemand, Ch. Fehrenbach, et les techniciens qui nous ont aidés. Nous soulignons avec plaisir la collaboration directe de A. Sellier, F. Gex, J. P. Lemonnier, D. Michet et L. Stoesel. Merci aussi à A. Sandage, R. Adgie et E. Fomalont qui nous ont communiqué leurs résultats.

Note

Ce texte résume l'essentiel d'un article soumis au journal *Astronomy and Astrophysics* **11**, 142 (1971).

Bibliographie

Adgie, R. L.: 1970, communication personnelle.
Burbidge, G. et Burbidge, M.: 1967, en *Quasi-stellar objects*, Freeman and Co, San Francisco.
Elsmore, B. et Mackay, C. D.: 1969, *Monthly Notices Roy. Astron. Soc.* **146**, 360.
Fomalont, E. B.: 1970, communication personnelle.
Lallemand, A., Canavaggia, R., et Amiot, F.: 1966, *Compt. Rend. Acad. Sci. Paris* **262**, 838.
Sandage, A.: 1970, communication personnelle.
Schmidt, M.: 1968, *Astrophys. J.* **151**, 393.
Walker, M. F. et Kron, G. E.: 1967, *Publ. Astron. Soc. Pacific* **79**, 551.
Wlérick, G.: 1969, *Adv. Electronics Electron Phys.* **28B**, 787.
Wyndham, J. D.: 1966, *Astrophys. J.* **144**, 459.

Discussion

Walker: (1) What emulsion was used to record the electronic images?; (2) What was the limiting magnitude of your one hour exposures and what was the half-intensity width of the stellar images?; (3) What is the probable error as a function of magnitude for a single observation on a one-hour exposure?

Wlérick: (1) We used Ilford G5 and K5 emulsions; these are already finer than Kodak II a-0, as may be seen in the pictures of the fields of 3C 49 but, as we plan to make longer exposures, we are considering using the finer Ilford L4 emulsion, as yours and Kron's group; (2) On the best night the width of the recorded images, at half intensity, varies from 1.4 to 2 arc sec. On these nights, I estimate that the limiting magnitude, in the blue, is $\geqslant 23$; (3) I may only say that, for $B \approx 21.0$, the error may be $\leqslant 0.10$ mag, in cases easier than 3C 173 (where we have an error of 0.15 mag). For $B \approx 19$, the error may be $\leqslant 0.04$ mag.

Hoyle: What can you say concerning the other seven sources?

Wlérick: The reduction is under way. In three cases, there is an object in coincidence. They are 3C 325, 3C 343 (for which I have heard that another group has also made the identification) and 3C 434. The fourth case, 3C 54, is an extended radio source. We see no object between the two components of the radio source but two objects lie in the vicinity of the NE component.

In the last three cases, no object appears: 3C 65 and 3C 107, with one hour exposures and 3C 238 with two hours exposure. This last object could be optically variable.

OPTICAL IDENTIFICATIONS OF COMPACT RADIO SOURCES

G. G. POOLEY

Cavendish Laboratory, Cambridge, Great Britain

Abstract. A study of twelve compact radio sources whose radio spectra show pronounced peaks at centimetre wavelengths has been made using the Cambridge one-mile radio telescope operating at a frequency of 5 GHz. For each of these sources we have measured accurate positions to closer than one arc sec and determined upper limits to their angular sizes.

A search for optical identifications using the prints of the Sky Survey has revealed probable candidates for eleven of the twelve sources, an unusually high success rate. Of the suggested identifications, three are galaxies (OC 328 = 4C 31.04, VRO 20.04.02, OQ 208), and eight are possible or confirmed QSOs (NRAO 140 = 4C 32.14 = OE 355, OI 315, P 0735 + 17 = VRO 17.07.02, OI 363, OJ 287 = VRO 20.08.01, OK 290, OR 103, DA 406). Details are given by G. M. Blake, *Astrophys. Letters* **6**, 201 (1970).

ATTEMPTS TO DETECT NEUTRAL HYDROGEN IN COMPACT OBJECTS

WILLIAM A. DENT

University of Massachusetts, Amherst, Mass., U.S.A.*

Abstract. An attempt was made to detect the redshifted H I line in absorption in the quasi-stellar sources 3C 273 and 1510 − 08 and in the N-galaxies 3C 120 and 3C 371. The Seyfert galaxies NGC 1068 and NGC 1275 were also searched for H I in absorption as well as in emission. Although no lines were detected, it was possible to place upper limits on the column density of H I in all these objects, as well as upper limits on the mass of H I in the Seyfert galaxies observed.

1. Introduction

The existence of intense, compact and frequently variable radio components in quasi-stellar sources, Seyfert galaxies, and N-type galaxies suggests the possibility of detecting neutral hydrogen in these objects by observing it in absorption against the intense continuum radio source present.

Following the suggestion of Bahcall and Ekers (1969), Shuter and Gower (1969) and Heiles and Miley (1970) have, without success, attempted to detect H I in absorption in a few distant $(z \sim 2)$ quasars which have optical absorption lines. This paper presents the results of a search for H I absorption in nearby variable radio sources where the probability of such a detection may be higher than for more distant objects.

Detection of an H I absorption line is more likely in those objects which are known to contain at least one variable radio component whose emission extends to frequencies near 1420 MHz. Such components, besides providing an intense background continuum source, have linear dimensions on the order of a few parsec or less and hence could be occulted by a H I cloud of comparable size or larger within the object.

From the observed differences between optical absorption and emission line redshifts the expected frequency of an H I absorption line would not necessarily be found at the frequency $1420.4/(1 + z_{em})$ MHz when only an optical emission line redshift is observed. However it is unlikely that as large a difference between emission and absorption redshifts would exist in Seyfert and N-type galaxies as in highly redshifted quasars.** The absorption redshift in 1510−08, one of the two quasars, observed in this paper, is 0.358 compared with an emission line redshift of 0.361 (Burbidge and Kinman, 1966). This difference corresponds to only about 1 MHz at the redshifted frequency of the 21-cm H I line. Thus it is felt that no absorption lines are likely to be found outside the actual frequency interval (up to 14 MHz) that was searched in each object.

* Contribution from the Four College Observatories, No. 89.
** Roberts (1970) has recently reported the existence of an H I absorption line in the nearby radio galaxy NGC 5128 that differs in radial velocity from that of the optical emission lines by only 23 km s^{-1}.

All except one of the objects observed (1510−08) have no observed optical absorption lines. The presence of absorption is not necessarily an important criterion since most of the strong resonances absorption lines seen in quasars are found at frequencies that are redshifted into the visible region in only high redshifted objects.

The observations were made with the 140-ft (43 m) antenna of the National Radio Astronomy Observatory using the 413 channel auto-correlation spectral line receiver at its maximum bandwidth of 10 MHz and with a resolution of 52 kHz. Three different parametric amplifier front-ends were used to observe sources in the range 1040 to 1420 MHz.

A load switched on-source, off-source observational procedure was used to remove first order instrumentally-produced spectral slopes. Second order spectral effects such as that due to the source continuum emission, uncertainties in the gain calibration resulting from extensive man-made interference outside the protected frequency bands, and radiated emission reflected from the surface of the disk were minimized as much as possible. The details of the observations and the above calibration procedures will be published elsewhere (Dent, 1971).

2. H I in Absorption

No obvious absorption lines were found in any of the objects studied. There may be a suggestion of a line near 1397 MHz in NGC 1275; however, the signal to noise ratio is not adequate to claim a positive detection. Observations with greater sensitivity are planned for this source.

Upper limits to the depth of the line, ΔT_L, and the ratio of line depth to continuum antenna temperature, $\Delta T_L/T_C$, are given for each source in Table I and represent the principle result of these observations. The quoted upper limits of ΔT_L are one-half the peak-to-peak noise fluctuations.

If the H I in the object is seen against the background of a continuum source, with

TABLE I

Source	Type of object	z (ref)	Frequencies searched (MHz)	ΔT_L (K)	$\Delta T_L/T_C$	N_{HI} (cm^{-2})[a]
NGC 1068 (3C 71)	Seyfert	0.0037 (1)	1410–1420	<0.075	0.052	<2 × 10^{21}
NGC 1275 (3C 84)	Seyfert	0.0176 (2)	1392–1403	≤0.157	0.042	≤2 × 10^{21}
3C 120	N-Gal	0.0334 (3)	1370–1382	<0.051	0.039	<2 × 10^{21}
3C 371	N-Gal	0.050 (4)	1348–1360	<0.070	0.093	<4 × 10^{21}
3C 273	QSS	0.158 (5)	1220–1234	<0.089	0.007	<3 × 10^{22}
1510−08	QSS	0.361 (6)	1040–1053	<0.068	0.080	<3 × 10^{23}

[a] Assuming $\Delta v \leqslant 100$ KHz; $T_S \approx 10^3$ K for Seyfert and N-Galaxies and $T_S \approx 10^5$ K for quasars.
References: (1) Walker (1968); (2) Burbidge and Burbidge (1965); (3) Burbidge (1967); (4) Sandage (1966); (5) Schmidt (1963); (6) Burbidge and Kinman (1966).

brightness temperature T_B, and if $T_B > T_S$, the spin (excitation) temperature of the H I, then the line will appear in absorption with the depth of a line, ΔT_L given by

$$\Delta v \Delta T_L / T_C = 2.85 \times 10^{-15} N_{\text{H I}} / T_S$$

where Δv is the effective line width in Hz, and $N_{\text{H I}}$ is the column density of H I in cm^{-2}. Thus the ratio $N_{\text{H I}}/T_S$ can be determined from the observational quantities on the left side of the equation.

Because the spin temperature of the neutral hydrogen in these sources is not accurately known, only crude estimates of the upper limit to the H I column densities can be made using a calculated value of T_S. However, since the contributions to T_S from Lyman α and 21-cm photons depend as r^{-2} upon the distance of the H I from the photon sources, it is also necessary to assume an average distance of any H I present from these emitting regions within the object. If we adopt a plausible distance of 10^4 pc; and use measurements of the radio continuum flux, estimates of the Lyα flux (Wampler, 1968) and expressions for T_S given by Bahcall and Ekers (1969); we obtain rough estimates of $T_S \approx 10^3$ K in Seyfert and N-type galaxies and $T_S \approx 10^5$ K in quasars.

Bahcall and Ekers (1969) have argued that the absence of absorption lines of O I λ1303.5 or N I λ1134.6, 1199.9 in the optical spectra of large redshifted quasi-stellar sources implies that the H I line (if it exists) must have line widths less than about 100 kHz. Although the O I and N I lines would not be redshifted into the visible in the objects studied in this paper, similar excitation and ionization conditions probably exist in these objects as well. Thus in the absence of a positive detection of a line, we will adopt a value of $\Delta v \leqslant 100$ kHz for the purposes of calculation.

Assuming this upper limit of Δv and the above estimates of T_S, upper limits to the column density of neutral hydrogen were calculated and are tabulated in the last column of Table I. It should be emphasized that because of the large uncertainties in estimating T_S, the quoted upper limits to $N_{\text{H I}}$ may be uncertain by more than an order of magnitude.*

3. H I in Emission

The 21-cm H I line will appear in emission from those regions of the object in which the spin temperature is greater than the background brightness temperature of the continuum emission. The H I line should therefore appear in emission everywhere in the object except directly in front of the small intense radio component. Since none of the objects observed here could be resolved with the 20' beamwidth of the antenna, only an effective antenna temperature, $\Delta T_L(\text{em})$, of the emission line integrated over the object could be observed. It is easily shown that an estimate of the total mass of H I in the object can be obtained from $M_{\text{H I}} = 3.1 \times 10^{-6} R^2 (\lambda^2)/A_e) \Delta T_L(\text{em}) \Delta v$ (solar

---

* It should be mentioned that the value of T_C used in the calculation is the total source antenna temperature consisting of contributions from the small intense components as well as emission from more extended regions. Thus depending on the location of the H I, it may be necessary to adjust further the limits of $N_{\text{H I}}$.

masses), where R is the distance to the source in parsecs, A_e is the effective area of the antenna, λ the wavelength, and Δv the effective line width in Hz.

In normal spiral galaxies Δv is a Doppler width due to the rotation of the galaxy, and is on the order of 200 km s^{-1} or about 1 MHz at 1420 MHz. Neutral hydrogen emission in these objects would be expected to have similar line widths since hydrogen with much higher velocities is likely to be almost completely ionized in these objects. The spectra were smoothed with a 500 kHz gaussian filter and examined for possible emission lines. Although no obvious lines were found, meaningful upper limits on $\Delta T_L(\text{em})$ were obtained for the two Seyfert galaxies. Upper limits on $\Delta T_L(\text{em})$ of 0.05 K and 0.06 K were obtained for NGC 1068 and NGC 1275 respectively. Assuming a Hubble constant of 100 km s^{-1} Mpc^{-1} and $\Delta v = 1$ MHz, these values yield upper limits to the masses of 1×10^9 $M_\odot$ and 3×10^{10} $M_\odot$ respectively. Because the upper limits to the masses are proportional to R^2 the remaining four more distant objects do not give meaningful upper limits to the mass of H<small>I</small>.

The most recent estimate (Walker, 1968) of the mass of the inner region (<2 kpc) of NGC 1068 is 6×10^9 $M_\odot$. An earlier study by Burbidge *et al.* (1959) which did not assume a non-rotational component of the gas obtained a higher figure of 2×10^{10} $M_\odot$ for the inner region. An estimate of the total mass of the entire galaxy would be at least a factor of 2 to 4 times greater than the mass obtained for the inner region (Roberts, 1969). Hence a lower limit to the total mass of NGC 1068 is about 4×10^{10} $M_\odot$. Thus the upper limit of 1×10^9 $M_\odot$ to the mass of neutral hydrogen in NGC 1068 (3C 71) obtained from these observations places an upper limit on the ratio $M_{\text{H}_\text{I}}/M_T$ of 0.025. This value is less than the average value of 0.05 (Roberts, 1969) for $M_{\text{H}_\text{I}}/M_T$ of Sb galaxies, the classification given to NGC 1068 by Humason *et al.* (1956) on the basis of its spiral structure.

Thus although the Seyfert galaxy NGC 1068 shows some similarities to normal spirals, it may be under-abundant in neutral hydrogen. This would not be surprising in view of the large flux of ionizing radiation likely to be emanating from the nucleus of this galaxy.

Acknowledgements

I would like to thank the staff of the National Radio Astronomy Observatory for making these observations possible. This work was supported in part by a Faculty Research Grant from the University of Massachusetts and by the National Science Foundation under grant number GP-14690.

References

Bahcall, J. N. and Ekers, R. D.: 1969, *Astrophys. J.* **157**, 1055.
Burbidge, E. M.: 1967, *Astrophys. J. Letters* **149**, L51.
Burbidge, E. M. and Burbidge, G. R.: 1965, *Astrophys. J.* **142**, 1351.
Burbidge, E. M., Burbidge, G. R., and Prendergast, K. H.: 1959, *Astrophys. J.* **130**, 26.
Burbidge, E. M. and Kinman, T. D.: 1966, *Astrophys. J.* **145**, 654.
Dent, W. A.: 1971, *Astrophys. J.* **165**, 451.

Heiles, C. and Miley, G. K.: 1970, *Astrophys. J. Letters* **160**, L83.
Humason, M. L., Mayall, N. U., and Sandage, A. R.: 1956, *Astron. J.* **61**, 97.
Roberts, M. S.: 1969, *Astron. J.* **74**, 859.
Roberts, M. S.: 1970, *Astrophys. J. Letters* **161**, L9.
Sandage, A.: 1966, *Astrophys. J.* **145**, 1.
Schmidt, M.: 1963, *Nature* **197**, 1040.
Shuter, W. L. H. and Gower, J. F. R.: 1969, *Nature* **223**, 1046.
Walker, M. F.: 1968, *Astrophys. J.* **151**, 71.
Wampler, J.: 1968, *Astrophys. J.* **144**, 443.

NEUTRAL HYDROGEN IN COMPACT GALAXIES

J. HEIDMANN

Observatoire de Paris, Meudon, France

In collaboration with Messrs P. Chamaraux and R. Lauqué we have observed five objects of compact type or with compact parts at Nançay with the 21-cm line of neutral hydrogen. Details will be found in *Astron. Astrophys.* **8**, 424, 1970.

II Zw 40 gives a well-defined line with nothing noticeable one beam East or West. I Zw 114 and 17 give more poorly defined lines which may possibly be absent though the agreement of optical and 21-cm radial velocities is in favour of their existences. Two are not detected.

All objects have low intrinsic luminosity. The three detected ones have sharp emission optical spectra and optical extensions.

Table I gives the main results:

TABLE I

Name	II Zw 40	I Zw 114	I Zw 17	Arp gal.	Barbon gal.
$M_H (10^8 \odot)$	2.3	1.0	<1.4	<1.8	<4.2
M_H/L_p	0.3	1.0	<12	<3.4	<1.8
$\sigma_H (\text{km s}^{-1})$	55	35	40	–	–

The hydrogen mass M_H is of the order of 10^8 solar masses as for low luminosity irregular galaxies.

The hydrogen mass to luminosity ratio M_H/L_p is about unity as for late type galaxies.

The rms radial velocity dispersion of neutral hydrogen σ_H is of the order of 50 km s^{-1}.

From the negative results obtained East and West of II Zw 40, its H I E–W width at half intensity is smaller than 3.8.

Using this upper limit, σ_H and the virial theorem, an upper limit of the total mass of II Zw 40 is 10^{10} solar masses. Hence its mass to luminosity ratio is smaller than 15 and its hydrogen mass to total mass ratio is larger than 0.02.

It is interesting to compare the H I velocity dispersions σ_H to the ones for stars σ_*. From its spectral resolution, Sargent gives 450 km s^{-1} as the upper limit for the emission lines in II Zw 40 and I Zw 114. Values of σ_* have been published for three other compacts by Minkowski and by Zwicky: 950, >700 and 370 km s^{-1}. The last value refers to NGC 4486 B which has a size and luminosity comparable to the objects which we observed; however its spectrum is a late absorption one.

If σ_* in II Zw 40 were as large as for NGC 4486 B, Kepler's law would lead to an H I spatial extent for II Zw 40 a few tens times larger than the compact part size. The situation would be comparable to the one for normal galaxies from the points

SEYFERT GALAXIES

B. M. LEWIS

Nuffield Radio Astronomy Laboratories, Jodrell Bank, Cheshire, Great Britain

Abstract. The four Seyfert galaxies with systemic velocities of less than 2000 km s^{-1} have been observed at 21 cm with the Jodrell Bank 250-ft telescope and a 5 MHz overall bandwidth setting of the 256 channel auto-correlation spectrometer. Relatively strong signals of 0.5 f.u. are observed in NGC 1068 and 4151, while less well-determined signals of ~ 0.25 f.u. and covering a wide velocity range are seen in both NGC 3227 and 4051. Measures of the systemic velocity, velocity width and H I mass are obtained, the H I mass/L being normal for galaxies of their Hubble type.

During an observing run in February and March, 1970, the 250-ft telescope was used to measure the H I content of 60 galaxies. The four northern Seyfert galaxies with systemic velocities of less than 2000 km s^{-1} were included in this study, and the preliminary data derived from fitting eye-estimates of the baselines are given here. The observing system, which consisted of a cooled paramp feeding the 256 channel auto-correlation spectrometer, resulted in an overall system noise temperature of ~ 100 K. Ten minute integrations on and 2 deg off source were subtracted to give a spectrum, the integration time being increased by stacking all observations made on the same day. Virgo A was used to give an absolute calibration of the temperature scale. Results obtained at Jodrell Bank for NGC 3344, 3718, and 6217 are about 10% larger than those given by other observers.

All the features seen in NGC 1068 and 4151 are strong and easily distinguished in every ten-minute spectrum, and in observations from different days. NGC 3227 is much weaker, being a rather broad low intensity profile, which is easily lost whenever the baseline is irregular. NGC 4051 is of intermediate strength, and may be over-emphasized here by the distorted baseline.

TABLE I

NGC	1068	3227	4051	4151
V_{opt}[a] (km s^{-1})	1080	1111	647	957
V_{21}[a] (km s^{-1})	1150 ± 20	1117 ± 30	807 ± 20	1107 ± 20
ΔV (km s^{-1})	340 ± 50	530:	440 ± 40	390 ± 30
$M_{HI} \times 10^9 M_\odot$	3.7	6.9	1.7	3.5
M_{HI}/L	0.065	0.13	0.16	0.11
Distance (Mpc)	10.8	16	8	10.8
i (deg)	37	51	44	50
Integ. Time (min)	50	40	60	35

[a] velocities with respect to the Sun

Table I lists the systemic velocity, velocity spread (ΔV) and H I masses determined from these profiles. The systemic velocity is defined as the median velocity in the range ΔV over which the signal is detected, and has an uncertainty of about 20 km s^{-1}.

Substantial differences exist between the optical systemic velocities listed in the Reference Catalogue of Bright Galaxies and the 21-cm velocity measures. However, Rubin and Ford (1968) found that in NGC 3227 the velocities determined from the emission lines in the nucleus are blueshifted by ~ 160 km s^{-1} with respect to the rest of the galaxy. Since the optical velocities are usually measurements made on the nucleus, it is likely that much of the difference between the two methods is due to the expansion of gas clouds in the nuclei of the Seyferts.

Rather a wide range of velocities is seen in each of the Seyferts. This might be expected to occur if there were very large masses of gas flowing out of the nuclear regions, or if the galaxies have large rotation velocities. It is possible to simulate the spectra of all these galaxies by assuming Sb-type rotation curves of moderate velocity dispersion and large rotation velocities. Since all Sb galaxies and in particular NGC 1068 and 3227 are seen from optical measurements to have observed rotation velocities of at least 200 km s^{-1} it is safe to assume that the velocity spread is due principally to the dynamics of each galaxy.

The hydrogen masses quoted here are likely to be a little optimistic as the base-lines are drawn by hand, and cannot be accurately determined for wide velocity profiles. An error estimate of $\pm 50\%$ is reasonable. Table I lists the masses together with the distance independent M_{HI}/L ratio. The luminosities are absorption-free estimates taken from Holmberg or the values of $B(0)$, after statistically adjusting to infinite radius given in the Reference Catalogue of Bright Galaxies. A mean M_{HI}/L ratio of 0.12 is found for the four Seyferts, almost exactly equal to the average ratio of 0.11 found for all 20 Sa–Sb galaxies whose HI masses are given in the literature (Roberts, 1969; Bottinelli *et al.*, 1970). Thus the Seyfert galaxies appear to have completely average gas contents for their Hubble type.

Three conclusions can be drawn from this preliminary observation:

(1) the Seyfert galaxies have velocity ranges which can be explained without assuming large expansion velocities generated in the nuclei;

(2) the optical systemic velocities are generally blue-shifted by expanding optically thick clouds in the nuclei;

(3) the Seyfert galaxies are quite normal in dynamics and gas content outside the nuclei.

References

Bottinelli, L., Chamaraux, P., Gouguenheim, L., and Lauqué, R.: 1970, *Astron. Astrophys.* **6**, 453.
Roberts, M. S.: 1969, *Astron. J.* **74**, 859.
Rubin, V. C. and Ford, W. K.: 1968, *Astrophys. J.* **154**, 431.

RADIO OBSERVATIONS OF NEUTRAL HYDROGEN IN FOUR SEYFERT GALAXIES

RONALD J. ALLEN

Kapteyn Laboratory, Groningen, The Netherlands

and

BERNARD F. DARCHY and ROBERT LAUQUÉ

Observatoire de Paris, Meudon, France

(To be published in *Astron. Astrophys.*)

Abstract. The 21-cm wavelength radiation from neutral hydrogen in NGC 1068, NGC 3227, NGC 4051 and NGC 4151 has been observed with the large radio telescope at Nançay, France. Since the angular sizes of these galaxies are of the same order as the telescope right ascension beam-width, no information on the angular distribution of the neutral hydrogen was obtained. However the radial velocity distribution of the total hydrogen (the 'integrated profile') of the whole galaxy was measured for each of the four galaxies. The hydrogen masses and total masses can be calculated from these profiles using simple models of galaxy shapes and rotation curves.

Optical spectra sometimes show evidence for explosive phenomena and radial outflow of gas in the central regions of Seyfert galaxies. We have examined the integrated radio profiles for indications of large-scale radial motions of neutral hydrogen in two ways. First, for all four galaxies observed, we compare the ratios of hydrogen mass to total mass with the values obtained from other galaxies (not Seyfert) of the same morphological type. Second, for these galaxies where the optical data are available, we compare the estimates of total mass obtained from the optical spectra with the estimates based on the width of the radio profile.

We conclude from these comparisons that the integrated profile of NGC 1068 is unusually broad. One possible interpretation which is qualitatively consistent with the optical data is that an appreciable fraction (about $\frac{1}{3}$) of the neutral hydrogen content of NGC 1068 is moving radially outward with velocities of about 200 km s^{-1} An indication of similar phenomena (although less extreme) is obtained for NGC 4051. The widths of the integrated profiles of NGC 3227 and NGC 4151 do not seem unusual.

Discussion of Papers Read by Lewis and Allen

Heidmann: Allen's argument is based on the use of the M_H/M_T ratio and this is a rather indirect way of comparison because this ratio is subject to uncertainty in distance, to cosmic dispersion of M_H and to cosmic dispersion of M_T. A more direct way is to use the 21-cm line width only, as we did for NGC 4151.

For NGC 1068, Lauqué showed me his records and when due account is taken of errors in inclination angle, the maximum rotational velocity V_m turns out to be 220–270 km s^{-1}. On the other hand the statistics of V_m vs morphological type by Gouguenheim (*Astron. Astrophys.* **3**, 281, 1969) show that for the Sb type, $V_m = 260$ km s^{-1}. Then NGC 1068 is rather on the low side and it does not appear necessary to invoke expansion of large H I masses in it.

I do not mean to say that the snow-plow effect which Oort and Minkowski used for supernovae does not exist. Walker's clouds may produce it, but this snow-plow effect will appear later, when enough H I has been pushed and when Seyfert activity may be no longer visible in the nucleus.

In my opinion this snow-plow effect could be invoked to explain asymmetrical H I distributions and even rings which would be asymmetrical distributions washed out by differential rotation. I made dynamical estimates which show this is possible.

Mrs Burbidge: You do not need to use the extrapolation factor of 5.13 for NGC 1068; the determination of mass out to the last easily observed Hα measurement, by Prendergast and my husband and myself, gave a mass of a few times 10^{10} $M_\odot$ (this remark also applies to the earlier paper by Dent).

Mrs Rubin: It appears that your ΔV increases with increasing optical faintness for the four galaxies. Could this just be a problem in defining the baseline for the weaker sources?

Lewis: NGC 1068 does seem to have the smallest velocity range ΔV and to be the closest of the Seyferts, while NGC 3227, the furthest away, has the largest velocity range. The difference is due mostly to the difference in apparent profile shape. NGC 1068 is more centrally peaked and its intensity shades away into the noise. It could indeed have a larger velocity width, though this is not seen in Allen's observations, which have a much better signal to noise ratio.

NGC 3227, however, in my observations, is of low intensity and very flat. The galaxy profile is delineated principally by the rapid decrease to zero signal at the extremities of the velocity range. Presuming the galaxy to have been detected, then the velocity range quoted is quite accurate. I do not think that any distance dependent effect is involved, other than the decrease it causes in the measured flux level.

Weliachew: I would like to point out that, in addition to the v^2 dependence of the derived total mass, there is a first power dependence with respect to the turn-over point radius which is not provided by the reported observations.

Heidmann: Following Weliachew's comment, Walker's work shows that in NGC 1068 the rotational velocity is 80 km s^{-1} at a 40 arc sec distance from the center.

On the other hand statistics of the ratio of turn-over radius to Holmberg optical radius give values from 0.5 to 1 for Sb galaxies. Then the turn-over radius in NGC 1068 should be about 150–300 arc sec. $V_m \gg 80$ km s^{-1} and Walker's optical observations cannot be invoked by Allen in his argument.

21-cm ABSORPTION IN BL LAC

M. H. COHEN

California Institute of Technology, Pasadena, Calif., U.S.A.

(To be published elsewhere)

Abstract. Observations with the Owens Valley Radio Observatory interferometer show that BL Lac has a 21-cm hydrogen absorption line, produced by a local intervening cloud. The line is centered on zero velocity (LSR), has a width of about 5 km s^{-1}, and shows an optical depth of 0.5. BL Lac is at galactic latitude $b = -10°$. No firm distance limit is possible, but such clouds are, typically, 100 pc apart. This observation is consistent with BL Lac being an extragalactic object. Other observations leading to a similar conclusion are of rotation measure (MacLeod and Andrew, 1968) and of reddening (Bertaud *et al.*, 1969).

References

Bertaud, Ch., Dumortier, B., Véron, P., Wlérick, G., Adam, G., Bigay, J., and Garnier, R.: 1969, *Astron. Astrophys.* **3**, 436.
MacLeod, J. M. and Andrew, B. H.: 1968, *Astrophys. Letters* **1**, 243.

THE LAW OF MOMENTUM CONSERVATION AND SOME PROBLEMS OF METAGALACTIC ASTRONOMY

I. S. SHKLOVSKY

Sternberg Astronomical Institute, Moscow, U.S.S.R.

Abstract. Because of the anisotropic character of the generation of relativistic particles and synchrotron emission of active galactic nuclei and QSS it should be expected that 'magnetoids', the plasma bodies which are responsible for such generation, will attain large momenta. Therefore, magnetoids should be pushed out of nuclei with high velocities. This mechanism can explain several puzzling effects which are observed in the Metagalaxy, e.g. the jet in NGC 4486, connections between some galaxies and QSS and the total positive energy of some clusters of galaxies.

During the past decade many facts were accumulated which evidently support a famous hypothesis advanced by V. A. Ambartsumian (1958) about ejection of large aggregates of matter from active galactic nuclei. According to the concept of V. A. Ambartsumian, the matter initially ejected by some explosion process from a galactic nucleus must be in some hypothetical superdense state and during some time evolve into galaxies. However, we do not agree with this concept because a consequent development of it would lead to the conclusion that the laws of 'usual' physics are not applicable to the description of the processes in galactic nuclei. It seems to us that today there is no reason for such a radical assumption.

In this report we will make an attempt to give a simple and natural explanation for some strange phenomena which were discovered in the Metagalaxy and which at first sight confirm the concept of V. A. Ambartsumian.

Let us summarize these facts (for details see Burbidge (1970)).

(1) V. A. Ambartsumian emphasized the fact that for galaxy clusters, as a rule, $E+\Omega>0$ so that they have positive total energy. In order to avoid this difficulty, many authors advanced the hypothesis that there are large amounts of unseen matter inside these clusters. However, these attempts were not successful. As a matter of fact, there do exist clusters of galaxies which are unstable expanding systems. The characteristic time of such an expansion is 10^8–10^9 yr, i.e. much less than the age of the Universe. From this fact follows inevitably the conclusion that the process of galaxy formation continues in our epoch. According to the point of view of V. A. Ambartsumian, galaxies are formed from the superdense matter ejected from active nuclei with high velocity. This hypothesis formally explains the observed total positive energy of a galactic cluster.

(2) There are chains and small groups of galaxies, e.g. Stefan's Quintet, Seyfert's Sextet, the Chain VV172, etc., in which one of the systems has an abnormally high velocity.

(3) Arp drew attention to the remarkable connection between the distribution of some galaxies and QSS over the sky. It is possible that in some cases there is a genetic

coupling between these objects. The impression was created that some QSS were 'shot out' from active nuclei.

Note that this circumstance cannot be considered as an argument in favor of the 'local' hypothesis of the origin of QSS. The observed very large red shifts should be explained presumably by the cosmological expansion while the peculiar velocities of QSS may be comparatively large.

(4) In addition to the above-mentioned arguments, we must add another one. This argument concerns the so-called 'jets' of some galactic nuclei, notably the famous jet in NGC 4486. The assumption that this jet was formed by ejection of clouds of relativistic plasma from the nucleus of the galaxy leads to great difficulties. First, the time of synchrotron energy losses is unacceptably short. Moreover, within the framework of this hypothesis there is no explanation for the observed absence of expansion of knots in this jet. Note that almost exactly in the direction of the jet is situated the radiogalaxy NGC 4374. The alternative hypothesis for the production of the jet in NGC 4486 is the supposition of ejection in some direction of compact, gravitationally-bound aggregates of matter, which are permanently generating relativistic electrons. This supposition is free from the above-mentioned difficulties. We encounter similar problems in the interpretation of very compact details which are observed in some extended radio sources (e.g. Cyg A).

The analysis of the results of observation leads to the conclusion that along with ejection of adiabatic expanding clouds of relativistic and nonrelativistic plasma from active nuclei of galaxies there takes place the ejection of compact, non-expanding aggregates of matter which are in a state of very high activity (i.e. there is a powerful generation of relativistic particles). We will show that this process is a natural consequence of the law of momentum conservation and current ideas concerning the nature of activity of galactic nuclei and QSS.

Everybody is accustomed to the fact that during activity of galactic nuclei extremely large amounts of energy are liberated, up to 10^{61}–10^{62} ergs. This energy is generated mainly in the form of relativistic particles. But, as far as we know, until recently nobody drew attention to the fact that the relativistic particles and photons of synchrotron radiation take away a very large momentum from the region of galactic nuclei. The reason for this is the very anisotropic character of generation of relativistic particles by nuclei of galaxies. As a rule this generation occurs in two opposite directions. This follows from the observed duality of most radio sources. The photon synchrotron emission also must be anisotropic. This follows from the analysis of outbursts of radio emission from QSS and Seyfert nuclei. According to this analysis, the value of $H_\perp$ is comparatively small (Kellermann and Pauliny-Toth, 1968). On the other hand, the equipartition condition demands that the total H should be comparatively large. Consequently, $H_\perp/H \ll 1$ and synchrotron emission must be anisotropic.

This result is valid also for the very powerful infrared emission of QSS and Seyfert nuclei (Shklovsky, 1970).

Evidently, the total loss of momentum by a galactic nucleus caused by generation of relativistic particles and synchrotron quanta should be

$$\Delta P = (2\pi/c) \int W(\theta) \sin\theta d\theta \tag{1}$$

where θ is the angle between the axis of symmetry of the source and any direction and $W(\theta)\sin\theta d\theta$ denotes the energy which was emitted in this direction during the whole time of activity of the nucleus. If the generation of relativistic particles and photons by galactic nuclei were exactly axially symmetric, we should have $\Delta P = 0$. However, in real conditions the case of exact axial symmetry does not exist. In any case the two components of a radio source are not similar in morphology and energy content. In some cases the two radio components and the 'parent' – the optically observed galaxy – are not situated along a straight line.

Consequently for our rough estimation we can assume, that

$$\Delta P \approx W/c \tag{2}$$

where W denotes the total energy generated during the active phase of a nucleus. If $W \sim 10^{61}$ erg, $\Delta P \approx 3 \times 10^{50}$ g cm s^{-1}.

According to the law of momentum conservation, the nucleus must obtain momentum $-\Delta P$ due to its energy generation. What are the consequences of this fact?

It seems to us that the most plausible hypothesis which explains the activity of QSS and galactic nuclei is the so-called 'magnetoid' hypothesis. According to this hypothesis, the source of activity of nuclei and QSS is connected with a large rotating magnetic plasma body or 'magnetoid'. Recently, Ozernoy (1971) undertook a very important and interesting development of this hypothesis.

It seems natural to suppose that the momentum $-\Delta P$ will be obtained by the magnetoid which is situated inside the nucleus of a galaxy. Denote the mass of magnetoid as M. Then we can write the following equations

$$W = \alpha M c^2, \tag{3}$$

consequently

$$\Delta P = \alpha M c = M v, \tag{4}$$

and

$$v = \alpha c \tag{5}$$

where $\alpha \approx 1\text{--}10\%$ denotes the efficiency of energy generation by the magnetoid. From this it follows that the 'back velocity' v of the magnetoid must be in the range between several thousand and several tens of thousand km s^{-1}. This is adequate for the explanation of the above-mentioned phenomena.

The present situation in magnetoid theory does not allow us to describe its evolution in a precise way. Evidently such an evolution should be possible in different directions depending on the initial conditions (value of the mass, angular velocity and the law of rotation, etc.). It may be expected that in some cases the magnetoid during its evolution will be fragmented into many condensations which will evolve into stars while in other cases the magnetoid will collapse. It may be expected that the masses of magnetoids which were ejected from nuclei of galaxies will cover a wide range e.g. from 10^3 to 10^{10} $M_\odot$. It is most probable that this process will be recurrent: after ejecting one magnetoid from the nucleus of a galaxy another one will be formed which will evolve in a more or less similar way. It is reasonable to assume that the time interval between successive ejections of magnetoids is proportional to their masses. The cause of formation of magnetoids in nuclei of galaxies may be, for example, accretion of interstellar gas (Shklovsky, 1962) or disintegration of the stars caused by collisions or close encounters.

It is essential for both cases that nuclei of galaxies may be considered as very deep potential boxes.

Thus, it is possible to explain naturally all the strange phenomena mentioned above. Let us consider, for example, the jet in NGC 4486. Each condensation of this jet can be considered as a small magnetoid with a more or less regular magnetic field having a different orientation. These condensations are similar to small quasars. The absence of the emission lines in the spectrum of the jet can be explained by the violet shift. According to Equation (5), we may suppose that the velocities of these magnetoids are near to $\sim 10^9$ cm s^{-1}.

Therefore, the age of condensations should be $\sim 10^5$ yr, while the mean interval between successive ejections of magnetoids from the nucleus of NGC 4486 must be several tens of thousands of years. From the total energy emitted by the jet during its lifetime ($\sim 3 \times 10^{42}$ erg) we can roughly estimate the mass of the magnetoids as 10^4 $M_\odot$.

It is natural to expect (by analogy with quasars) that optical emission from the condensations of the jet may be variable. In addition we can expect variability of polarization from the condensations. Evidently the duration of activity of the nucleus of NGC 4486 may be much longer than 10^5 yr.

Therefore, it may be expected that in NGC 4486 there will be many comparatively old magnetoids which are essentially less active than in the jet. It is not impossible that such objects may be identified with the so-called 'globular clusters' which are observed in NGC 4486. In (Shklovsky, 1968) we drew attention to a very strange phenomenon: most of the 'globular clusters' in NGC 4486 are situated in a quadrant with the axis coincident with the direction of the jet. In the case of 'true' globular clusters it is natural to expect spherical symmetry in their distribution.

Thus we see that the application of the law of momentum conservation to some of the problems of metagalactic astronomy will open new and far-reaching possibilities. In conclusion, we should emphasize that our attempt at analysis of this problem is a very rough and preliminary one, but further development of this idea may give new and interesting results.

References

Ambartsumian, V. A.: 1958, in *Structure and Evolution of the Universe*, Solvay Conference, Brussels, p. 241.

Arp, H. C.: 1967, *Astrophys. J.* **148**, 321.
Burbidge, G. R.: 1970, *Ann. Rev. Astron. Astrophys.* **8**, 369.
Kellermann, K. I. and Pauliny-Toth, I.: 1968, *Astrophys. J. Letters* **152**, L169.
Ozernoy, L. M.: 1971, this volume, p. 290.
Shklovsky, I. S.: 1962, *Astr. Zh.* **39**, 591.
Shklovsky, I. S.: 1968, *Astr. Zh.* **45**, 919.
Shklovsky, I. S.: 1970, *Astr. Zh.*, in press.

THEORETICAL CONSIDERATIONS OF COMPACT OBJECTS

L. WOLTJER

Columbia University, New York, U.S.A.

Abstract. The rotation of massive objects with magnetic fields may provide the highly efficient means of energy conversion required to explain the high luminosity of compact objects. Several variations of the 'massive rotator' model are discussed.

The two main difficulties in our understanding of compact objects concern the energy requirement and the radiation mechanism. With regard to the latter various models involving synchrotron and Compton radiation, with cutoff due to self absorption or cyclotron turnover have been discussed. The main difficulty is caused by the sharply peaked strong infrared radiation observed in several objects. Thermal dust models for the infrared are attractive because we know from the intensity ratios of certain emission lines that dust is present in the nuclei of some Seyferts. However, the rapid time variations in the infrared, which were reported here, appear to stretch the dust models beyond the breaking point. If so a non-thermal mechanism is called for and it seems likely that, whatever may be the detailed process, much of the radiation derives from energetic particles.

In Table I we have assembled estimates of the energy requirements and time scales for some representative objects. The values listed are on the conservative side. For example recent results by Kleinmann and Low (1970) may indicate that the far infrared emission of 3C 273 could be as high as 10^{49} erg s^{-1}, while the results reported here

TABLE I

Luminosity L, time-scales and energy requirements E for some objects

	L (erg s^{-1})	(yr)	E (erg)
3C 273	10^{47}	$10^{5.5}$	10^{60}
NGC 1068	$10^{46.5}$	10^{8}	10^{62}
Cyg A	$10^{44.5}$		(10^{60}) [a]

[a] equipartition energy for electrons.

by Dr Neugebauer show that some QSOs emit at least 10^{48} erg s^{-1} already in the more accessible parts of the infrared. The time-scales are quite uncertain. In the case of 3C 273 all we have is the light travel time to the tip of the jet, while for the Seyferts the fact that 1% of the brighter galaxies belong to this class only establishes a minimum total duration (not necessarily continuous) of 10^{8} yr. Inspecting the table we see that energy yields of 10^{6} $M_{\odot}$ c^{2} – partly in the form of relativistic electrons are required for radio galaxies and QSOs, while for the Seyferts the long term yield is no less than $10^8 \, M_\odot \, c^2$. While the figures in Table I represent modest estimates for the more energetic objects it should be kept in mind that the luminosity functions of QSOs and radio galaxies are quite steep. For example for the optical radiation of the QSOs Schmidt (1970) finds a luminosity function $n(L) \propto L^{-2.2}$. Either the formation rate of the fainter objects is much larger than that for bright ones or the fainter objects live much longer.

The time variations impose severe constraints on the spatial extent of the objects. Although counter examples can be constructed, it seems unlikely that the emission comes from regions with linear dimensions *much* larger than the velocity of light multiplied by the characteristic time. Both in the QSOs and in NGC 1068 time scales of the order of a day appear to be involved, indicating emission regions no more than 10^{16} cm across. Direct interferometric observation of a radio component in M87 with a diameter less than 10^{17} cm appears to confirm these very small sizes. Probably the region in which the energy is primarily released is even smaller. The presently available data do not yet allow one to conclude that the same emission region is responsible for all variations, but the very fragmentary polarization data which tend to show some reproducibility perhaps suggest that this is the case.

Not much is known about the total mass available in the inner nuclei of galaxies, but some useful upper limits exist. In NGC 1068 Burbidge *et al.* (1959) obtained an upper limit of $3 \times 10^9 \, M_\odot$. From Dr King's remarks here it would seem that the limits in some giant ellipticals are of the same order, while the rotation curve for our own galaxy seems to be incompatible with the presence of a central mass much in excess of $1 \times 10^9 \, M_\odot$.

Comparing the available mass with the total energy release in NGC 1068 we conclude that a conversion efficiency of several percent of Mc^2 seems to be achieved. At least in the radio galaxies much of the energy seems to go into relativistic electrons. Inspecting the older mechanisms we are extremely doubtful that a Fermi-mechanism, supernova-shell expansion or star collisions could have the required efficiency, particularly for the electrons.

In the case of the Crab Nebula pulsar we have an object of very small size which converts rotational energy with very high efficiency (20% or so) into energy of relativistic electrons, presumably through the effect of a strong magnetic field in the pulsar. It seems tempting to scale the Crab Nebula mechanism to objects of much greater mass. Investigations of massive rotating objects with magnetic fields have recently been made by several investigators (Morrison, 1969; Cavaliere *et al.*, 1969; Woltjer, 1970; Fowler, 1970; and with somewhat different emphasis by Ozernoy, 1969; and Piddington, 1970; also Lynden-Bell, 1969). Closely related objects were discussed earlier by Hoyle and Fowler (1963) who referred to these objects as supermassive stars, spinars, magnetoids or massive rotators. A rough outline of the contents of some of these investigations follows.

Suppose we have a non-relativistic cold object which is supported mainly centrifugally (angular velocity Ω) and which has a magnetic field B. We treat the object as if it were a sphere of uniform density with radius R. From the virial theorem we find that mechanical equilibrium requires that

$$\Omega^2 R^3 = 3GM/5$$

If we assume that the electromagnetic stresses take away angular momentum, J, in the same way as it is thought to occur in pulsars (Pacini, 1967; Ostriker and Gunn, 1969; Goldreich and Julian, 1969) we have an expression of the following form

$$\frac{dJ}{dt} = \frac{2}{5} M \frac{d(R^2\Omega)}{dt} = -\text{const}\,\Omega^3 R^6 B^2.$$

If magnetic flux, Φ, is conserved during the evolution of the body we have

$$\Phi = BR^2 = \text{const}$$

and combining these expressions

$$\Omega = \frac{\Omega_0}{(1 - t/\tau_0)^{1/2}}, \quad \tau_0 = \text{const}\,\frac{M}{\Phi^2 \Omega_0^2},$$

with Ω_0 referring to $t=0$. If we assume that the energy is ultimately converted into particles and radiated with an efficiency s we obtain for the luminosity L:

$$L \propto s\Phi^2 \Omega^4 = \frac{L_0}{(1 - t/\tau_0)^{4/3}}$$

with $L = L_0$ at $t = 0$. We note that in case of a steady state population this luminosity law corresponds to a steep luminosity function, $n(L) \sim L^{-7/4}$, even if all objects have the same value of M and Φ.

In order to fix the values of M and Φ some assumptions have to be made. Morrison assumes that the 200 day period of 3C 345 can be identified with the rotation period while Cavaliere et al. let Φ correspond to that in a typical region of the interstellar medium. Fowler (whose object is not cold but an $n=3$ polytrope) also obtains Ω from the 200 day period but interprets the additional 50 day time scale as a pulsation period. These approaches lead to values of $M \approx 10^9 M_\odot$. We ourselves have considered it unlikely that the evolution terminates before R is of the order of the Schwarzschild radius R_{sch} and therefore take the most luminous objects to have $R \approx R_{sch}$. Somewhat smaller masses ($10^8 M_\odot$), radii of $10^{13} - 10^{14}$ cm and values of B near 10^6 G are obtained, corresponding to rotation periods near 10^4 sec. We then interpret the bursts with a time scale of 200 days as perhaps more similar to the 'wisps' in the Crab Nebula. Clearly in these discussions a better knowledge of the fastest time variations in QSOs would be most useful. The very active and luminous QSOs may well be in their later evolutionary phases and the activity for detecting rotation related regularities would present itself in the more stable objects without major fluctuations. At the same time the amplitudes of the more regular variations may be quite small. Again the case of the Crab Nebula is instructive, the pulsar related variations in the integrated light being only of the order of 1%. Because in the massive rotators differential rotation may be expected to occur the situation might be even less favorable.

The above discussion of the massive rotators is quite primitive. In some phases the objects may well be hot – supported partly by radiation pressure – as a consequence of nuclear processes and a substantial amount of light could be thermal (Hoyle and Fowler, 1963). If the objects are cold they are likely to be more in the shape of thin disks – like those studied relativistically by Salpeter and Wagoner. In that case instabilities probably will result in fragmentation. If the fragments are very small the disk might look more like a collection of pulsars – as in the model proposed by Rees. If in the inner parts a collapsed object, (black hole), has been formed and if angular momentum from infalling gas is transported out magnetohydrodynamically the situation would be that envisaged by Lynden-Bell with a steady flow into the black hole. And finally even if the rotators are formed they are unlikely to be of uniform density and differential rotation is expected. As a result the poloidal magnetic fields would be twisted up, resulting in much instability as discussed by Morrison and Ozernoy.

The formation of bodies of large mass in the centers of galaxies presents great difficulties. Spitzer (1970) has studied a model where repeated cycles of star formation and stellar mass loss lead to a nucleus of high stellar density. Once such a region is formed, stellar dynamical relaxation and finally star collisions, become important and will result in a large increase of the central density. During the stellar collisions much thermal energy may be radiated. However the effects of angular momentum remain to be fully understood. Direct condensation of intergalactic gas may be an alternative.

Acknowledgement

This research has been supported in part by AFOSR under contract AF 49 (638) 1358.

References

Burbidge, E. M., Burbidge, G. R., and Prendergast, K. H.: 1959, *Astrophys. J.* **130**, 26.
Cavaliere, A., Pacini, F., and Setti, G.: 1969, *Astrophys. Letters* **4**, 103.
Fowler, W. A.: 1970, in R. D. Davies and F. G. Smith (eds.), 'The Crab Nebula', *IAU Symp.* **46**, 364.
Goldreich, P. and Julian, W. H.: 1969, *Astrophys. J.* **157**, 869.
Hoyle, F. and Fowler, W. A.: 1963, *Monthly Notices Roy. Astron. Soc.* **125**, 169.
Kleinmann, D. E. and Low, F. J.: 1970, *Astrophys. Letters* **159**, L165.
Lynden-Bell, D.: 1969, *Nature* **223**, 690.
Morrison, P.: 1969, *Astrophys. Letters* **157**, L73.
Ostriker, J. P. and Gunn, J. E.: 1969, *Astrophys. J.* **157**, 1395.
Ozernoy, L. M.: 1969, *Soviet Astron.* **12**, 901.
Pacini, F.: 1967, *Nature* **216**, 567.
Piddington, J. H.: 1970, *Monthly Notices Roy. Astron. Soc.* **148**, 131.
Schmidt, M.: 1970, in Proceedings, Vatican *Semaine d'Etude* on Active nuclei in Galaxies, to be published.
Spitzer, L.: 1970, in Proceedings, Vatican *Semaine d'Etude* on Active nuclei in Galaxies, to be published.
Woltjer, L.: 1970, in Proceedings, Vatican *Semaine d'Etude* on Active nuclei in Galaxies, to be published.

Discussion on Papers by Shklovsky and Woltjer

Arp: I have looked at the distribution of globular clusters around M87 several times and have never noticed any strong relationship to the line of the jet. I would be very disappointed if I had missed such an asymmetry.

The objects which *are* lined up very well with the line of the jet and counterjet are the E galaxies in the neighborhood of M87. These E galaxies have masses which are orders of magnitudes greater than those of the globular clusters which Shklovsky suggests are ejected. The Shklovsky suggestion seems observationally, however, a step in the right direction.

Baum: Some years ago I investigated the distribution of the small bodies, supposedly globular clusters, surrounding M87. My purpose was to compare their radial distribution with the luminosity profile of the main body, but I would have noticed a pronounced asymmetry if it were present, as asserted by Shklovsky. I did not notice any such asymmetry but shall be interested in re-examining my data with this possibility in mind.

Barnothy: You have mentioned Morrison's rotating model as one of the more plausible models of QSOs. I have investigated the efficiency of the synchrotron mechanism which is used in this model to transform the rotational energy into optical radiation and found that its efficiency is lower by a factor 10^{24} than needed. During my correspondence with Dr Morrison he suggested two additional assumptions, but even these would leave a discrepancy by a factor of 10^9.

Woltjer: There appears to be no difficulty in converting rotational energy efficiently into relativistic electron energy. With a suitable magnetic field much of this energy can be converted into infrared and (possibly indirectly by inverse Compton) optical radiation.

Felten: With regard to Shklovsky's idea (remarkable how much response is drawn by a qualitative suggestion which is specific and somewhat individual), one would not expect that most of the relativistic particles would end up in the blobs which represent their recoil. One might think, for example, that they had mostly flown off 10 or 100 times as far in the opposite direction. We should think about where these electrons are and whether they can be observed to test the model. Has Shklovsky discussed this?

Woltjer: No. However it seems to me that the electrons might again be deflected by magnetic fields in the galaxy; if so, the interstellar matter in which these fields are anchored would ultimately absorb the momentum.

Smith: An observational comment may be of interest, regarding your suggestion that 10^4 s should be some kind of lower limit to the time scale of significant brightness variations to be seen in massive core objects. Angione has examined a number of QSOs photo-electrically for rapid variations; the fastest found was 3C 454.3, with changes of 30–40% in times of several hours, although in most cases the changes were of the order of, or less than, a few per cent per day.

Ozernoy: If, firstly, the physical nature of quasars and quasarlike phenomena in galactic nuclei is the same and differs mainly in the scale of energy output from magnetoids and if, secondly, the

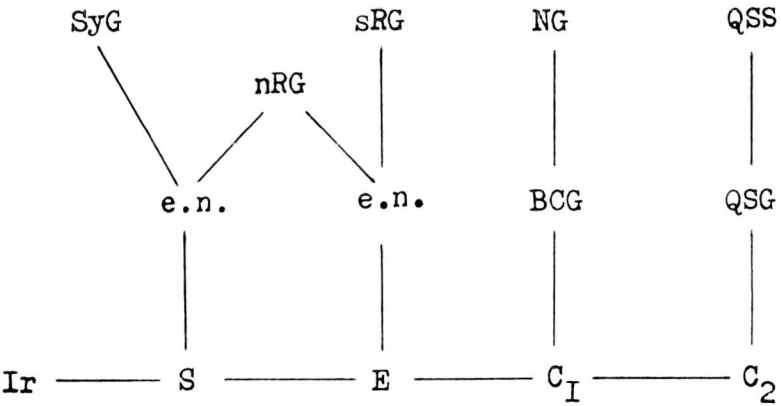

Fig. 1. The extended Hubble sequence of galaxies.

harmonic mean time for the active phase is about 10^8 yr, one can derive theoretically the directions of evolution of different kinds of galaxies. These directions are represented as an extended Hubble sequence of galaxies (see Figure 1).

Here Seyfert galaxies (SyG), strong radio galaxies (sRG), N-galaxies (NG) and quasi-stellar radio-sources (QSS) are regarded as a 'ceiling' of excited states of galaxies of the types spiral (S), elliptical (E) and two kinds of very compact galaxies designated C_1 and C_2 respectively. These galaxies (S, E, C_1 and C_2) are considered as 'ground states' of galaxies.

Blue compact galaxies (BCG) and quasi-stellar galaxies (QSG) are the intermediate excited states' analogous to 'excited nuclei' in normal galaxies, where compact sources of non-thermal radiation exist as described in the paper by Dr Ekers.

The galaxies in the ground state are not connected genetically and differ by initial conditions, that is mainly by the total mass and angular momentum. The vertical lines point out the only possible directions of the excitation of galactic nuclei. For example, the transition of QSS to RG or vice versa is unlikely to be expected. The ground states of QSS and NG are special compact galaxies, and it may be expected that their spatial densities nearly coincide with those for BCG and QSG if a mean harmonic life-time of NG and QSS is about 10^8 yr. If this time is closer to 10^6 yr, these densities are 10^2 greater than those for BCG and QSS.

An excited state of a galaxy should be periodically repeated (with a characteristic interval of about 10^8 yr) during the whole life of a galaxy. The decrement of this process is determined by a gradual change of conditions for a magnetoid reproduction.

Dr Morgan kindly informed me at this Symposium that the theoretical scheme presented (and described in detail in Astronomical Circular of the U.S.S.R. No. 581) is very similar to his classification obtained from an observational point of view.

THE LARGE-SCALE VARIATIONS OF
QUASI-STELLAR OBJECTS

W. H. McCREA

University of Sussex, Falmer, Great Britain

The proposition for consideration is:

A QSO is an approximately standard object: it has an approximately standard history consisting mainly of an increase of optical brightness to a very high peak, and a subsequent decline. I suggest that the history has a relevant duration of about 10^6 yr, and that the total increase of brightness is by a factor about 10^3, while the mean brightness is of the order of 1% of the peak brightness (so that the peak is rather sharp).

The following corollaries would result:

(1) The energy requirement of an individual QSO would be less than that usually assumed (a few percent);

(2) The spread of intrinsic brightness of QSOs would be due mainly to the variation;

(3) QSOs would have a particular greatest intrinsic brightness, the same for all distances;

(4) Since, by hypothesis, the peak brightness is very high, it is plausible that QSOs with very large redshift be detectable. This need not demand some QSOs of exceptionally great energy;

(5) The variation would tend to obscure the effect of distance upon apparent magnitude, and so to mask the Hubble relation, but we might get a significant relation from the brightest QSOs at each redshift.

(6) Most optical properties would be statistically the same for all QSOs at the same stage in their history. If QSOs at redshifts z_1 and z_2 show any similar variabilities to which we assign characteristic times T_1 and T_2, then we should expect to find

$$T_1/T_2 = (z_1 + 1)/(z_2 + 1) \qquad (1)$$

(7) Since the calculation of intrinsic properties from the observations is model-dependent, then, conversely, we could seek a model that gives the same intrinsic luminosity-function for all values of the redshift, z.

Equation (1) would provide a powerful test of the cosmological, or at any rate the Doppler, character of z. For if the line-shift were, say, gravitational, it is unlikely that variations in the continuous emission would be subject to the same 'shift'. The test could be applied in a quite crude fashion. We could take all the light curves, such as those discussed by Harlan J. Smith and by T. D. Kinman, and plot those for QSOs having, say, $z>1$, on half the time scale for those having $z<1$ and see if the two lots are more similar than when they are plotted on the same time scale.

The corollaries 5, 6 and 7 may provide ways of exploiting the large z-values of QSOs for cosmological tests: hitherto not much success has been achieved in this respect.

1. Evidence in Support of the Proposition

It seems to be generally agreed that a QSO has an active life of the order of 10^6 yr. So QSOs must be continually appearing and disappearing, i.e. rising to the typical QSO brightness and then fading. It is economy of hypothesis to accept this as the main reason for the great spread of intrinsic luminosities.

On the $q_0 = +1$ Friedman model, the intrinsic brightness of the brightest QSOs is indeed nearly constant. In particular, I find that the object 4C 05.34 recently found by C. R. Lynds and Derek Wills to have $z = 2.877$ has almost exactly the same absolute luminosity as 3C 273 on this model.

There are the well-known cases of 3C 446 and 5C 2.56 (recently discovered by E. M. Burbidge) that show great variations in brightness so that at any rate some of the spread in brightness must be due to variations.

It has always seemed significant that the energy required by a QSO was calculated to be not a great deal more than what could be produced by known energy-sources. If by allowing for variations we find that the energy requirement is appreciably reduced, the fact that the requirement can then be met by known sources appears to be an argument in favour of such behavior.

2. Remarks

There is nothing basically new in these ideas. The discussion about density evolution vs luminosity evolution is essentially the same. However, it seems opportune to call attention to some of the foregoing considerations.

The suggestions made here are simple and tentative, but some of the consequences would follow even were the main proposition only very approximately valid. Also there may be complicating factors, e.g. a QSO may have several main peaks of brightness, and there might be directional properties.

The radio properties of a QSO are probably secondary consequences of the main mechanism, and so these properties may not be accommodated in a simple discussion. One would be inclined to suppose that the radio properties may be associated more with the later part of the life of a QSO.

Discussion

Mackay: It may be necessary to consider QSO lifetimes considerably in excess of 10^6 yr. There are now data on a number of QSOs whose radio components are separated by great distances from the optical object and which appear to have ages of the order of 10^6 yr. However, in these cases the QSO is still optically bright – perhaps only one magnitude fainter than the brightest QSOs – and in some of these sources another component is observed coincident with the optical QSO.

McCrea: As I have mentioned in the written account, I should suppose that there may be more than one outburst in certain QSOs. Consequently, in the case mentioned by Dr Mackay the extended features may be associated with a much earlier outburst.

A 'SINGLE ELECTRON' SYNCHROTRON RADIATION MODEL AND THE QUASI-STELLAR OBJECTS

D. F. FALLA and A. EVANS
The University College of Wales, Aberystwyth, Great Britain

Abstract. In this synchrotron radiation model, the essential feature is that the basic form of the theoretical continuum is a characteristic of the energy evolution of a single electron. The model may be applied to sources containing high magnetic field regions where electrons have radiation lifetimes short compared with typical observation times; the electrons are generated in condensations of material which also amplify magnetic fields to values required for the emission of optical synchrotron radiation. A more specialized version of the model can provide a description of the QSOs that is consistent with their cosmological interpretation.

The optical radiation continua of the QSOs may be represented by $F(v) \propto v^{-n}$. The usual explanation for a spectral index of any given value is that it results from synchrotron radiation, in a uniform magnetic field, by electrons having a power-law energy spectrum with an index directly related to n. We discuss here a theoretical model in which the particular values $n = \frac{1}{3}$ and $n = 1$ can be given a completely different interpretation; the essential feature of the model is that the basic form of the theoretical continuum obtained is a characteristic of the energy evolution of a single electron (Falla, 1970).

The synchrotron radiation power spectrum, for an electron of energy γmc^2 in a magnetic field H, is described by the characteristic frequency

$$v_c = (3eH/4\pi mc) \gamma^2. \tag{1}$$

For x defined as v/v_c, the power spectrum is given approximately by

$$P(v) \propto H x^{1/3}, \quad \text{for} \quad x \leq \tfrac{1}{3}, \tag{2a}$$

and

$$P(v) \propto H \exp(-ax^{2/3}), \quad \text{for} \quad x > \tfrac{1}{3}, \tag{2b}$$

where a is a constant. The expression for the electron energy as a function of time,

$$\gamma = \gamma_0/(1 + \beta H^2 \gamma_0 t), \tag{3}$$

where γ_0 represents the initial electron energy and β is a constant, gives the radiation half-life for the electron as

$$t_{1/2} = (\beta H^2 \gamma_0)^{-1}. \tag{4}$$

The corresponding evolution of the synchrotron power spectrum can be represented by the variation of characteristic frequency with time,

$$v_c = (v_c)_0/(1 + t/t_{1/2})^2 \tag{5}$$

where $(v_c)_0$ is the value of v_c for $\gamma = \gamma_0$.

D. S. Evans (ed.), External Galaxies and Quasi Stellar Objects, 285–289.
All Rights Reserved. Copyright © 1972 by the IAU.

It is normally assumed that v_c does not change appreciably over a given time interval Δt, which we take here to be the period of observation, so that the radiation continuum can be derived by combining the synchrotron power spectrum with the electron energy spectrum. We examine here the opposite situation, for which a significant change in v_c occurs during the relevant time interval and thereby causes an appreciable evolution of the synchrotron power spectrum. We suggest that for the change in v_c to be significant, $\Delta t \gtrsim t_{1/2}$. From the above equations, $t_{1/2}$ can be expressed in terms of H and v_c for the observed radiation. In the radio region ($v_c \sim 10^9$ Hz), a magnetic field $H \sim 10^{-4}$ G gives $\Delta t \gtrsim 10^6$ yr: clearly, evolutionary effects are insignificant for an electron radiating in this region. In the optical region, however, larger magnetic fields may be expected to occur. For the QSOs 3C 273 and 3C 446, as discussed by Burbidge and Burbidge (1967), the existence of magnetic fields $H \sim 10$–100 G is required if these objects are to be given a cosmological interpretation, and if synchrotron radiation is to dominate that by the inverse Compton process. We find that these values of H give $\Delta t \gtrsim 10^2$–10^3 sec, which are time intervals that are comparable with photographic plate exposure times, so that in this case our condition for a significant evolutionary effect is fulfilled. The same applies to the low-energy X-ray region, for which magnetic fields of the same orders of magnitude give $\Delta t \gtrsim 1$–10 sec, which are time intervals that are certainly exceeded in X-ray photon detection.

For electrons in high magnetic fields, the radiation mean free paths are short compared with the estimated dimensions of the objects: electron generation *in situ* is therefore required. We consider pion decay to be the principal source of these electrons. Pions of low energy ($\lesssim 100$ MeV) produce decay electrons with a most probable energy given by $\gamma_0 = 75$; these electrons radiate in the optical region if magnetic fields $H \sim 10^4$–10^5 G are available, and have $t_{1/2} \ll 1$ s.

We consider, for one of these decay electrons, an evolving power spectrum of the

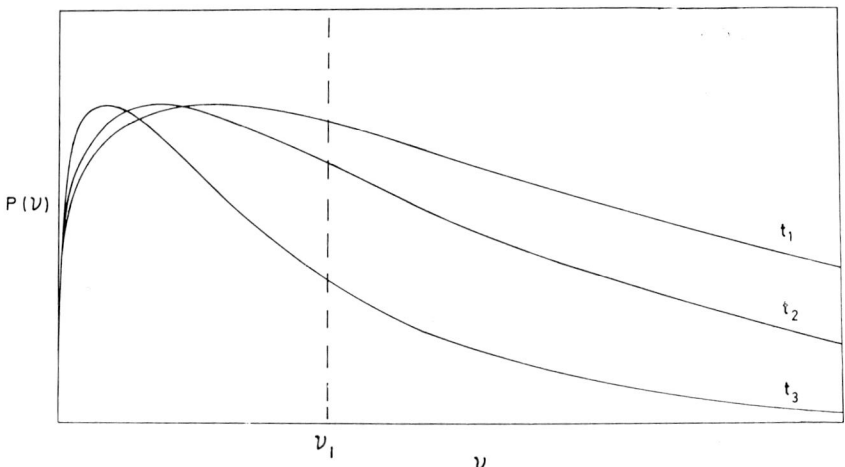

Fig. 1. Power spectra at times t_1, t_2, and t_3 from Equations (2).

form (2), in which v_c changes with time according to (4) and (5): the evolutionary effect is illustrated in Figure 1, in which the power spectrum is shown for successive moments of time t_1, t_2, and t_3. The total radiation continuum has been derived by taking the frequency v as an independent parameter and at each of its values, (for example, v_1 in Figure 1), computing the integral of the evolving power spectrum over the whole radiation lifetime of the electron. Figure 2(a) shows the radiation continuum

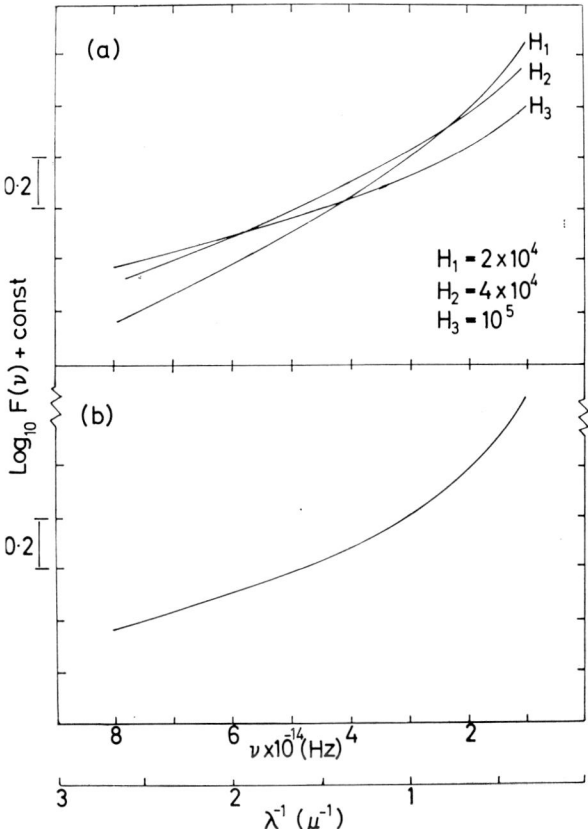

Fig. 2. (a) Single electron radiation continua. (b) Locus of maxima for optimum H.

for the optical and near infrared regions, computed for $\gamma_0 = 75$, for several magnetic field values in the range from 2×10^4 G to 10^5 G: each curve gives the radiation continuum produced by a single electron.

If we perform the integration by an analytical method we find that, provided that γ_0 is sufficiently high, (a condition that can be expressed by the inequality $H\gamma_0^2 \gg 10^9$ G for the optical region, and which is easily fulfilled for the high values of H taken), the integral can be written in the simple asymptotic form

$$F(v) = 2.63 \times 10^{-10} (vH)^{-1/2} \text{ erg Hz}^{-1}. \tag{6}$$

If the above condition for γ_0 is not satisfied then the expression (6) is only an approximation; in this case the more general formula used in obtaining the curves of Figure 2(a) has to be taken.

The requirements for the validity of the theory described are that the electron should be confined to a uniform magnetic field region, and be observed for the whole of its radiation lifetime. Furthermore, the theoretical radiation continuum does not depend upon the initial energy of the electron: this has the important consequence that the spectral index n is independent of the initial energy spectrum of the electrons.

For a complete radiation source it is necessary to integrate the computed radiation continua over an appropriate magnetic field distribution. In principle, any distribution of high magnetic fields could be taken and its parameters adjusted to fit the observational value of n: this approach would be similar to that of Hoyle and Burbidge (1966) to the radio spectrum of 3C 273, but with the difference that for high magnetic fields the electron energy spectrum would no longer be important. We consider now only one particular case, – a flat distribution, in which all magnetic field values are equally probable. Inspection of the curves in Figure 2(a) reveals that for any frequency there exists an optimum H value at which the computed flux is a maximum; the locus of these maxima is the upper envelope of the curves and is shown in Figure 2(b). We take the envelope plotted in Figure 2(b) as being an approximate representation of the total radiation continuum; for the optical and near infrared regions, this curve can be described by the spectral index $n \approx 1$. We suggest, therefore, that this value of n has a special significance: it could indicate that we are observing cosmic systems in which the different high magnetic field values occur with equal probability, and in which the radiating electrons have synchrotron power spectra for which evolutionary effects are appreciable.

High magnetic fields may be produced by the localized condensation of material initially associated with fields of much lower intensity (Ginzburg, 1964; Hoyle, 1969). Dyson's concept of a QSO, in which there occurs a random succession of local gravitational collapse events (Dyson, 1968), might be relevant to our model if magnetic field amplification were also included.

In our model, the electrons are generated in the interactions of high-energy protons incident upon the condensed material. Their rate of generation from a given proton flux is approximately proportional to the mass of the condensation, and does not depend upon the magnetic field contained by it; for the situation where the mass and magnetic field are completely uncorrelated, the suggested flat magnetic field distribution would be obtained. We would expect this type of model to apply to all cosmic systems containing a flux of high-energy protons, together with condensations of material where there are high magnetic fields and where electron production can occur.

For the particular case of the QSOs, a slightly more specialized version of the model is required. Burbidge and Burbidge (1967) have concluded that "if the QSOs are assumed to be at cosmological distances then their magnetic fields must be very large; the relativistic particles must be generated or accelerated *in situ*; and the very large number of subregions... must be phased together". Regarding this third requirement, we see that if these subregions are identified with the condensations that we have considered here, then the necessary correlation of the radiation from the subregions may be obtained by locating them around a well-defined source, (perhaps a central physical object), from which, as suggested by Hillier (1966), the interacting protons are emitted. Alternatively, following Dyson's random local collapse model, we might suggest that each condensation itself becomes a source of fast protons at some stage of its lifetime. This more specialized version of our model, in either of the two forms suggested, would seem to provide a consistent physical description of the QSOs that has all the properties demanded by their cosmological interpretation.

References

Burbidge, G. R. and Burbidge, E. M.: 1967, in *Quasi-Stellar Objects*, Freeman, London and San Francisco.
Dyson, F. J.: 1968, *Astrophys. J. Letters* **154**, L37.
Falla, D. F.: 1970, *Astrophys. Letters* **6**, 77.
Ginzburg, V. L.: 1964, *Dokl. Akad. Nauk.* **156**, 43.
Hillier, R. R.: 1966, *Nature* **212**, 1334.
Hoyle, F.: 1969, *Nature* **223**, 936.
Hoyle, F. and Burbidge, G. R.: 1966, *Astrophys. J.* **144**, 534.

A PROBABLE MECHANISM OF REPEATED EXPLOSIONS OF COMPACT OBJECTS

L. M. OZERNOY

*P. N. Lebedev Physical Institute, Academy of Sciences of the U.S.S.R.,
Moscow, U.S.S.R.*

Abstract. This paper considers a magnetoid model for nuclei of compact objects in which repeated explosions are caused by instability in twisted magnetic fields. The derived frequency and energy of individual explosions agrees with observations.

Lately more and more theorists and observers have inclined to the opinion that galaxy and quasar nuclei contain, in addition to stars, a magnetoid, i.e. a supermassive plasma body, the equilibrium and stability of which are determined by internal motions of a rotational type and magnetic fields. The elementary magnetoid theory, developed several years ago (Ozernoy, 1966; 1968), offered the possibility of explaining semi-quantitatively the two fundamental properties of quasars and quasar-like phenomena in galactic nuclei, that is, the powerful energy output from a small volume and the variability of this radiation. Recently we have obtained new results concerning three relevant topics: (a) magnetoid structure and evolution (Ozernoy and Usov, 1971); (b) the mechanism of repeated explosions (Ozernoy and Somov, 1971); and (c) the role of magnetoid explosions in the formation of stars and star systems (Ozernoy, 1971). These results give a more complete theoretical picture of quasi-stellar phenomena.

My purpose is to describe briefly the mechanism of repeated explosions, which probably operates in the nuclei of galaxies and quasars leading to ejection of clouds and jets of relativistic plasma.

The idea of this mechanism is the following: Let us imagine a supermassive rotating star, connected with the surrounding plasma by a magnetic field, inclined to the axis of rotation. As a result of the rotation the magnetic field is twisted. When rot **H** increases up to a definite threshold, an explosive instability occurs near the zero line of the magnetic field. Part of the magnetic field energy is transformed into kinetic energy of the current sheet plasma, ejected with up to relativistic velocities. Meanwhile the dissipated magnetic field is regenerated by continued twisting, which ensures the recurrence of explosions. The origin of the explosions is connected with the interaction of the rotation and the magnetic field. That is why I call such explosions *magneto-rotational*.

Now I would like to dwell on the most important features of the mechanism of magneto-rotational explosions (for more details, see Ozernoy and Somov, 1971).

The formation of the zero line of a magnetic field together with the corresponding current sheet is necessary for the appearance of such an instability. Let us consider the change of the geometry of the outer magnetic field in the course of contraction of the cloud with the initially homogeneous magnetic field perpendicular to the rotation axis. Let the rotation initially be small and insignificant. As shown in Figure 1, the first zero line of the X-type which has the form of a circle appears when contraction by a factor of $\sqrt{3}$ has already taken place. Figure 2 shows that the dissipation of the magnetic field takes place near the zero line and as the contraction continues this dissipation diminishes the number of force lines, going to infinity.

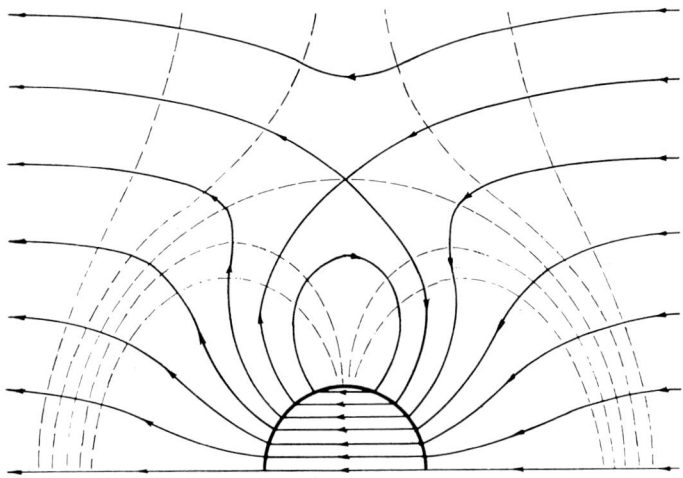

Fig. 1. The external magnetic field of the contracting cloud (solid lines) and the corresponding magnetic equipotentials (dashed lines).

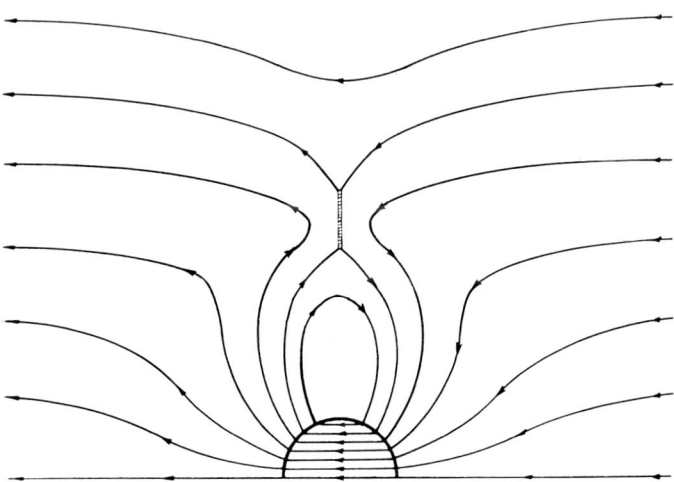

Fig. 2. The current sheet in the vicinity of the zero line of a magnetic field.

The process of cloud contraction is non-homologous, because the inner parts contract quicker than the outer ones. This leads to a more rapid increase of the radial component H_r compared with the transverse one H_θ. As a result, as shown in Figure 3, in the course of contraction the field acquires a predominantly radial character.

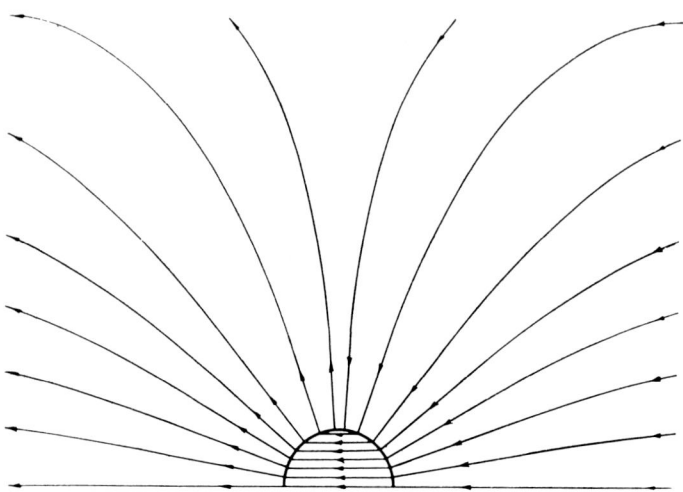

Fig. 3. The quasi-radial field forming due to the non-homologous character of a contraction.

The cloud contraction is brought to a stop by the growth of the pressure gradient and by the rotation, leading to the formation of a stationary magnetoid. In the intermediate stage of transition from contraction to equilibrium one or more bounces take place. As was suggested independently by Piddington (1966) and Ozernoy (1966, 1968) these bounces can lead to explosions producing clouds of relativistic electrons with a total energy characteristic of radio galaxies. The energy released is the magnetic energy which has been increased during the gravitational contraction. This is why I call such explosions *magneto-gravitational* as distinct from magneto-rotational which appear in the later stages of equilibrium.

When a cloud reaches its equilibrium state, a stationary magnetoid is formed. Subsequent contraction of the magnetoid is quasi-static with the velocity determined by the loss of angular momentum.

At this stage of quasi-stationary equilibrium, rotation becomes important for the amplification of the magnetic field. During contraction because of conservation of angular momentum, the cloud atmosphere is forced into differential rotation given by

$$v_\varphi = \Omega R^2/(r) \sin \theta \qquad (r > R), \tag{1}$$

where R is the radius of the cloud, r is the distance from the center, Ω is the angular velocity and θ is a polar angle. Such differential rotation will lead to a twisting of the quasi-radial field, so that a toroidal component, H_φ, will appear:

$$H_\varphi = -4\pi N (R/r)^4 H_R \sin \theta \qquad (r > R), \tag{2}$$

where N is the number of revolutions. The topology of the twisted magnetic field near the rotational axis is shown in Figure 4.* The curvature of the force lines on one

---

* It is of interest that the resulting magnetic field (which is parallel to the equator near the surface and roughly perpendicular in outer layers) is very similar to that inferred by Kinman (1970) from his observations of some variable quasars.

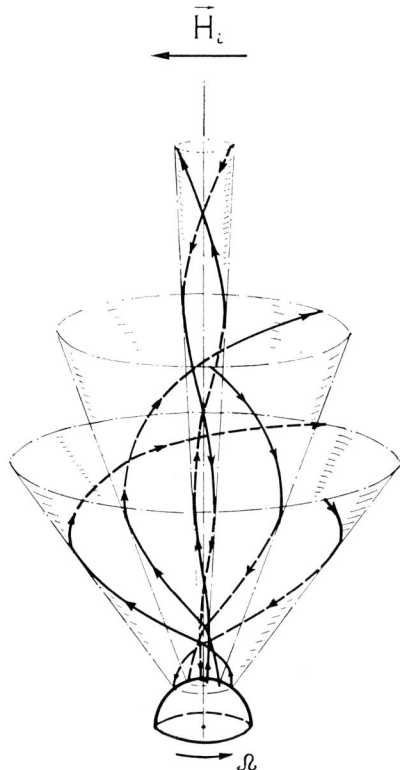

Fig. 4. The twisted magnetic field near the axis of rotation.

meridian decreases with approach to the rotational axis. Near the axis two force lines meet, oppositely directed and forming the zero line. Such a geometry offers the possibility of transforming the surplus magnetic energy, accumulated due to the rotation, into two symmetric ejections of accelerated particles, oppositely directed along the rotation axis. The acceleration may be conditioned by the mechanism of dynamic dissipation of a magnetic field (Syrovatsky, 1966). When the twisting of the magnetic field exceeds some threshold, the magnetic field near the zero line cannot be screened by quasi-stationary currents, but penetrates through the plasma into the zero line and prolongs the electric field $\mathbf{E} \sim \mathbf{H}$, which accelerates the current sheet plasma up to relativistic velocities.

The energy released by an explosion is of the order of the magnetic energy contained in the cylindrical canal surrounding the zero line. The radius of the canal

$$r_k \approx v_A \tau \qquad (3)$$

Here r_k is a cumulative radius, i.e. the radius within which the 'raking' of the magnetic field lines take place in the direction of the zero line as a result of the cumulative effect, and τ is the interval, during which the picture of the magnetic field is restored by the rotation after the previous explosion.

For the energy of the explosion a simple formula is obtained

$$\mathscr{E} \approx 10^{-2} H_{\varphi,e}^2 R^3 (r_k/R)^4 \qquad (4)$$

where $H_{\varphi,e}$ is a toroidal magnetic field at the equator. The time interval between the successive explosions is determined by the expression

$$\tau \approx 0.3 (v/v_A)^{4/3} N^{-1/3} T \qquad (5)$$

where v is the rotational velocity, v_A is the Alfvén velocity in the magnetosphere, and $T = 2\pi (R^3/GM)^{1/2}$ is a rotation period in the regime of the plasma outflow from the equator.

Though the appearance of explosions is connected with the rotation, their periodicity does not coincide with the rotational period T, because the restoration of the dissipated magnetic energy by the rotation may be going on, generally speaking, both slower and quicker than for one revolution. When N is small, $\tau \gg T$. With the growth of the number of revolutions, the value of τ/T decreases as $N^{-5/3}$ and becomes much less than unity, when the rotoidal field is large enough. The energy of explosions also decreases with time: $\mathscr{E} \sim N^{-2/3}$. Thus, magneto-rotational explosions become more 'feverish' and 'degenerate' with time.

As an example we shall estimate the parameters of explosions in the quasar nuclei. Let a gas conglomerate begin contraction with the following parameters:

$$R_i \sim 10 \text{pc}; \quad \varrho_i \sim 10^{-16} \text{ g cm}^{-3}; \quad H_i \sim 3 \times 10^{-6} \text{ Oe}.$$

As a result, a magnetoid is formed with mass $M \sim 3 \times 10^9 M_\odot$ and with the following main parameters:

$$L \sim 5 \times 10^{47} \text{ erg s}^{-1} \ R \sim 10^{17} \text{ cm}, \ T \sim 10 \text{ y}.$$

These parameters are close to those for the nucleus of the quasar 3C 273.

In order to predict the parameters of the explosions, the theory uses two items of observational data: the age of the active phase, and the Alfvén velocity in the magnetosphere. These are derived from the movements in the emission line region, surrounding the nucleus (Greenstein and Schmidt, 1964) and from fluctuations of its optical variability (Ozernoy and Chertoprud, 1967). They are:

$$t \sim 10^4 \text{ y}, \quad v_A \sim 3 \times 10^8 \text{ cm s}^{-1}.$$

Now we can find theoretically $N \sim 10^3$, $H_R \sim 3 \times 10^{-1}$ Oe, $H_{\varphi,e} \sim 3 \times 10^3$ Oe, $\tau \sim 3$ yr, $r_k \sim 3 \times 10^{16}$ cm, $\mathscr{E} \sim 10^{54}$ erg. These values prove to be in a rather good agreement with the observations available, including the time interval between successive radio bursts and the energy of a single burst calculated from the picture of radio variability.

The frequency spectrum, time variations and polarization of the flux of synchrotron radio emission from two clouds of relativistic electrons originating in a burst and flying apart in opposite directions with relativistic velocities, at the same time expanding, are calculated in the paper by Ozernoy and Sazonov (1969).

In conclusion I would note briefly some additional points.

(1) According to the dynamic dissipation theory, electrons of high energies, up to

$E \sim 10^{22}$ eV, radiating in the X-ray and γ-ray regions, should be generated during a burst. Estimates of γ-fluxes are given by Shklovsky (1970).

(2) The cumulative mechanism of explosions proposed follows a pulse regime. However, a magnetoid serves also as a source of continuous ejection of relativistic particles generated by the plasma turbulence.

(3) The dynamic dissipation leads to the acceleration of the current sheet plasma up to relativistic energies. The total mass of this relativistic plasma is, of course, too small. But there is no need to search for a new source of energy for the acceleration of thermal plasma clouds. The same relativistic particles may serve as such a source. By collisions with a magnetized thermal plasma cloud they can accelerate this cloud with their pressure up to high velocities.

(4) This last mechanism for acceleration of gas masses could, probably, explain the high velocities of the dwarf companions of some galaxies (for instance, in chains) as a result of their acceleration at the protogalaxy stage caused by the activity of a magnetoid in the parent galaxy.

References

Greenstein, J. L. and Schmidt, M.: 1964, *Astrophys. J.* **140**, 1.
Kinman, T. D.: 1970, this volume, p. 164.
Ozernoy, L. M.: 1966, *Astr. Zh.* **43**, 300.
Ozernoy, L. M.: 1968, in *Problems of Stellar Evolution and Variable Stars*, Symposium, Moscow, 24–27 Nov. 1964, (in Russian), "Nauka" Pub. House, p. 140.
Ozernoy, L. M.: 1971, *Astr. Zh.* **48**, in press.
Ozernoy, L. M. and Chertoprud, V. E.: 1967, *Astron. Zh.* **44**, 537.
Ozernoy, L. M. and Sazonov, V. N.: 1969, *Astrophys. Space Sci.* **3**, 395.
Ozernoy, L. M. and Somov, B. V.: 1971, *Astrophys. Space Sci.* **11**, 264.
Ozernoy, L. M. and Usov, V. V.: 1971, *Astrophys. Space Sci.*, in press.
Piddington, J. H.: 1966, *Monthly Notices Roy. Astron. Soc.* **133**, 163.
Shklovsky, I. S.: 1970, *Astron. Zh.* **47**, 742.
Syrovatsky, S. I.: 1966, *Astron. Zh.* **43**, 340; *Zh. Eksperim. Teor. Fiz.* **50**, 1133.

Discussion

Wray: At Northwestern University a student of mine, Harry Heckathorn, has carried out a comprehensive study of the velocity field of the ionized gas in M82. Our results, obtained with the 40-in. telescope of the Lindheimer Astronomical Research Center, are consistent with earlier observations by E. M. Burbidge, and by Lynds and Sandage, at points observed in common. We find that, out to the limits of our observations, some 45 arc sec from the major axis, the ionized hydrogen is both expanding and rotating. Since it seems likely that the mechanical energy dominates the magnetic field in the regions we have observed, it would appear that we may have evidence for the twisting of any magnetic field which might be present, much in the way you require. It would be interesting to see if your theory may indeed be applicable to the problem of M82 and, if so, it would be interesting to see if there are any parameters common to both theory and our observations which could be investigated for agreement or disagreement, as the case may be.

ON THE EVOLUTION OF QUASARS AND THEIR REMNANTS

P. KAFKA

Max-Planck-Institut für Physik und Astrophysik, Munich, Germany

Abstract. The evolution of massive cores in the post-quasar phase is considered.

Obviously galaxies are able to produce nuclear condensations of about 10^8 (or more) solar masses within a light-month (or less). If we disregard the possibility that such nuclei are 'retarded cores' in universal expansion, or places where "new matter is poured into our world", we are faced with the problem, how gravitation manages to defeat rotation. It is clear, that angular momentum must be transported from inner to outer regions, such that the former can shrink, while the latter expand. Hence, we have to start with more total mass than we finally need.

If the original accumulation of mass is already fragmented into stars, the desired transport must be undertaken by their gravitational interaction. The stellar-dynamical 'evaporation' of stars from a cluster makes it shrink approximately along lines RM^{-2} = const in the mass-radius diagram. Thus the process might produce quasar-like nuclei in galactic centers. It was proposed and discussed by various authors. (e.g. Gold, 1965; Spitzer and Saslaw, 1966; von Hoerner, 1968). The shrinking time predicted by classical stellar dynamics is far too long and only at the desired final state does it become as low as the age of the universe. However, since the influence of a broad mass function and the effect of binary formation have been discussed as possibilities to lower the time scale, it seems that the stellar dynamical process may not yet be excluded.

If the original large mass is *not* fragmented into stars, or if a dense interstellar material is left (or newly produced), friction and turbulence can transport angular momentum outward. More efficient, however, will be the winding up of magnetic fields in a rotating disk of gas. This mechanism was discussed by various authors (cf. Kardashev, 1964; Piddington, 1964, 1967; Ginzburg and Ozernoy, 1964; Pikelner, 1965; Ozernoy, 1966, and contribution to this symposium). More detailed examination is in progress, also on a stellar scale, where the mechanism will be important for supernova explosions.

Now let us assume that one of the mechanisms produced a condensation of several 10^8 solar masses within only $\frac{1}{10}$ pc. The question is then, how this object liberates energy in the form of fast particles very efficiently.

Let us first look at the fragmented situation: As indicated in Figure 1, stellar collisions become important at the density considered. Coalescence leads to more massive stars which rapidly burn out their nuclear fuel and suffer gravitational instability. In other close encounters stars may be torn and will shed much matter into interstellar space. (The same will happen in the collapse events themselves, due to the presence of magnetic fields and rotation.) The resulting situation, with frequent local collapses embedded in a dense medium, would certainly lead to quasar-like phenomena (Colgate, 1967; Dyson, 1968). It is very important that we have learned from current pulsar theory (Pacini, 1968; Ostriker and Gunn, 1969; Goldreich and Julian, 1969) how energetic particles may be produced very efficiently. If a neutron

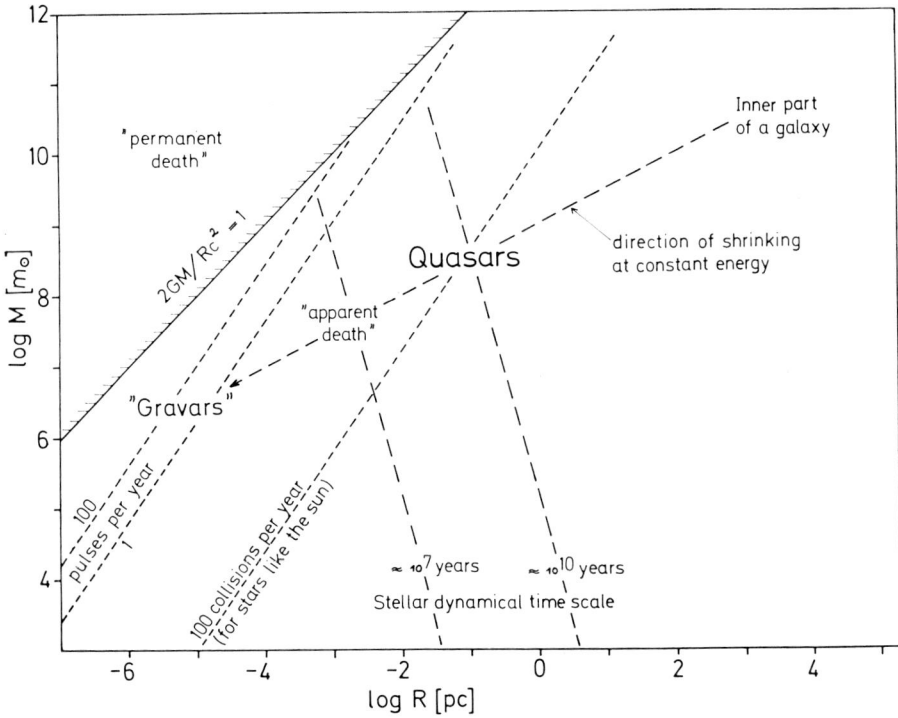

Fig. 1. Mass-radius diagram of clusters of stars (at the right-hand side of the line labelled '100 collisions per year') or black holes (at the left-hand side).

star is left after collapse, a considerable fraction of total emission will be concentrated within the first few days. It seems likely, however, that a similar pulse must arise in the formation of a 'rotating black hole', which cannot remain magnetically connected with its neighborhood, and must 'radiate away' the magnetic field.

Many details of final 'Magneto-rotational-gravitational' collapse are still not understood, but it seems more and more likely that there is no severe problem with the conversion of enough gravitational potential energy to fast particles, if rotation and magnetic fields are cooperating with gravitation.

A quasar model of this 'already fragmented' type would not account for periodic activity. Whoever is convinced by observations of such periodicities, must discard this hypothesis or else introduce some coherence of the whole cluster which allows for influence of its overall pulsation or rotation on the rate of local collapses.

If the compact object is still unfragmented, it might work as the now fashionable giant pulsar-like machine (Morrison, 1969; Cavaliere *et al.*, 1969; Fowler at the IAU-Symposium No. 46 at Jodrell Bank; Woltjer's review at the present symposium), or the magneto-turbulent 'magnetoid' (Ozernoy and Chertoprud, 1967). If it can remain coherent even through its final states, the remnant will be a single giant black hole (cf. Lynden-Bell, 1969), probably of the Kerr-type (Bardeen, 1970).

However, it seems likely that fragmentation cannot be avoided (cf. Dyson, 1968). In this case, even if part of the catastrophic activity took place during a coherent stage, local instabilities would finally take over and leave us with a (flattened) cluster of collapsed objects, as in the first case.

As long as there are normal stars left, they will collide with each other or with collapsed stars. After some time all stars are collapsed or have been thrown out. As most of the collapsed stars are expected to be heavier than a few suns, they will be (Kerr-type) black holes, and not neutron-stars. Their cross-section is about 10^{10} times smaller than before collapse, and hence collisions no longer occur. The only detectable activity might come from the accretion of the dense interstellar medium. This process also brakes the collapsed star and induces a shrinking of the cluster in addition to the stellar dynamical 'evaporation'. But even the latter works now on a time-scale of several 10^9 yr (neglecting the possibly accelerating modifications mentioned above). Hence, even when most of the gas within the cluster has been swallowed by its members, and the object seems dead, contraction still must go on, and finally the remaining cluster members come so close to each other, that collisions of black holes begin to occur. Fusion or close passage of black holes, however, is certainly the source of strong gravitational radiation. This may appear a rather speculative idea, but a glance at the diagram shows that a quasar remnant will inevitably pass through the critical region, where collisions of black holes become dominant, *if* fragmentation into stars occurred at any time during the earlier evolution. "For shortness and for fun" the author called such dense clusters of black holes 'gravars' (Kafka, 1969, 1970). During their short life they emit a considerable fraction of their rest energy in pulses of gravitational radiation which could fit the results reported by Weber (1969, 1970), if remnants of a quasar-like stage were situated in the nucleus of our own galaxy. (In the simplest model the collision rates, indicated in the diagram by the lines for 1 and 100 pulses per year, do not depend on the mass of the fragments but only on the total mass M.) In a real quasar remnant one might expect some subclustering, such that instead of a single very massive cluster many smaller ones would cross the 'gravar' region, causing strong fluctuations in the gravitational radiative activity.

During the stage of apparent death and in the final 'gravar' phase the quasar emnant must supply its mother-galaxy with a total mass of about 10^8 $M_\odot$ in the form of black holes. Immediately after the exhaustion of the quasar activity their ejection-velocity would be of the order of several thousand km s^{-1}. In the gravar phase they are thrown out nearly at the velocity of light. When they come to large distances from the cluster, their velocity will on the average have dropped to low values. For some 'tail' of the distribution, however, extreme velocities may survive.

At the very end of such a quasar remnant, this 'sling mechanism' as well as the fusion of member-black-holes and the gravitational radiation compete in the destruction of the cluster. Whatever is left must collapse in a single black hole, but very likely this has only a small fraction of the original quasar's mass. All the rest became redistributed in the mother-galaxy in the form of collapsed stars or has been radiated away in pulses of gravitational radiation.

Hence, even if astronomers should find that there are no dark massive objects in the nuclei of galaxies (cf. King's contribution), this would not necessarily mean that there have never been any. If fragmentation occurred, a quasar will not leave much.

References

Bardeen, J.: 1970, *Nature* **226**, 64.
Cavaliere, A., Pacini, F., and Setti, G.: 1969, *Astrophys. Letters* **4**, 103.
Colgate, S. A.: 1967, *Astrophys. J.* **150**, 163.
Dyson, F. J.: 1968, *Astrophys. J. Letters* **154**, L37.
Ginzburg, V. L. and Ozernoy, L. M.: 1964, *Zh. Eksper. Teor. Fiz.* **47**, 1030 (*Soviet Phys. JETP* **20**, 689).
Gold, T.: 1965, in *Quasi Stellar Sources and Gravitational Collapse*. Proceedings of the first Texas Symposium on Relativistic Astrophysics.
Goldreich, P. and Julian, W. H.: 1969, *Astrophys. J.* **157**, 869.
Von Hoerner, S.: 1968, *Bull. Astron. Paris, Ser. 3*, **3**, 147.
Kafka, P.: 1969, *Mitt. Astron. Ges.* **27**, 134.
Kafka, P.: 1970, *Nature* **226**, 436.
Kardashev, N. S.: 1964, *Astron. Zh.* **41**, 807 (*Soviet Astron.* **8**, 643).
Lynden-Bell, D.: 1969, *Nature* **223**, 690.
Morrison, P.: 1969, *Astrophys. J. Letters* **157**, L73.
Ostriker, J. P. and Gunn, J. E.: 1969, *Astrophys. J.* **157**, 1395.
Ozernoy, L. M.: 1966, *Astron. Zh.* **43**, 300 (*Soviet Astron.* **10**, 241).
Ozernoy, L. M. and Chertoprud, V. E.: 1967, *Astron. Zh.* **44**, 537 (*Soviet Astron.* **11**, 428).
Pacini, F.: 1968, *Nature* **219**, 145.
Piddington, J. H.: 1964, *Monthly Notices Roy. Astron. Soc.* **128**, 345.
Piddington, J. H.: 1967, *Monthly Notices Roy. Astron. Soc.* **136**, 165.
Pikel'ner, S. B.: 1965, *Astron. Zh.* **42**, 3 (*Soviet Astron.* **9**, 1).
Spitzer, L. and Saslaw, W.: 1966, *Astrophys. J.* **143**, 400.
Weber, J.: 1969, *Phys. Rev. Letters* **22**, 1320.
Weber, J.: 1970, *Phys. Rev. Letters* **24**, 276.

Discussion

Question: Would a black hole, flying through the interstellar medium, be observable?

Kafka: Probably not. Some aspects of this problem have been discussed by Salpeter (E. E. Salpeter: 1964, *Astrophys. J.* **140**, 796), however, especially for high velocities it should be re-examined, if the evidence for evaporation of black holes from the galactic center would grow with a confirmation of Weber's results.

SUPERMASSIVE DISKS*

R. V. WAGONER** and E. E. SALPETER

Cornell University, Ithaca, N.Y., U.S.A.

Abstract. The ultimate source of energy in compact objects is investigated. The model is a highly flattened body of mass $\gtrsim 10^4 \, M_\odot$, supported by centrifugal force in two directions and by gas and radiation pressure in the third. The properties of this model are compared with those nearly spherical supermassive stars introduced by Hoyle and Fowler.

We would like to discuss some work, done partly in collaboration with James Bardeen (Bardeen and Wagoner, 1969; Wagoner, 1969; Salpeter and Wagoner, 1971), which has been motivated in part by a desire to understand the nature of the ultimate source of the energy produced by the compact objects we have discussed. The type of source

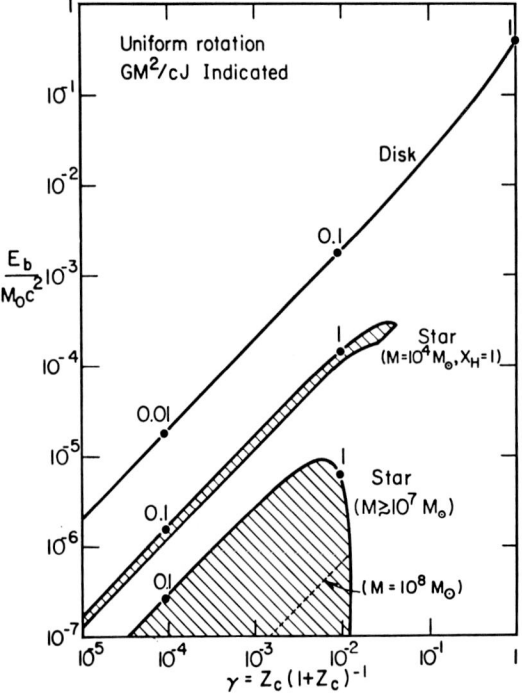

Fig. 1. A comparison of the fractional binding energies of supermassive stars and thin (low entropy) disks as a function of central redshift z_c.

we shall consider is a highly flattened body of mass $M \gtrsim 10^4 \, M_\odot$, supported by centrifugal force in two directions and by gas and radiation pressure in the third. We

---

* Supported in part by the National Science Foundation (GP-9621).
** Alfred P. Sloan Foundation Research Fellow.

shall compare the properties of such supermassive disks with those of the nearly spherical supermassive stars introduced by Hoyle and Fowler (1963a, b) some seven years ago.

The assumption of uniform rotation is made for mathematical simplicity, but it must be stressed that there are reasons for believing that differential rotation must occur, so that our results probably represent only a qualitative indication of the properties of real disks. However, this is sufficient for our purposes.

The major point of this talk is illustrated on Figure 1, which indicates the domains of quasi-static evolution of uniformly rotating supermassive stars and disks in the fractional binding energy, relativity parameter plane. Here z_c is the redshift of the center of the body, and the points indicate various values of the dimensionless rotation parameter GM^2/cJ, which, by comparison, has the value $\sim 10^{-3}$ in typical spiral galaxies.

A supermassive star with $GM^2/cJ \lesssim 1$ evolves by losing photons until it begins to shed mass at the upper boundary of the shaded region, whereas if $GM^2/cJ \gtrsim 1$, the star eventually becomes unstable to gravitational collapse at the lower boundary of the region (not shown for $M \gtrsim 10^7 M_\odot$). For the mass range of most interest to us, $M \gtrsim 10^7 M_\odot$, the star can reach a binding energy of $E_b \approx 10^{-5} M_0 c^2$ and a central redshift $z_c \approx 10^{-2}$. As Fowler (1966) has noted, however, the introduction of differential rotation allows these limits to be increased somewhat.

Let us compare some properties of a supermassive star (with $M \gtrsim 10^7 M_\odot$) at the onset of shedding with those of a disk of the same rest mass and angular momentum. We find that the relativity parameter increases by $\sim 10^2$, the binding energy increases by $\sim 10^4$, the radius decreases by $\sim 10^2$, and the central pressure increases by $\sim 10^8$ in going from the star to the disk. The fractional binding energy $\to 0.37$ and $GM^2/cJ \to 1$ as the central redshift of the disk approaches infinity. The basic reason for these differences in properties is the fact that the stars are supported by radiation pressure in nearly neutral equilibrium, while the disks are supported by rotation. Although the binding energy of the disks does not reach a maximum at finite z_c, it is not known whether they are in fact stable against overall gravitational collapse, however. Of course, the well-known smaller scale Newtonian instabilities can be present as well. Nevertheless, it is seen that large rotational energies can at least in principle be made available for conversion into high energy particles and their accompanying radiation through pulsar-like mechanisms (Cavaliere *et al.*, 1969; Morrison, 1969; Cavaliere *et al.*, 1970), for instance.

Note that the disk evolves radially by changes in J/M^2, whereas the loss of photons only leads to increased flattening with slightly increased binding. In Figure 2 is shown the relation between rotation period and angular momentum J for disks of fixed mass, and the corresponding domains for stars. Unlike the stars, the disks rotate faster as they lose angular momentum. The period of a disk can also be much shorter than that of a star of the same mass.

The properties discussed up to now have been independent of β, the ratio of gas pressure to total pressure in the disk, as long as it is thin. In Table I are included the lifetime $E(4p \to {}^4\text{He})/L$, with a normal CNO abundance, yields a maximum hydrogen-burning mass of $\sim 2 \times 10^{10}\ M_\odot$, as compared with $3 \times 10^6\ M_\odot$ for a uniformly rotating supermassive star. It is also seen that the Kelvin-Helmholtz flattening time E_b/L can be greater than the nuclear burning lifetime of 3×10^6 yr. The disk will flatten at constant central temperature until the gas pressure dominates, at which point it begins to cool.

Finally we apply this model to the QSO 3C 345, where it is assumed that the light variations correspond to a local rotation period of 200 days, although of course the

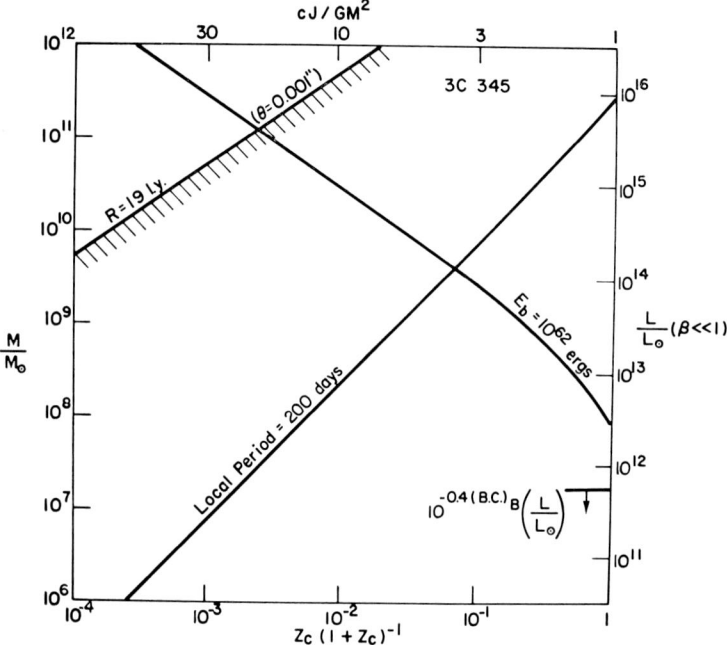

Fig. 4. Curves of constant period, binding energy, and radius of uniformly rotating disks. The values (or upper limits) chosen are related to observations of the variable QSO 3C 345. The upper limit on the amount of thermal radiation seen through the B filter is also indicated.

evidence is not compelling at present (Kinman et al., 1968; Morrison, 1969). In Figure 4 is shown the locus of disks having this period on the mass parameter, relativity parameter plane. Also shown is a curve of constant binding energy, taken to be near the upper limit of that thought necessary. It is seen that the limit on the radius, set by the radio interferometric result of Kellermann et al. (1968), easily allows the solution $M \sim 4 \times 10^9\ M_\odot$ and $z_c \sim 0.1$ obtained from the assumed period and estimated binding energy. (In fact, the resulting radius is 0.02 light years.) As has been pointed out by Fahlman (1970), no such solution exists for uniformly rotating supermassive stars. However, such a solution requires rather high luminosity, $L \sim 10^{14}\ L_\odot$. This can be compatible with the observational upper limit shown if either: (a) the peak of the spectrum is well away from the visible region, so that the bolometric correction to the observed B magnitude $(BC)_B$ is large; or (b) the gas pressure dominates, so that the luminosity is no longer so high.

References

Bardeen, J. M. and Wagoner, R. V.: 1969, *Astrophys. J. Letters* **158**, L65.
Cavaliere, A., Pacini, F., and Setti, G.: 1969, *Astrophys. Letters* **4**, 103.
Cavaliere, A., Morrison, P., and Pacini, F.: 1970, *Astrophys. J. Letters* **162**, L133.
Fahlman, G. G.: 1970, *Astrophys. J. Letters* **160**, L87.
Fowler, W. A.: 1966, *Astrophys. J.* **144**, 180.
Hoyle, F. and Fowler, W. A.: 1963a, *Monthly Notices Roy. Astron. Soc.* **125**, 169.
Hoyle, F. and Fowler, W. A.: 1963b, *Nature* **197**, 533.
Kellermann, K. I., Clark, B. G., Bare, C. C., Rydbeck, O., Ellder, J., Hansson, B., Kollberg, E., Hoglund, B., Cohen, M. H., and Jauncey, D. L.: 1968, *Astrophys. J. Letters* **153**, L209.
Kinman, T. D., Lamla, E., Ciurla, T., Harlan, E., and Wirtanen, C. A.: 1968, *Astrophys. J.* **152**, 357.
Morrison, P.: 1969, *Astrophys. J. Letters* **157**, L73.
Salpeter, E. E. and Wagoner, R. V.: 1971, *Astrophys. J.* **164**, 557.
Wagoner, R. V.: 1969, *Ann. Rev. Astron. Astrophys.* **7**, 553.

Discussion

Pachner: I am not sure whether the simplifying assumption of a uniform rotation does not substantially disturb the physical situation in rotating bodies. Since the paper of Raychaudhuri published fifteen years ago it is well-known that any deviation from an isotropic rotation increases the possibility of a gravitational collapse, even in the bodies without the uniform distribution of matter.

Wagoner: I am not familiar with all the details of the paper of Raychaudhuri. What is isotropic rotation? In any case, I know of no one who has investigated the stability against overall gravitational collapse of rapidly rotating, relativistic, isolated objects.

A MATTER-ANTIMATTER MODEL FOR QUASI-STELLAR OBJECTS

AINA ELVIUS
Stockholm Observatory, Sweden

Abstract. Observations of quasi-stellar objects and radio galaxies indicate that the total energy radiated from such objects is so large that the most likely source of energy is annihilation.

The demand for symmetry in the universe between ordinary matter and antimatter indicates that there must be equal quantities of the two kinds of matter in every galaxy. From this it seems likely that a galaxy is born as an ambiplasma body, in which separation of matter from antimatter leads to reasonably stable configurations.

The violent events observed in quasi-stellar objects are then interpreted in terms of collisions between stars of opposite kinds of matter. Such collisions are expected to occur frequently in very young galaxies with a high stellar density in the nucleus. Most of the gamma-radiation released in the annihilation will be absorbed in the gases of the colliding bodies, causing strong heating and violent explosions. Strong ionizing radiation and expanding gas clouds will give rise to the observed optical line emission. Expanding clouds of light ambiplasma will emit synchrotron radiation.

1. Introduction

The present paper will be concerned mainly with the explosive events occurring in quasi-stellar objects and in the active nuclei of some types of galaxies. General ideas concerning the presence of antimatter in galaxies and the energy release in quasi-stellar objects and in different types of radio galaxies due to the annihilation of matter and antimatter have been outlined in an earlier paper (Alfvén and Elvius, 1969). A 'symmetric cosmology' was adopted, assuming the presence of equal amounts of antimatter and ordinary matter in the universe and in every individual galaxy.

As antimatter cannot be well mixed with ordinary matter without being annihilated in a short time at the densities of interstellar gas in a typical galaxy, the formation of the galaxy from a thinner ambiplasma must have been accompanied by a separation process. Some processes have been suggested (Alfvén and Klein, 1963; Alfvén, 1965), which may cause a separation, resulting in many small cells of pure matter and pure antimatter, later growing into larger regions in which stars can form. Unseparated parts of the ambiplasma will be annihilated. In the present paper a stage of evolution of the galaxy is considered in which the stars have already been formed in great numbers.

2. Outline of Model

We assume that the phenomena of quasi-stellar objects occur in certain young galaxies in which the concentration of stars in the central part is unusually high. The observed activity within very small regions of QSOs and of some galaxy nuclei makes such an assumption seem plausible, and similar ideas have been adopted by a number of authors working with various models (e.g. Field, 1964; Colgate, 1969; Low, 1970). In the very dense nucleus collisions between stars will be frequent. On the average, in our model every second collision will occur between bodies of opposite kinds of matter which leads to violent explosions. Colliding stars need not be of the same size. Typically a small star or a body of substellar size will fall into a bigger star with a velocity of the order of 1000 km s^{-1}. When the bodies collide, annihilation will start immediately, but in order that the velocity of the falling body may be reversed, a fairly large quantity of matter must be annihilated. This requires thorough mixing of antimatter with matter. As such a mixing is a relatively slow process, the smaller body will penetrate the bigger star well below its surface, before most of the annihilation takes place. Most of the gamma-radiation from the annihilation process will then be absorbed in the stellar gas, causing strong heating and shock waves. The consequence will probably be a violent explosion with a release of very large energies, toward which also nuclear reactions in the distorted stars may contribute. Only in favourable cases will the star of smaller mass be entirely annihilated. In less efficient collisions where only a small part of the colliding masses annihilates, the energy due to the explosions of the stars may reach the same order of magnitude as the annihilation energy itself.

3. Energy Requirements

We adopt the hypothesis that the quasi-stellar objects are very distant and therefore very luminous. According to recent observations by Kleinmann and Low (1970) the infrared flux from the QSO 3C 273 might be as high as several times 10^{48} erg s^{-1}. This flux may be used as a plausible upper limit for this object, because it is based on an extrapolation of the intensity of infrared radiation actually observed toward longer wavelengths, when it was assumed that the infrared spectrum of 3C 273 is very similar to the spectra recorded for the nucleus of our galaxy and for NGC 1068.

Proton-antiproton annihilation produces neutrinos, gamma-rays, and relativistic electrons and positrons with energies around 100 MeV. If the unobservable energy going into neutrinos is not included in the calculations, we find that the annihilation of one solar mass per year will release, on the average, 3×10^{46} erg s^{-1}. The annihilation of matter and antimatter in a QSO required to explain the total energy radiation would thus be approximately in the range from around one solar mass to about a hundred solar masses per year.

4. Comparisons of Model with Various Observed Properties

A. SIZE OF SOURCES

Interferometer measurements (Kellermann *et al.*, 1968) as well as the shape of the radio spectra (Kellermann and Pauliny-Toth, 1969) and observations of rapid intensity variations by numerous investigators, all indicate that the sources in QSOs are often very small and that in several cases more than one source may exist in a QSO. This is in good agreement with our model. The remnants of colliding and exploding stars are certainly small sources and the model is necessarily characterized by strong variations in the energy output. Several sources will usually be active at the same time in the nucleus of a QSO, although they will typically be of different strength and in different stages of evolution.

B. OPTICAL CONTINUUM

The energy released in the annihilation will be degraded by a number of complex processes. Intense local heating of the gas as well as ejection of hot plasma from the exploding stars may give rise to the bright ultra-violet and blue continuum observed in QSOs. Probably synchrotron radiation will contribute even at such short wavelengths because of strong magnetic fields connected with the remnants of the exploding bodies. As such fields must be quite strong to give optical synchrotron radiation, the life-time of the relativistic particles must be correspondingly short. This might explain the rapid light variations often observed in QSOs. Optical variations with longer time scales may be due to the variations in thermal radiation and to variations in the mean activity.

The abovementioned observations of strong and in some cases variable infrared radiation from the QSO 3C 273 and from the nuclei of some Seyfert galaxies and our own galaxy, were interpreted by Low (1970) in terms of a new kind of infrared sources, which he called irtrons. Although there are certain similarities between Low's model and our own, the two models are based on entirely different cosmological hypotheses. Low assumes that matter and antimatter are created in the nuclei of galaxies, whereas Alfvén and myself assume that the galaxies condensed out of a thin ambiplasma. Instead of collisions between mysterious irtrons we speak of collisions between stars of opposite kinds of matter. In both models the annihilation of matter and antimatter will give rise to electrons and positrons with energies around 100 MeV. If moving in magnetic fields of the order of 100 G, such particles emit synchrotron radiation with the peak in the flux distribution around 4×10^{12} Hz, as observed by Kleinmann and Low for some objects. This radiation will show rapid variations. As mentioned before, such strong fields must be of very small dimensions. In our model they are thought to be confined to a small volume around the exploding stars.

At least in objects where the infrared radiation does not show any rapid intensity variations, part of the infrared energy may come from large dust clouds which are heated to the appropriate temperature by the intense radiation in shorter wavelengths.

C. OPTICAL LINE EMISSION

Expansion and cooling of the hot plasma ejected from the exploding stars will lead to the formation of gas clouds emitting broad spectral lines. The existence in the same spectrum of lines originating in gas clouds of very different densities or temperatures is in agreement with our model in which the spectrum will typically be composed of the radiation from several regions, not only inner and outer shells surrounding one explosion but also remnants of many explosions occurring at various times. In addition there will be optical line emission from much larger regions of gas excited by radiation at higher frequencies from the intense source.

D. POLARIZATION OF LIGHT

The variable polarization of light observed in several QSOs (Kinman *et al.*, 1966; Kinman *et al.*, 1968; Elvius, 1968) may be explained as follows. When a smaller body penetrates the outer parts of a star of the opposite kind of matter and is partly annihilated, the explosion will most likely be asymmetric and produce a massive jet of hot plasma, ejected radially from the star. Electron scattering of light from the intense source as well as optical synchrotron radiation may give highly polarized radiation from such a jet. Each new explosion could result in a new more-or-less polarized component of radiation. Thus, sudden changes in the degree of polarization as well as in the position angle could accompany major increases in the intensity of optical radiation.

E. RADIO RADIATION

In the annihilation process considered in our model, a large fraction (17%) of the energy goes into relativistic (100 MeV) electrons and positrons. Although some of these particles may be annihilated immediately, the rest will form a cloud of light ambiplasma, which may leave the star in the form of a jet. This cloud will later continue to expand, as it moves away from the star where it was formed.

Soon after the explosion, however, the relativistic particles of the light ambiplasma will spiral in a magnetic field related to the exploding star. As was mentioned above, such a field may be many orders of magnitude stronger than the general field of the young galaxy. On the other hand it has a very small volume. Thus it seems possible to account for the emission of rather energetic synchrotron radiation without assuming impossibly large energies stored in the magnetic field of the whole QSO.

Our model, therefore, predicts young radio emitters that are extremely small as compared to typical radio sources. Since, we believe, the collisions and explosions take place mainly in the nucleus of the young galaxy, the angular separation of young radio source components should also be quite small, even when the components are due to independent explosions within the same galaxy. The particles will probably, in a later stage, spiral around the lines of force in a larger-scale galactic field, emitting synchrotron radiation at radio wavelengths. Frequencies are probably high in the early stages after the cloud forms, and the energy maximum will shift toward longer wavelengths as the cloud expands and moves away from the active region. We have seen beautiful examples of such shifts demonstrated by several radio astronomers at this symposium. The clouds of relativistic electrons and positrons which comprise the light ambiplasma will drift along the magnetic lines of force. It is possible that in many galaxies a dipole-like magnetic field of large dimensions may exist, along which the light ambiplasma may drift in two opposite directions (each cloud containing both electrons and positrons). In this way the double radio sources which are so often observed may be formed. Immediately following a period of strong activity in the nucleus the new radio source will be concentrated in the nucleus. As time goes on, the two main components of the source become more separated and also become expanded.

It should perhaps be pointed out that the extended clouds of relativistic particles typical of strong radio galaxies and of some QSOs (which are not among the youngest) according to our model did not originate in a single explosive event but contain the light ambiplasma ejected in many collisions during a period of high activity.

5. Origin of Elements

The problem of how the elements were synthesized constitutes a challenge to all cosmologies, and so far none has been able to provide a very satisfactory explanation. If quasi-stellar objects are considered to be very young, the theory of element production by ordinary thermonuclear reactions in stars runs into difficulties because the quasar absorption spectra show the presence of heavy elements. This means that these elements must have been produced no later than the epoch of the quasi-stellar phenomenon.

We have already seen that the violent fluctuations in quasi-stellar object emissions require that stars or starlike bodies of one kind of matter are hit by bodies of the opposite kind of matter. Especially if the hitting bodies are not too big, they will penetrate a star to some depth and cause an explosion at the time of annihilation. A great release of energy will greatly heat a large part of the star and cause a number of shock waves. It is conceivable that heavy elements are produced under such conditions by processes which, in some respect, are similar to those thought to take place in the interior of normal stars and in connection with supernova explosions. The gases ejected from the explosion should then contain newly produced heavy elements. It is likely that some emission and absorption lines in quasars are produced by gases ejected in this manner, when they have cooled down at some distance from the source.

References

Alfvén, H.: 1965, *Rev. Mod. Phys.* **37**, 652.
Alfvén, H. and Elvius, A.: 1969, *Science* **164**, 911.
Alfvén, H. and Klein, O.: 1963, *Arkiv Fysik* **23**, No. 19.
Colgate, S. A.: 1969, *Physics Today* **22**, 27.
Elvius, A.: 1968, *Lowell Obs. Bull.* **7**, 55.
Field, G. B.: 1964, *Astrophys. J.* **140**, 1434.
Kellermann, K. I., Clark, B. G., Bare, C., Rydbeck, O. E. H., Ellder, J., Hansson, B., Kollberg, E., Höglund, B., Cohen, M. H., and Jauncey, D. L.: 1968, *Astron. J.* **73**, S 101 and *Astrophys. J. Letters* **153**, L209.
Kellermann, K. I. and Pauliny-Toth, I. I. K.: 1969, *Astrophys. J. Letters* **155**, L71.
Kinman, T. D., Lamla, E., and Wirtanen, C. A.: 1966, *Astrophys. J.* **146**, 964.
Kinman, T. D., Lamla, E., Ciurla, T., Harlan, E., and Wirtanen, C. A.: 1968, *Astrophys. J.* **152**, 357.
Kleinmann, D. E. and Low, F. J.: 1970, *Astrophys. J. Letters* **159**, L165.
Low, F. J.: 1970, *Astrophys. J. Letters* **159**, L173.

MATTER-ANTIMATTER ANNIHILATION AS AN ENERGY SOURCE IN SEYFERT GALAXIES

G. STEIGMAN and P. A. STRITTMATTER

Institute of Theoretical Astronomy, Cambridge, Great Britain

(To be published in *Astron. Astrophys.* **11**, 279 (1971))

Abstract. The extreme infrared luminosities of some Seyfert galaxies place severe requirements on the efficiency of the energy source. It is attractive to suggest that matter-antimatter annihilation supplies the necessary energy source; efficiencies of up to $\sim 50\%$ are, in principle, obtainable. However, there is a price to pay for such an explanation: gamma rays and neutrinos.

In a typical nucleon-antinucleon annihilation roughly $\frac{1}{3} Mc^2$ is released in electron-positron pairs, $\sim \frac{2}{3} Mc^2$ in ~ 3 gamma rays and $\sim Mc^2$ in ~ 3 electron-neutrinos and ~ 6 muon-neutrinos. As a result, if the observed infra-red power is to be derived from the energy in electron-positron pairs, then the flux of gamma rays would be 10^2–10^3 times the upper limits to the gamma ray flux. It is, of course, possible to account for the absence of the gamma rays by insisting that they be absorbed in the source. The neutrinos, however, will not be stopped and hence provide the possibility of testing the annihilation hypothesis.

We have computed the spectrum of neutrinos produced in annihilation. Assuming the product of the space density and infrared luminosity of Seyfert galaxies varies with redshift (z) as: $L(z)n(z) = L_0 n_0 (1+z)^m$ we have computed the flux of μ-neutrinos contributed by all Seyfert galaxies out to a given red-shift for $m=3$ and $m=6.5$ (strong evolution). Further, assuming the "3K" and "0.3 mm" background radiation fields to be caused by a burst of "Seyfert type" objects at appropriate redshifts, we have again computed the expected μ-neutrino flux. When these results are compared with present limits on the flux of μ-neutrinos at the Earth, as determined from experiments performed by several groups deep underground, it emerges that the predicted flux is comparable to or greater than present upper limits. Thus, annihilation probably does not supply the infrared sources in Seyfert galaxies with the energy they require.

Discussion of Papers Read by Elvius and Steigman

Kellermann: I think it is unlikely that the observed radio emission can be explained as being from annihilation electrons. In order for 100 MeV electrons to radiate at millimeter wavelengths, magnetic fields of about 1 G are required, whereas the radio data indicate that the fields are about 10^{-4} G. Also, in a 1 G field, the lifetime for millimeter radiation is only about one month, again contrary to the data for most radio sources.

Longair: Evolution laws of the form $(1+z)^\beta$, where $\beta \simeq 6$, are primarily derived from observations of radio sources and refer only to the most powerful classes of radio source. Seyfert galaxies do not belong to the class and they cannot exhibit such powerful effects. This only refers to the radio properties, of course, and it is possible that they behave quite differently in the infra-red. Obviously, counts of Seyfert galaxies in the infra-red is the only way of getting a clue about this.

Steigman: If one wishes to explain the background at about 0.3 mm as a burst of Seyfert-type infrared emitters at a redshift $z \simeq 2.5$ whose energy is derived from annihilation then the flux of muon-neutrinos is about three times the present upper limits.

Allen: I suppose that if the gamma-rays produced in the matter-antimatter model are absorbed in the Seyfert galaxy itself, the result must eventually be ionization of the interstellar gas in the galaxy. Is it possible that this disagrees with the apparently normal neutral hydrogen content of the four Seyfert galaxies which I presented this morning?

Steigman: If the gamma rays are stopped then the ionization produced causes trouble. In particular the free-free emission is probably too large and if the absorbing medium is too dense the electrons and positrons will lose their energy to ionization losses rather than synchrotron losses.

Felten: It did not look as if the non-evolutionary case on your graph violated the Reines condition very strongly. Can you make such a firm conclusion in that case?

Steigman: The case $m = 3$ does not provide a strong test of the annihilation hypothesis since the luminosity and/or space density may be in error.

Ozernoy: Very large γ-fluxes produced by annihilation in a matter-antimatter model for quasi-stellar and other compact objects may turn out to be mortal for these models. Indeed, as was shown recently by V. L. Ginzburg and myself (*Astrophys. Space Sci.*, in press), the photo-disintegration of nuclei due to photo-nuclear reactions must lead to significant differences of chemical abundance from normal in emission-line regions of quasars if the gamma-ray output with energy 20–30 MeV is as large as, say, 10^{46} erg s^{-1}. The absence of appreciable distortions in chemical composition of quasars may give useful restrictions on models of the central part of a quasar and, in particular, indicates the effective mixing of a plasma in its nucleus. But in the matter-antimatter model any mixing may lead only to additional annihilation in which about $\frac{1}{3}$ of the energy is released in the form of gamma-rays. Therefore this model seems to be inconsistent with a normal chemical abundance in quasars.

TRANSIENT ANNULAR STRUCTURES IN EXPLODING GALAXIES*

J. L. SĚRSIC**

Observatorio Astronómico, Córdoba, Argentina

Abstract. The explosive events going on in the central parts of some galaxies are related to a very high mass concentration. As an explosion is actually a drastic rearrangement of the concerned masses with energy release, the binding energy of the central core will change and, correspondingly, its effective gravitational mass. A test particle far from the nuclear region, although within the galaxy, will be moving accordingly in a variable-mass Newtonian gravitational field.

On the other hand the observations suggest that explosions in galaxies have axial symmetry, so we are concerned with the global properties of the motion of a particle in a variable mass axisymmetric gravitational field. In order to get rid of the mass variation a space-time conformal transformation is made, which, after imposing some not very restrictive conditions, leads to a conservative potential in the new variables. This new potential has additional terms due to the elimination of the variable mass. The equations of motion in the new variables provide the motion of the test particle *relative* to an expanding or contracting background which depends on the choice of the transformation and the law of the mass variability. The problem is, at this point, formally similar to Hill's. It is possible to write an equation for the relative energy (a generalization of Jacobi's integral) and also to define surfaces of zero relative velocity for the infinitesimal particle. The general topological properties of these surfaces require singular points along the symmetry axis (analogous to the collinear Eulerian points) and also a dense set in a circumference on a plane perpendicular to the symmetry axis (analogous to the Lagrangian points). The latter one is the main feature characterizing the topology of the zero relative velocity surfaces. Even when we lift some of the restrictive conditions, the Lagrangian ring preserves its properties, as for example, the one of being the only region where zero-velocity curves and equipotentials coincide when the configuration evolves in time (in the transformed space-time).

It is easy to understand that the topology of the surfaces is kept when we reverse the transformation and go back to physical space-time. If the dust, gas or stars in the system has definite upper limits for its Jacobian constants, spatial segregation of them will arise, as is the case in radio-galaxies such as NGC 5128, NGC 1316, etc. where ringlike dust structures are observed.

---

* To appear in extenso in the Proceedings of the São Paulo, Symposium on Celestial Mechanics, September 1969, Reidel, Holland.
** Under contract with the Consejo Nacional de Investigaciones of Argentina.

COSMOLOGICAL INFORMATION FROM GALAXIES AND RADIO GALAXIES

J. V. PEACH

Department of Astrophysics, Oxford, Great Britain

Abstract. An account is given of recent developments in the derivation of the value of the Hubble parameter applicable to regions beyond the local anisotropy. Recent observations relevant to interpretations of the magnitude-redshift diagram for cluster galaxies are discussed, in an attempt to assess uncertainties in the value of the deceleration parameter.

1. Introduction

Some twenty years ago Hubble (1951) in a review of the prospects for observational cosmology outlined what he believed to be the most promising methods of choosing observationally among the available world models. These were, firstly, to determine the local mean density of matter and the local law relating velocity to distance, and, secondly, to determine whether or not there are systematic changes with distance in either of these data. The search for a variation of density with distance through the extension of galaxy counts to successively fainter magnitude limits has not been actively pursued. Robertson's (1955) criticism of the possibilities of the method was followed by Sandage's (1961b) calculations, which indicated that at the limit of the Hale telescope the differences between the predictions of the models would be much less than observational errors and further uncertainties due to the unknown effects of clustering, the broad luminosity function of field galaxies and incalculable evolutionary effects.

The central line of development in classical optical cosmology has been in the more precise formulation of the local velocity-distance relation and the evaluation of the Hubble parameter H_0, together with the search for second-order effects in this relation and estimates of the deceleration parameter q_0. Indeed, except for some preliminary investigations which have been made into the use of the angular diameter-redshift relation applied to galaxies and clusters of galaxies, the redshift-magnitude (m–z) relation is the only cosmological test to have been exploited since Hubble wrote.

Since that time it has of course slipped from its position at the centre of the subject as a whole, partly due to the exploitation of radio-source counts and to the discovery of the isotropic background radiation, both of which offer the possibility of sampling at greater redshifts than are accessible to optical observers, but also partly due to the extreme difficulty of the observation of magnitudes and velocities at large enough distances so that the differences between the interesting models become larger than the observational errors. The main reason for persevering with the m–z relation is that one can get in principle a single conclusion among the possible models, if one can calculate the numerous corrections needed to interpret the observations. Furthermore, these corrections are easier to calculate for the test objects of the m–z relation than are, for example, the evolutionary effects associated with the interpretation of the radio-source counts.

This paper does not aim to give an exhaustive account of work in these fields. Critical reviews of methods for H_0 determination have been given recently by Tammann (1969) and van den Bergh (1970a, b), and progress in the photometry of cluster galaxies and radio galaxies has been reported on several times by Sandage (1966a, b, 1968a, b). We have restricted ourselves to a very brief account of recent developments in the derivation of a value of H_0 applicable to regions beyond the local anisotropy, and to a somewhat more extended discussion of recent observations relevant to the interpretation of the m–z diagram for cluster galaxies, in an attempt to assess the uncertainties in the value of the deceleration parameter.

2. Basic Formulae and Their Application

For the zero-pressure Friedman models with zero cosmological constant the relation between the bolometric magnitude of a test object m_{bol}, its absolute magnitude M and the redshift z is (Mattig, 1958)

$$m_{bol} = 5 \log \frac{c}{H_0 q_0^2}$$

$$\times \{q_0 z + (q_0 - 1)[\sqrt{(2q_0 z + 1)} - 1]\} + M + 25 \qquad (q_0 > 0) \qquad (1)$$

The Hubble parameter H_0 and the deceleration parameter q_0 are defined as

$$H_0 \equiv \dot{R}_0/R_0 \quad \text{and} \quad q_0 \equiv - R_0 \ddot{R}_0/\dot{R}_0^2$$

where R is the time-dependent scale factor in the Robertson-Walker line element, and subscript zero indicates that the quantities are evaluated at the present cosmic time. The deceleration parameter is related to the mean density ϱ_0 by

$$\varrho_0 = \frac{4\pi G}{3 H_0^2} q_0 \qquad (2)$$

where G is the gravitational constant, and with the spatial curvature k/R_0^2 by

$$kc^2/R_0^2 = H_0^2 (2q_0 - 1) \qquad (3)$$

Expanding (1) in powers of z gives the m–z relation derived without use of the field equations by Heckmann (1942) and Robertson (1955)

$$m_{bol} = 5 \log (cz/H_0) + 1.086 (1 - q_0) z + \cdots + M + 25 \qquad (4)$$

For the steady-state model the Hubble parameter is time-independent; thus $R(t) = A \exp(Ht)$ and consequently $q_0 = -1$. The exact m–z relation for this model is

$$m_{bol} = 5 \log [cz (1 + z)/H_0] + M + 25. \qquad (5)$$

Given a set of test objects with small dispersion in M and a guarantee that M is time-independent one could measure m_{bol} and z and use the relation (1) to derive q_0; the precision of the determination will depend on the dispersion in M and on the redshift interval sampled. Clearly larger redshifts will have higher weight in the solution for the higher order terms. To determine H_0 on the other hand, one must find the modulus of one of the test objects for which z is small enough for the second (q_0-dependent) term in (4) to be negligible, and yet high enough to be not only much larger than the galaxian peculiar velocities of a few hundred km s^{-1}, but also large enough for it to be outside the local anisotropic velocity field, so that it will reflect the motion of the presumably isotropically expanding substratum unperturbed by the gravitational interactions within the local supercluster.

3. The Value of H_0 Outside the Local Supercluster

Since the preliminary mapping of the anisotropy of the local velocity field it has been obvious that it is no longer possible to base a value for the Hubble parameter appropriate at large distances, $H_0(\infty)$, on the modulus and velocity of the Virgo cluster, until the perturbation of the isotropic flow at the Virgo cluster is known. De Vaucouleurs' (1958) model of the anisotropy indicated a reduced expansion rate at Virgo such that $H_0(\infty)/H_0$ (Virgo) $=1.58$; a more recent kinematic model based on more velocities gives 1.35 ± 0.15 (de Vaucouleurs and Peters, 1968). Although the present spread in quoted values for H_0 (from about 50 to 120 km s^{-1} Mpc^{-1}) is larger than the error that would arise from neglecting this anomalous expansion velocity, it is clearly an effect that must be investigated. We describe in this section two methods that have been applied to derive $H_0(\infty)$; the use of the redshift-magnitude plot for brightest cluster galaxies and the luminosity function of rich clusters of galaxies.

Sandage (1968c) has found the modulus of M87 in the Virgo E-cloud by using Racine's (1968) luminosity function for the M87 globular clusters. The brightest of these has $B=21.3$. The brightest globular in M31 is taken to be B282 with $B=15.01$, and the M31 apparent blue modulus is taken as $(m-M)_{AB}=24.84$ from the Cepheid photometry of Baade and Swope (1963), using Sandage and Tammann's (1968) calibration of the Cepheid P–L curve. Then $M_B = -9.83$ for B282 and the apparent modulus for M87 is $(m-M)_{AB}=31.1$, if its brightest globular is of the same absolute magnitude. Now NGC 4472, which according to Sandage's (1970) photometry is the brightest Virgo cluster member, has an apparent magnitude to an isophote of 25 mag. arc sec^{-2} of $B=9.42$. Thus $M_B=-21.68$ for NGC 4472. The m–z curve for brightest cluster members has a dispersion of only $\simeq 0.3$ mag. If NGC 4472 is now assumed a typical brightest cluster member, it can be used to calibrate the m–z plot at distances with velocities greater than 4000 km s^{-1} and outside the effects of the Supercluster, and $H_0(\infty)$ can be based on the mean line through the cluster points. This gives $H_0(\infty)=75$ km s^{-1} Mpc^{-1} with 950 km s^{-1} for the E-cloud velocity, whereas H_0(Virgo)$=64$ km s^{-1} Mpc^{-1} as NGC 4472 falls below the mean cluster line, presumably due to its anomalous velocity. Thus $H_0(\infty)/H_0(\text{Virgo}) = 1.17$. Assessing all the errors of the method, excluding possible errors in the distance scale in the Local Group, suggests that these values could be in error by ± 25 km s^{-1} Mpc^{-1}.

De Vaucouleurs (1970) has reexamined this use of brightest globular clusters as distance indicators, and has shown a correlation between the absolute magnitude of the brightest globular cluster M_{gc} and the absolute magnitude of the parent galaxy M_{gal} for seven Local Group members. As the brightest calibrating galaxy is M31 some extrapolation is needed to cover gE galaxies. Assuming his calibration and a measured B magnitude for M87 of 9.0, he uses Sandage's calibration of the brightest globular to find the modulus and absolute magnitude for M87. Treating this as a first approximation he uses the relation between M_{gc} and M_{gal} to find a second approximation to M_{gc} and a revision to the modulus, and this process is repeated until the value of M_{gc} converges. The total effect is to produce a change in M_{gc} for the brightest globular in M87 of 0.5 mag. As he also finds M87 brighter than NGC 4472 by 0.25 mag., following Sandage's procedure in entering the m–z diagram leads to the revised value of $H_0(\infty) = 50$ km s^{-1} Mpc and negligible difference between this and $H_0(\text{Virgo})$. Clearly much more work is necessary on the luminosity function of globular cluster populations of galaxies, and on the possible correlation of M_{gc} with the size of this population, before the method becomes more reliable, but it offers the possibility of a modulus to the gE galaxies of the Coma cluster, whose globular clusters will be accessible at about 24 mag.

A different approach has recently been made by Abell and Eastmond (1968), who have used the form of the cumulative luminosity functions of rich clusters of galaxies as a distance indicator. By matching the discontinuity in the luminosity functions of the Coma and Corona Borealis clusters to the Virgo cluster, they found the difference in modulus to be 4.7 mag. and 7.2 mag. respectively. Using Sandage's modulus of M87 of 31.1, this gives moduli of 35.8 for Coma and 38.3 for the Corona cluster. With their mean velocities of 6866, and 21651 km s^{-1} respectively one has $H_0(\infty) = 47$ km s^{-1} Mpc^{-1}. Once again one has bypassed the Virgo anisotropy.

These two methods indicate some of the inaccuracies in present Hubble parameter determinations and give some indication of the possible spread in values of H_0; an age of the universe, H_0^{-1}, of as much as 20×10^{10} yr or as little as 5×10^{10} yr cannot be excluded by present data.

4. The Determination of q_0. Generalities

As the intrinsic dispersion in M for field galaxies is so great as to mask any cosmological effects over the accessible redshift range, Hubble (1936) suggested the use of cluster galaxies as test objects, because for these it would be possible both to select from a fixed part of the luminosity function, viz., its bright end, and to get better space penetration owing to the high luminosity of brightest cluster galaxies. The first attempt to derive q_0 by this method was by Humason *et al.* (1956) (subsequently referred to as HMS) who measured magnitudes in P and V for the 1st, 3rd, 5th and 10th brightest galaxies in 18 clusters spanning the distance from the Virgo cluster to the Hyades cluster (0855+0321) at $z=0.2$; these magnitudes were then combined to give a synthetic brightest galaxy. This combined magnitude showed a standard deviation of only 0.3 mag when the data were fitted to Equation (4) and gave $q_0 = 2.5 \pm 1$. The intrinsic dispersion in the absolute magnitudes is clearly smaller, as part of the observed dispersion must be due to measuring error, patchiness in galactic absorption and possible peculiar velocities of the clusters measured. Baum (1957, 1961a, b) later extended the measurements as far as $z=0.46$ using an eight colour photometric system which obviated the need of a K-correction and gave the redshifts non-spectroscopically from the shift of the continuum through the measuring bands. This work remains of very considerable importance, and Baum's photometry of the brightest members of the clusters 0024+1654, 1448+2617 and 1410+5224 remains the only photometry for cluster galaxies with $z > 0.2$. Baum did not analyse his observations in detail, only indicating that a straight line in the m–z plane ($q_0 = 1$) seemed an adequate description of the data; σ_M for these measures is 0.20 mag., significantly smaller than the HMS result, but perhaps due to the small number of data points. The available data have recently been extended by Sandage's (1970) photometry of about forty cluster galaxies with $z \leqslant 0.2$ in B and V.

Radio galaxies share the property of high optical luminosity with brightest cluster galaxies, and the high proportion of emission line objects makes redshift measurement easier. Not only are their luminosities closely similar, but many radio galaxies are brightest members of clusters. There are problems of interpretation connected with them that do not occur in the interpretation of normal ellipticals, namely, the question of the K-correction in the presence of line emission and possible non-thermal radiation, the evolutionary properties over the light travel time, and the observational problem of how to treat the photometry of the so-called 'dumb-bells'. The magnitude dispersion of field radio galaxies is about 0.5 mag., so that over the presently accessible range of z they are of much less weight than brightest cluster galaxies, although possibly avoiding some of the selection effects associated with the latter.

5. The Determination of q_0. Assessment of Systematic Errors

The search for the second-order term is still in a very preliminary stage and it is impossible at present to give a value for q_0 in which any great degree of confidence can be placed. This is due not simply to the comparatively small redshift range of the m–z data but also to the possibility of large systematic errors in these data. This section considers some of the corrections that must be made to the observed magnitudes to free them from effects irrelevant to the problem in hand, and attempts to assess the size of possible systematic errors and their consequential effect on our knowledge of q_0.

A. K-CORRECTION

A K-correction is applied to the measured apparent heterochromatic magnitudes to compensate for the redshift of the energy curve $I(\lambda)$ of the observed galaxy, which sweeps different parts of $I(\lambda)$ measured in the rest frame of the galaxy through the fixed pass-bands of the observer's photometer. Effects due to the decrease in the observed photon energy and the decrease in the photon arrival rate are already incorporated in the m–z relation (1). There are two terms in the K-correction. The first arises from the narrowing of the photometer pass-band in the rest frame of the galaxy by a factor $(1+z)$; this term is wavelength-independent and simply increases the apparent magnitude by $2.5 \log(1+z)$. The second term is due to the fact that the radiation received by the observer at wavelength λ is that emitted by the galaxy at wavelength $\lambda/(1+z)$; this effect is clearly wavelength-dependent and its sign and magnitude depend on the gradient of $I(\lambda)$ over the wavelength region of interest. We write the two terms of the K-correction as

$$K = 2.5\log(1+z) + 2.5\log \frac{\int_0^\infty I(\lambda)\, S(\lambda)\, d\lambda}{\int_0^\infty I(\lambda/(1+z))\, S(\lambda)\, d\lambda}. \qquad (6)$$

Here $I(\lambda)$ is the flux per wavelength interval and $S(\lambda)$ is the photometer response function. For the evaluation of the second term we need $I(\lambda)$ and an assurance that $I(\lambda)$ is sufficiently homogeneous among giant E and S0 galaxies that we can base a universal K-correction on the detailed observation of $I(\lambda)$ for a few nearby objects. The pass-bands $S(\lambda)$ are typically broad ($\Delta\lambda/\lambda \cong 0.1$) so that the $I(\lambda)$ required is the continuum intensity distribution.

A first approximation to the K-correction for gE galaxies, apart from Hubble's (1936) assumption that they could be represented by a 6000 K black body, was made by HMS. They used an $I(\lambda)$ based on the 6-colour photometry of M32 by Stebbins and Whitford (1948). With the benefit of hindsight it is now clear that this was unfortunate for two reasons. Firstly, M32 is bluer than a typical gE, being about 0.3 mag. brighter in the ultraviolet. Secondly, as first suggested by de Vaucouleurs (1948) and confirmed by later scanner observations (Code, 1959), the broad bands of the 6-colour system had inadequate spectral resolution to give a sufficiently accurate representation of the sharp drop in the $I(\lambda)$ curve near 3900 Å. Accordingly $I(\lambda)$ in this region appeared too smooth and the K-correction for the P or B bands systematically too small. It was this $I(\lambda)$ and K-correction which gave rise to the Stebbins-Whitford effect; the colours of distant ellipticals appeared redder by as much as 0.3 mag. at $z=0.13$ than would have been predicted from the M32 curve. The presence of the Stebbins-Whitford effect and the possibility of evolutionary effects in colour of this magnitude confused the discussion of the HMS photometry where the problem was left essentially unresolved. A certain confusion still remained when Code's (1959) and Oke's (1962) scans of the Virgo elliptical NGC 4374, which has normal $U-B$ and $B-V$ colours for a gE, were subsequently used by Sandage (1966) to recalculate K_B and K_V.

There was a small systematic difference between the measured $I(\lambda)$ curves, largely due to the different absolute calibrations used in the spectrophotometric reductions. The Oke curve showed no Stebbins-Whitford effect when the predicted run of $B-V$ colour with z was compared with observed colours out to $z=0.17$, while there was an excess reddening of 0.2 mag. present with the Code curve.

The situation is now more satisfactory as a result of the work of Oke and Sandage (1968) and Whitford (1970). Oke and Sandage's work is based on scans with an exit slit of about 50 Å of the central 10″ to 12″ of a number of nearby giant E and S0 galaxies. The $I(\lambda)$ curves for these, based on Oke's (1964) calibration of Vega with a small modification in the ultraviolet, appeared identical from 3375 Å to 5840 Å to within the error estimate for each object (± 0.07 mag. for $\lambda > 4000$ Å and ± 0.13 mag. for $\lambda < 4000$ Å), and the mean for all of the observed galaxies was in turn identical with $I(\lambda)$ for the central region of M31, which is known with a standard deviation of only ± 0.02 mag. at all wavelengths. K_B and K_V were calculated using the $I(\lambda)$ for M31 for $\lambda < 5840$ Å and for the less well determined region to the red of 5840 Å from scans of NGC 3379. It was realised that because of the colour gradient known to exist in ellipticals (de Vaucouleurs, 1960; Tifft, 1963) this K-correction would not necessarily be appropriate to photometry with aperture to diameter ratios A/D_0 greater than those used in the spectrophotometry ($\simeq 0.08$), but systematic effects due to this were thought to be less than 0.04 mag. in K_V for $z < 0.2$.

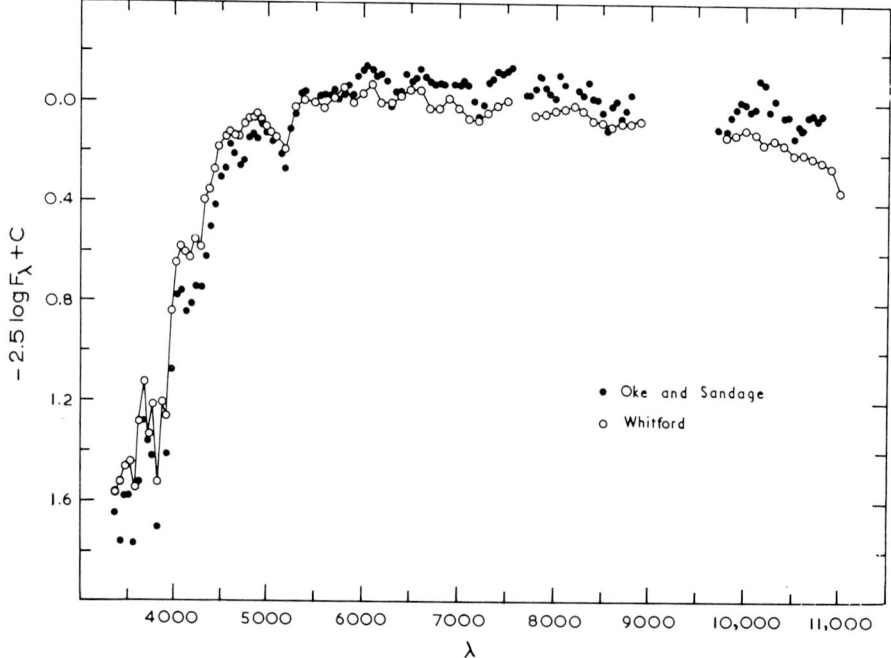

Fig. 1. Intensity distributions in magnitudes per unit wavelength interval for giant ellipticals from the observations of Oke and Sandage (1968) and Whitford (1970). The significantly bluer $I(\lambda)$ with Whitford's larger A/D_0 ratio should be noted.